Patrick Horster (Hrsg.)

Chipkarten

Grundlagen, Realisierungen,
Sicherheitsaspekte, Anwendungen

DuD-Fachbeiträge

herausgegeben von Andreas Pfitzmann, Helmut Reimer, Karl Rihaczek
und Alexander Roßnagel

Die Buchreihe DuD-Fachbeiträge ergänzt die Zeitschrift DuD – Datenschutz
und Datensicherheit in einem aktuellen und zukunftsträchtigen Gebiet, das für
Wirtschaft, öffentliche Verwaltung und Hochschulen gleichermaßen wichtig ist.
Die Thematik verbindet Informatik, Rechts-, Kommunikations- und Wirtschafts-
wissenschaften.
Den Lesern werden nicht nur fachlich ausgewiesene Beiträge der eigenen
Disziplin geboten, sondern auch immer wieder Gelegenheit, Blicke über den
fachlichen Zaun zu werfen. So steht die Buchreihe im Dienst eines interdiszi-
plinären Dialogs, der die Kompetenz hinsichtlich eines sicheren und verantwor-
tungsvollen Umgangs mit der Informationstechnik fördern möge.

Unter anderem sind erschienen:

Hans-Jürgen Seelos
Informationssysteme und Datenschutz
im Krankenhaus

Wilfried Dankmeier
Codierung

Heinrich Rust
Zuverlässigkeit und Verantwortung

Albrecht Glade, Helmut Reimer
und Bruno Struif (Hrsg.)
Digitale Signatur &
Sicherheitssensitive Anwendungen

Joachim Rieß
Regulierung und Datenschutz im
europäischen Telekommunikationsrecht

Ulrich Seidel
Das Recht des elektronischen
Geschäftsverkehrs

Rolf Oppliger
IT-Sicherheit

Hans H. Brüggemann
Spezifikation von objektorientierten
Rechten

Günter Müller, Kai Rannenberg,
Manfred Reitenspieß, Helmut Stiegler
Verläßliche IT-Systeme

Patrick Horster (Hrsg.)

Chipkarten

Grundlagen, Realisierungen, Sicherheitsaspekte, Anwendungen

ISBN-13: 978-3-322-89204-1   e-ISBN-13: 978-3-322-89203-4
DOI: 10.1007/978-3-322-89203-4

Vorwort

Plastikkarten in Form einer Kreditkarte sind längst zu unentbehrlichen Helfern des täglichen Lebens geworden – nicht nur, wenn wir telefonieren, Bargeld abheben, zum Arzt gehen oder nach einem Essen bargeldlos zahlen. Plastikkarten mit und ohne Chip – sichtbar oder unsichtbar – erfreuen sich zunehmender Beliebtheit und werden daher in zahlreichen Anwendungen eine breite Akzeptanz erfahren. Dabei hat sich die Kartentechnologie in den letzten Jahren entscheidend verändert, in vielen Fällen reden wir nicht nur von der simplen Chipkarte, wir sprechen von der SmartCard und der CryptoCard. Beide Karten sind zwar weiterhin Chipkarten, die eine ist aber mit einem Prozessor, die andere zudem mit einem Crypto-Controller versehen.

Mit der zunehmenden Komplexität der Chipkarten haben sich auch die Einsatzbereiche geändert. Während die ersten Chipkarten lediglich Speicherfunktionen mit und ohne Sicherheitslogik realisieren konnten, sind moderne SmartCards in der Lage, komplexe und zugleich sicherheitsrelevante Aufgaben flexibel zu erfüllen. Dies wird durch den Einsatz spezieller CryptoCards weiter verstärkt, durch sie werden innovative Ideen wie digitale Signaturen und dadurch auch Electronic Commerce und Electronic Banking erst ermöglicht.

Als gemeinsame Veranstaltung der GI-Fachgruppe 2.5.3 Verläßliche IT-Systeme, des ITG-Fachausschusses 6.2 System- und Anwendungssoftware, der Österreichischen Computer Gesellschaft und TeleTrusT Deutschland ist die Arbeitskonferenz Chipkarten nach den Veranstaltungen Trust Center und Digitale Signaturen die dritte einer Reihe, die sich einem speziellen Thema im Kontext der IT-Sicherheit widmet.

Die Schwerpunkte der in diesem Band behandelten Themen stellen sich wie folgt dar. In einer Einführung werden Anforderungen, Eigenschaften und Anwendungen überblicksartig beschrieben, neue Wege für SmartCard-Anwendungen aufgezeigt und die verifizierbare Sicherheit multiapplikationsfähiger SmartCards dargestellt. Bereits hier wird klar, daß Sicherheit durch Chipkarten und Sicherheit von Chipkarten die beherrschenden Themen der Beiträge sind. Von besonderer Bedeutung ist zudem der Aspekt des Datenschutzes und der damit einhergehenden betroffenenorientierten Technikbewertung. Patientenkarten und Chipkarten eingesetzt in einem regionalen klinischen Tumorregister sind davon ebenso betroffen wie individualisierte Kommunikationsumgebungen mit kontaktlosen Chipkarten.

Kontaktlose Chipkarten sind für viele Anwendungen als ideales Medium anzusehen, sie erfordern aber besondere Sicherheitsmaßnahmen und eine Aufklärung der Benutzer, um letztendlich sichere und vertrauenswürdige Applikationen zu gewährleisten. Im Kontext kontaktloser Chipkarten werden ein spezieller Mikrokontroller und das Problem sicherer Transaktionen betrachtet. Dabei kommt der Kryptographie eine besondere Rolle zu.

Sicherheitsrelevante Chipkartenprojekte können zumeist ohne kryptographische Verfahren nicht realisiert werden, wobei Hardwarerealisierungen, Algorithmen auf der Basis elliptischer Kurven und die Schlüsselerzeugung auf einer Chipkarten zunehmend an Bedeutung gewinnen. An konkreten Beispielen werden die wesentlichen Designkriterien des Hardwareentwurfs einer CryptoCard, die Implementierung elliptischer Kurven, die Realisierung zugehöriger effizienter Algorithmen und Konzepte zur dublettenfreien Schlüsselgenerierung vorgestellt. Werden so erzeugte Schlüssel zur Verwirklichung digitaler Signaturen eingesetzt, so kann der Forderung des Gesetzes zur digitalen Signatur, daß jeder Benutzer einen sicheren und einmaligen Signaturschlüssel besitzt, Rechnung getragen werden.

Ein bedeutendes und anspruchsvolles Anwendungsgebiet für Chipkarten ist im Umfeld digitaler Signaturen zu erkennen. Hier müssen spezielle CryptoCards ihre Sicherheit und Wirksamkeit unter Beweis stellen. In aktuellen Beiträgen wird behandelt, wie Signaturgesetz und Signaturverordnung dazu beitragen können, daß Chipkarten als Signier- und Verifizierkomponente eingesetzt werden können. Außerdem wird gezeigt, wie digitale Signaturen zur Sicherheit im Internet ihren Beitrag leisten können.

Wie Chipkarten in offenen und geschlossenen IT-Systemen gewinnbringend eingesetzt werden, wird in verschiedenen Anwendungen detailliert betrachtet. Die vorgestellten aktuellen Chipkartenprojekte geben dabei auch einen Einblick in Probleme, die bei der praktischen Umsetzung entstehen. So wurden etwa bei der Einführung von chipkartenbasierten Studentenausweisen besondere Erkenntnisse gewonnen, die in Erfahrungsberichten dargestellt sind.

Die zudem behandelten Themen, Java für Chipkarten, Chipkarten im Internet und Chipkarten im Trust Center, spiegeln wider, daß Chipkarten Multitalente mit multifunktionalen Eigenschaften sein können. Mit der zunehmenden Akzeptanz und Verbreitung gewinnen aber auch neue Aufgaben an Bedeutung. Chipkarten und Chipkartensysteme müssen verwaltet und betrieben werden, zudem muß in zahlreichen Anwendungen die Qualität der eingesetzten Chipkarten eine angemessene Sicherheit garantieren, wozu geeignete Sicherheitsüberprüfungen durchzuführen sind.

Die vorgestellten Beiträge spiegeln die Kompetenz der Autoren in eindrucksvoller Weise wider. Bedanken möchte ich mich bei allen Autoren, sowie bei Dagmar Cechak, Peter Schartner und Petra Wohlmacher für ihre Unterstützung bei der technischen Aufbereitung des Tagungsbandes und den umfangreichen Vorbereitungen. Mein Dank gilt weiter allen, die bei der Ausrichtung der Konferenz mitgeholfen und zum Erfolg beigetragen haben, den Mitgliedern des Programmkomitees, A.Beutelspacher, J.Borchert, A.Driessen, D.Fox, M.Horak, P.Kraaibeek, G.Meister, R.Posch, H.Reimer, P.Sonntag, B.Struif, J.Swoboda, M.Waidner, F.Weikmann, M.Welsch, T.Wille und P.Wohlmacher, und dem Organisationskomitee M.Horak, P.Kraaibeek, J.Swoboda und P.Wohlmacher, wobei die Mitglieder aus München unter Mithilfe von L.Heckmann die schwere Aufgabe der lokalen Organisation auf sich genommen haben.

Ich hoffe, daß die Arbeitskonferenz Chipkarten zu einem Forum regen Ideenaustausches wird und mit der Auswahl der Beiträge ein Einblick in die Lebendigkeit der Forschungs- und Entwicklungstätigkeit eines bedeutungsvollen, innovativen und sicherheitsrelevanten Gebiets der Informatik - der Chipkarten und Chipkartentechnologie - vermittelt werden kann.

Patrick Horster
pho@ifi.uni-klu.ac.at

Inhaltsverzeichnis

VIII

Smart Cards
Requirements, Properties and Applications

Klaus Vedder · Franz Weikmann

Giesecke & Devrient GmbH
Prinzregentenstr. 159, D-81677 München
{klaus.vedder,franz.weikmann}@gdm.de

Abstract

Smart cards play an increasing role as 'active' security devices. Due to its microcomputer and programmable memory, a smart card can cater for the specific needs of the environment it is used in. Smart cards allow the secure handling and storage of sensitive data such as user privileges and cryptographic keys as well as the execution of cryptographic algorithms. They are secure tokens by means of which a user can be identified and authenticate a computer system or communication network and vice versa. This paper provides a comprehensive introduction into the features of chip cards, the principals of their operating system, their life-cycle and the standards governing them. It also includes a brief discussion of major applications and an outlook on the future development.

1 Introduction

"Smart Cards. The ultimate personal computer." is the title of a book by J. Svigals [15] one of the first treatise of this subject. Since its publication in 1985 smart cards have gone a long way towards achieving this claim. What are smart cards? Smart cards are a specific type of chip cards. These are cards, usually made of plastic, containing a chip. Depending on the properties and features of this chip and its 'carrier' we distinguish the following types, which we will discuss briefly to achieve a common understanding of the terminology.

Memory cards contain non-volatile memory and allow 'free' reading and, in many instances, writing or updating of data stored. Writing usually refers to changing a 0-bit/byte to a 1-bit/byte (or vice versa), while updating is erasing the contents of the memory cells followed by writing. Such cards are used instead of magnetic stripe cards as they are more reliable and offer far more memory. Not all types of memory allow the erasure of data, a feature which is a basic requirement in many applications.

Intelligent memory cards contain a security logic in addition to the non-volatile memory. This allows the introduction of security attributes for reading and writing data. A memory zone may be secret (the data is used for card internal purposes only), public or sensitive. The latter means that it is accessible only after the presentation of a correct "personal feature" of the user. This is in most cases a Personal Identification Number (PIN) consisting of 4 to 8 digits. The PIN is protected against trial and error attacks by a 'false-presentation-counter'. After a specified number of consecutive false entries the security logic blocks the non-public data against any further access. The new generation of telecommunication chips (for pre-paid telephone applications) also include hardware algorithms for challenge-response mechanisms for the authentication of the card by the system to increase the security against cloning.

Smart cards are chip cards where the chip is a microcomputer with programmable memory.

Super smart cards are smart cards with an integrated key board, a display and solar cells or a battery. Due to its complexity and high price, this type of card has not advanced much beyond field trials.

A *contactless card* can be any of the above. Its name is derived from the way the data is transmitted between the chip and the InterFace Device (IFD) or the Card Accepting Device (CAD); the standardised names for chip card readers. In the last three years the standardisation of such cards has made enormous progress (see chapter 6 below).

Hybrid cards usually refer to chip cards which have two or more interfaces. Such an interface can be a magnetic stripe, contacts, contactless, or an optical memory. Most present-day chip cards use contacts for the transmission of power and data.

2 Interfaces and Dimensions

In this chapter we discuss the interfaces and the dimensions of chip cards or integrated circuit(s) cards (IC cards) with contacts. The dimensions and locations of their contacts are laid down in Part 2 of the International Standard ISO/IEC 7816. This standard [11], which is jointly published by the International Organization for Standardization (ISO) and the International Electrotechnical Commission (IEC), is the basic reference for such cards.

In addition to the eight contacts, the card can be equipped with a magnetic stripe or be embossed. Embossing and magnetic stripe shall be on opposite sides. The reader is referred to references [9], [10] and [11-2] for details of the specifications of these two features. While the current version of ISO/IEC 7816-2 allows contacts and magnetic stripe to be on the same side, this possibility has been excluded in the revision of this standard which is expected to be published later this year. Figure 1 below thus shows the only possibility for a card which provides all three interfaces. Such cards are typically cobranding cards comprising, for instance, a credit card and a telecommunication function.

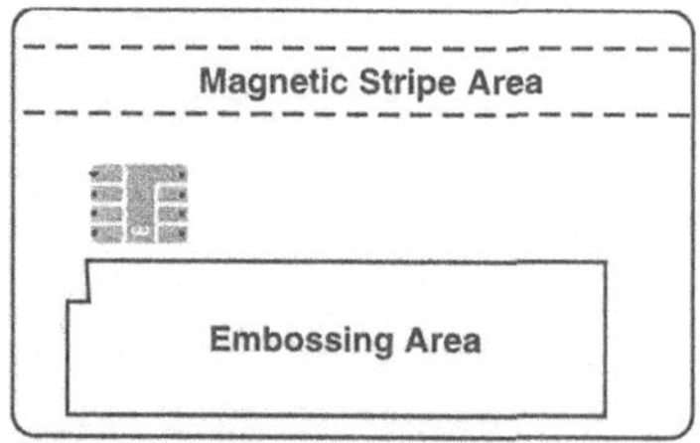

Fig. 1. Interfaces

2.1 Dimensions

The 'standard' identification card or ID-1 card [8] is the size of a credit card. Figure 2 gives the approximate dimensions of this card and the location of contacts as well as height and width of the so called Plug-in card. This ID-000 card with a height of 15 mm and a width of 25 mm has first been specified by the European Telecommunications Standards Institute (ETSI) for GSM, the Global System for Mobile communications, where about half the number of cards have this format [5]. They are mainly used in mobile phones which are too small to accept an ID-1 card.

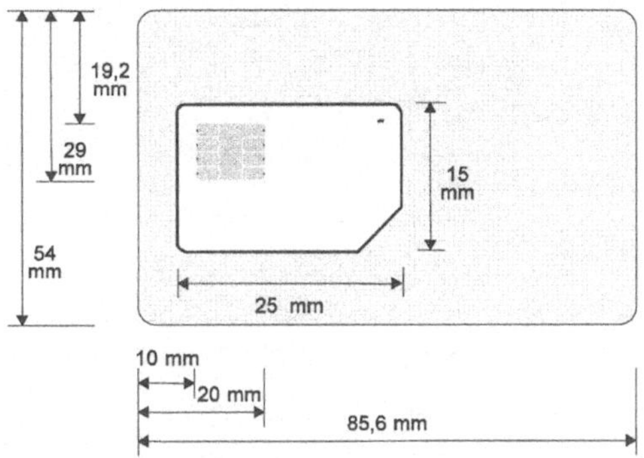

Fig. 2. Dimensions

A third type, the so-called "Mini Card" or ID-00 card is 66 mm in width and 33 mm in height [1]. It has the same location of contacts with respect to the upper and the left edge as the ID-1 card. This allows the same card reader, if so designed, to accept either card. Both new formats can be thought of as obtained from an ID-1 card by cutting away excessive plastic.

2.2 Electrical Interface

Figure 3 shows a layout for a contact area providing all 8 contacts C1 to C8 specified in [11-2] (the logo of the card manufacturer is etched into the surface). The contacts C4 and C8 are reserved for future use by ISO/IEC and are often not provided.

The supply voltage Vcc is specified as 5 V ± 10% in [5]. GSM also specified a 3 Volt interface to improve battery life and thus stand-by and operation times of mobile phones [6]. Both properties have been incorporated in the revision of ISO/IEC 7816-3 [11-3]. Development within GSM has already started on mobile phones using an even lower voltage than 3 V. This will also be reflected on a new standard for GSM cards operating at 1.8 V and lower. ISO/IEC has also approved a new work item standardising low voltage interface on a general level. One of the major problems facing such specification is the task of achieving backwards compatibility between "new" and "old" phones and "old" and "new" cards. Dual voltage cards supporting 1.8 V and 3 V will most certainly not work at 5 V and may, if no precautions are taken, even be destroyed. It is expected that chips which require only 1.8 V at the interface will be available around the turn of the millennium.

Contact C2 is used for the reset signal of the card. The baudrate for the answer to reset (ATR) is equal to the frequency supplied on contact C3 divided by 372. The divisor of 372 is due to a quartz commonly used to drive the chip and which supplies about 3.57 MHz resulting in a baudrate of 9600 bits/sec. During the ATR the baudrate can (within certain limits) be increased or decreased for the subsequent communication. It is also possible to change other transmission parameters or to select a protocol with different features to the default protocol known as T=0 (see paragraph 3.2). During reset the chip shall support 1-5 MHz.

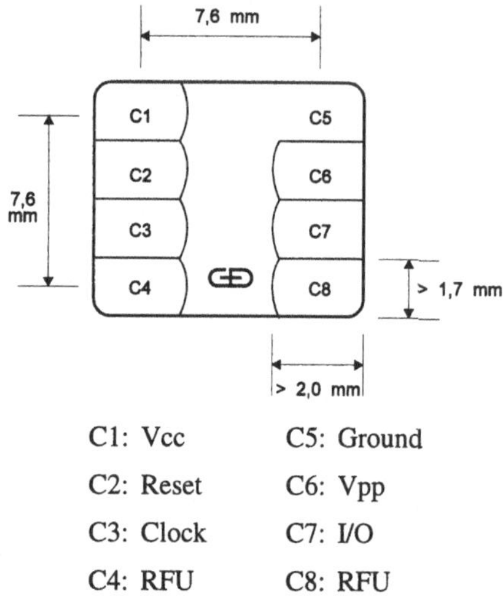

C1: Vcc   C5: Ground

C2: Reset   C6: Vpp

C3: Clock   C7: I/O

C4: RFU   C8: RFU

Fig. 3. Contact layout and assignment of the contacts

Contact C6 is used to supply the programming voltage Vpp for the non-volatile memory. An external programming voltage is needed if the chip has no internal charge pump to derive it from the supply voltage. The change from EPROM to EEPROM technology (see below) during the last years is the reason that contact C6 is no longer used.

Contact C7 is the only port for the transmission of data.

3 Performance

The performance of a smart card depends to a great extent on the chip and the protocol run between the card and the interface device.

3.1 The microcomputer

Bearing in mind, the thickness of the card (0.76 ± 0.08 mm [8]) and the torsion and bending it will be subjected to during its life, only special purpose chips can be used. Today's microcomputers are single chip solutions which means that all the parts shown in figure 4 are integrated onto a single piece of silicon. This has some obvious advantages from the points of embedding, reliability and security. Connecting the chip to the contact area (the bonding of the chip) and embedding this module (or stamp) into the plastic carrier is a specialised profession.

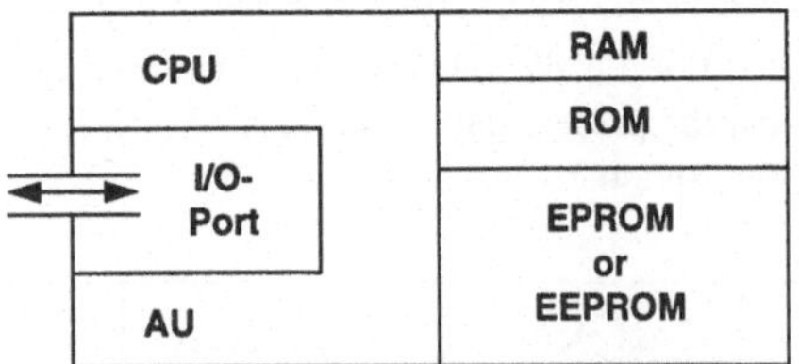

Fig. 4. A smart card microcomputer

The Central Processing Unit (CPU) comprises an 8 bit controller and is in most instances a variant of a 6805 (e.g. Motorola, SGS-Thomson), a 8051 processor (e.g. Philips, Siemens) or a manufacturer specific CPU (e.g. Hitachi). As memory is sparse, programming is usually done in assembler taking into account the specific properties of the CPU, its instruction set and the available memory. As an example let us consider the Data Encryption Standard (DES) [2] which is often used for message or entity authentication (i.e. the authentication of the card to the system or vice versa). This algorithm requires approximately 800 bytes of memory, giving a processing time of about 10 msec at 5 MHz for a 64 bit block. In the last few years the programming language C is often used if the application does not require too much memory (the overhead of programming in C is estimated at 30 %). It should be noted that many card manufacturer, who in most cases also do the software development, have their own operating system(s).

The *Read Only Memory* (ROM) is mask programmed. The photographic mask containing the customer specific ROM code is one of several layers used in the production process of the chip (see [13] for details). This has some important consequences for the design and the quality control of the software. The ROM code should only contain programs and data which are not specific to a card but are the same for a large number of cards and which are constant during the life of the card. A change of the ROM code requires a new mask. It takes several months from the completion of the software to the first die becoming available.

The ROM code typically contains the operating system of the card, the transmission protocol(s) and commands, the security algorithms and the software for the application. The card specific (secret) keys needed for the execution of the security functions are contained in the non-volatile memory.

The *Random Access Memory* (RAM) is a volatile memory the contents of which are lost when the power supply is switched off. It is used by the CPU as a buffer for storing transmission data and as a very fast access memory for the intermediate results (workspace) produced during the execution of an algorithm. Reading or writing a byte takes a few microseconds, which is a magnitude of 1,000 times faster than writing a byte into the (E)EPROM.

The non-volatile programmable memory of present-day cards consists of *Electrically Erasable Programmable Read Only Memory* (EEPROM). This type of memory allows a minimum of 100.000 update (i.e. erase/write) cycles. They derive internally the required programming voltage of about 18 V from the supply voltage Vcc. First generation smart cards had *Erasable Programmable ROM* (EPROM) with an external source for the programming voltage Vpp. As the contents of the EPROM can only be erased by UV light, every cell can be programmed just once by the CPU. The use of EPROM cards was thus limited to applications which do not require a frequent update of the memory.

One important boundary condition is the surface area of the chip which should not exceed 25 square millimetres. This and the present day technology (0.6 to 1.2 μm HCMOS) explain the size of the memory available for smart card chips (table 1).

Memory	typical	maximum	factor chip area
ROM	8 - 16 kByte	32 kByte	1
EEPROM	2 - 8 kByte	16 kByte	4
RAM	128 - 256 Byte	512 - 680 Byte	16

Table 1. Memory size

The "factor" gives a rough estimate for the silicon area used by such a cell of memory compared with the area needed for a ROM cell. This explains why RAM is sparse.

New non-volatile technologies such as Flash EEPROM and Ferro Electrical RAM (FRAM or FeRAM) will have quite an impact on the performance of smart cards. Flash EEPROM uses less area than EEPROM but can only be updated a few thousand times. Flash memory provides a good alternative to ROM, especially for the rapid production of application prototypes or different versions of operating systems. FRAM is probably the more exciting one because it has a write cycle similar to the RAM (200 ns) and maintains stored data for the same duration as EEPROM (up to 10 years). The features of these three types of non-volatile memory are shown in table 2.

The non-volatile memory is organised in so-called pages which are physically group memory cells. A page usually consists of between 4 and 64 bytes which can programmed (write/erase) in "parallel". This substantially reduces the programming time. In most cases it is also possible to program a specific byte of a page.

Technology	Write/Erase Cycles	Write / Write+Erase Time
EEPROM	10^4-10^6	1.75ms/3.5ms
Flash EEPROM	10^3-10^4	1.5ms/3.5ms
FRAM	10^{10}	200ns

Table 2. Features of Memory Technologies

Added Units (AU) provide special functions which could otherwise not be handled at all or at least not within an acceptable time. A typical example is a coprocessor in the form of extra hardware for the execution of modular arithmetic. This is, for instance, needed for the execution of public key algorithms. Other examples of AUs are timers, Universal Asynchronous Receiver/Transmitter (UART), Memory Management Units (MMU) and hardware security logic.

3.2 The transmission

For the exchange of data between a smart card and an interface device two protocols have been standardised [11-3]. They are denoted by T=0 and T=1, respectively. Both of them are asynchronous, half duplex protocols. The main difference between the two lies in their handling of data and the OSI reference model.

The character transmission protocol T=0 [11-3] can be characterised as a (byte-oriented) protocol of the first generation when computing power and RAM of the chips were fairly limited. Checking the parity of a byte immediately after having received it and requesting its retransmission is the only error correction. As this can not be achieved by 'standard' hardware a special UART is needed in the IFD. There is no clear separation of the transport and application layer which, for instance, makes it impossible to encrypt the header of a command. Nor is it possible to transmit data in both the request and the response of one command. This causes some overhead, for instance, during any authentication process, since data has to be exchanged between the card and the interface device. A fair amount of overhead is also unavoidable if the data to be transmitted is larger than the buffer in the RAM. The data have to be sent byte after byte until the buffer can cope with all remaining bytes. These can then be requested by using a special acknowledgement for the last byte received prior to this.

The byte-oriented transmission protocol T=0 is, however, less complex than the block protocol T=1. The standardisation of the latter was finalised in 1992. T=1 respects the OSI (Open System Interconnection) reference model and data may be sent in both a request and the response. As its name suggests, data can be handled (and transmitted) in "blocks" and the error check is carried out on a block of data.

3.3 The throughput

Optimising an algorithm for a smart card is an act of balancing speed against memory. For the speed of the algorithm is only one of several factors which determine the performance of the card. Input, calculation and output are sequential operations. Each byte of user data is sandwiched by a start bit preceding it and by a parity bit and two stop bits following it. So the transmission of each user byte requires the transmission of 12 bits. This yields a throughput of at most 3200 user bits per second at a baudrate of 9.600 bits/sec and an infinitely fast algorithm. This does not take into account any overhead caused by the transmission protocol or the buffering of data.

As an example consider the DES, which acts on blocks of 8 bytes. Each block requires the transmission of 96 bits which takes 10 msec each way. Table 3 gives some (theoretical) values for the throughput of the card depending on the speed of the algorithm (given in both msec/DES block and bits/sec).

algorithm	40	20	10	5	msec/block
algorithm	1.600	3.200	6.400	12.800	bits/sec
smart card	1.066	1.600	2.132	2.560	bits/sec

Table 3. Throughput

One can see from table 3 that the speed of the algorithm has only a minor effect on the throughput once the transmission of a block takes about as much time as the calculation. Other means to increase the throughput are a higher baudrate (e.g. 115 kbit/sec) and the use of only one stop bit, which is possible in T=1.

4 Smart Card Operating System

Operating systems are used on different digital hardware platforms – from smart cards, pocket calculators to organisers, PCs and mainframes. Broadly speaking, the term operating system, for which there is no general definition, refers to a group of system programs required for operating a computer.

The operating system provides defined functions to the user, who in turn need not know anything about the computer hardware. The user can run and program applications almost completely independently of the hardware. The operating system takes care of controlling and organising memory media (RAM, hard disks, CDs, etc.) and of programming processors. It allows the hardware manufacturer to incorporate many technological developments without there being a need to change functions of the operating system.

What has been said about operating systems for "large-scale" computers also applies to operating systems for smart cards, the ultimate personal computer. A smart card operating system and its basic tasks can be characterised as follows:

- miniature operating systems requiring memory capacity of a few kBytes;

- administration of security hardware based on single-chip microcontrollers;

- single processing systems;

- machine-interfaces in the form of a serial interface;

- secure data storage;

- support of cryptographic methods for authenticating system components and enciphering/ deciphering data;

- provision of IT-security for applications.

The operating system and the smart card hardware have to protect themselves against the following three basic threats for the duration of their entire life cycle (see 4.4 below):

- loss of confidentiality by spoofing or unauthorised information about programs and data such as keys, PINs and user data;

- loss of integrity by manipulation or unauthorised modification of information such as identification number and counters of electronic purses;

- loss of availability by unauthorised withholding of information or system functions.

4.1 Software standardisation

The basis for all industry specifications and thus the most important smart card standard is "ISO/IEC 7816, Identification cards – Integrated circuit(s) cards with contacts" [11]. The following list describes the major topics of some parts of this International Standard which are relevant for smart card operating systems:

- ISO/IEC 7816-3: Transport protocols T=0 and T=1;

- ISO/IEC 7816-4: File organisation, commands, historical bytes and secure messaging;

- ISO/IEC 7816-5: Definition and registration of the Application IDentifier (AID);

- ISO/IEC 7816-6: Definition of data objects;

- ISO/IEC 7816-7 (Draft): Commands for Structured Card Query Language (SCQL) for database application;

- ISO/IEC 7816-8 (Draft): Security functions for Public key procedures and extended PIN functions;

- ISO/IEC 7816-9 (Draft): Personalisation commands and extended file handling commands;

- ISO/IEC 7816-11 (Draft): Access conditions and security attributes.

Even if smart card operating systems have been implemented in accordance with these International Standards, it does not mean that they are compatible with each other. This is to some extent due to the options provided for in the standards. Specifications are thus required to select particular items from the standard to implement compatible smart card operating systems (of different manufacturers) for one and the same application. One such specification is GSM 11.11. Furthermore, some applications as, for instance, electronic purses require functionality not defined in ISO/IEC 7816. They need, therefore, "private-use-commands". Today, approximately 60 % of the functions of a smart card operating system are private use. In 1994 the leading credit card organisations EUROPAY, MasterCard and VISA (EMV) began to draft a specification for smart cards in international payment systems based on ISO/IEC 7816. In June 1996, version 3 of the EMV specifications has been published [3]. This allows the implementation of the basic command set for compatible credit card systems based on smart cards.

4.2 Structure of smart card operating systems

The flow of information between an interface device and a smart card occurs via transport protocols in the form of command-response pairs. In most cases the interface device or the application (terminal, PC, etc.) has the role of the master, i.e. the commands will be generated and processed by the IFD. In those cases, where the smart card represents the application (e.g. in the case of security modules or in specific applications within the GSM SIM Application Toolkit [7]), the roles may be reversed (see figure 5).

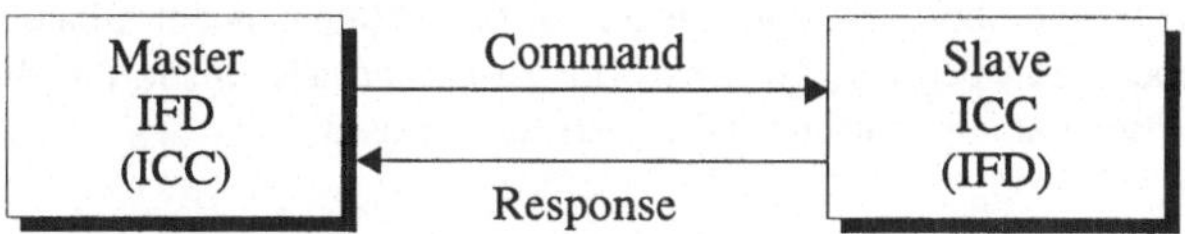

Fig. 5. Information transaction

In accordance with the OSI 7-layer model the information transaction can be divided into three protocol sections:

- physical layer (layer 1);

- data transmission protocols (layer 2);

- information protocols, for command and response data (layer 7).

Smart card operating systems are divided into modular functional units called managers – just like computer operating systems. According to reference [18] an information transaction may run through the following managers (see figure 6):

- The *Transport Manager* controls and secures the data transmission using asynchronous, half duplex transport protocols. These can be the already mentioned protocols (T=0, T=1) or a national protocol (all national protocols no matter how different they may be are denoted by T=14).

- The *Secure Messaging Manager* covers the cryptographic protection of the transmission by enciphering or deciphering information and/or by checking information for authenticity.

- The *Logical Channel Manager* is required for accessing an application if several applications are "open" simultaneously. This semi-multitasking is required, for example, in the case of a bank application transferring money to an electronic purse.

- The *Command Manager* verifies command syntax. In some operating systems it also controls the command processing protocol by using the state machine(s) of an application.

- The *Security Manager* is in charge of object access control, in particular for keys. It controls, for instance, the states of the state machine which are depending on the successful or unsuccessful execution of the identification of the user and of the authentication mechanisms.

- The *File Manager* administers the different file categories and supports the different file types.

- The *Memory Manager* administers the entire memory organisation, e.g. installation, applications and files. It calculates checksums, repairs defective file structures and takes care of the administration of the available memory.

- *Functions*. Among these are, for example, mathematical operations for specific cryptographic functions. In some chips they are supported by hardware. These functions may be part of the whole system or just part of one or more applications.

The smart card operating systems of the STARCOS® family (**Smart Card Chip Operating System**) [14] incorporate these managers as part of the system concept. Since these universal, application-independent operating systems offer structured administration they are suitable for administering several applications and issuers on one card.

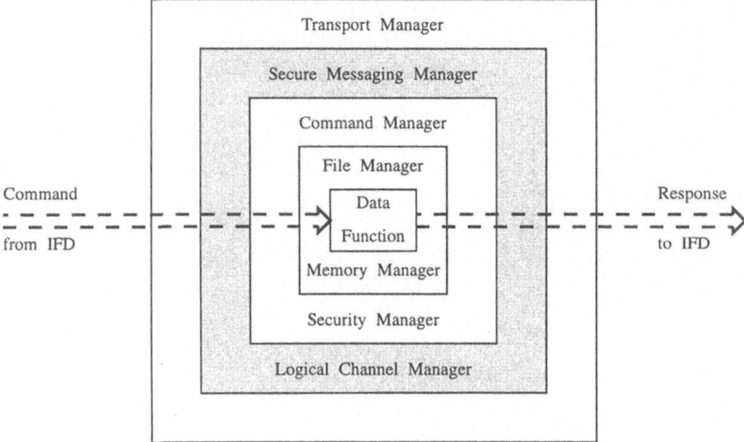

Fig. 6. Structure of a smart card operating system

4.2.1 File organisation

A smart card operating system administers files in directories – similar to conventional PC operating systems. One distinguishes the following components within the file organisation which one can visualise as having a "tree-structure" (figure 7):

- Master File (MF): the root of the file system (tree);

- Dedicated Files (DF): a DF contains substructures and the actual application (DFs are also called "directories");

- Elementary Files (EF): an EF contains application information or data.

The MF is usually selected implicitly after the card has been reset. It may also contain an application, for instance on a mono-application card. In the case of multiapplication cards each application is contained in a separate dedicated file or in further subdirectories. An application is selected by using an application identifier (AID), which can be registered on both international and national level with ISO/IEC [11-5], or by a (fixed) file identifier consisting of two bytes. Application identifiers have the advantage that the interface device does not have to know the file identifier for the application and that, as a registered AID (a so-called RID) is unique world-wide, several applications can easily be combined on a card. In the case of (fixed) file identifiers a collision can arise if two applications which use the same file identifier are to be combined on one card. Data which is used for all applications in the card (for example administrative and general security information such as serial number, keys, PIN) and data concerning the administration of the card life cycle are stored in the master file. This information is, for instance, used by the operating system for the creation of a new application. Application specific control information and files are stored in a DF containing the application. They are separated logically and, in some cases, physically from other applications contained in different DFs. The administration of rights for DFs arranged in a "vertical" structure is always performed by the "parent" DF i.e., the DF which resides "one level up". A dedicated file can administer several dedicated files which are "one level below". An example would be a financial card where the same service provider is responsible for a payment transaction application and credit, debit and electronic purse functions.

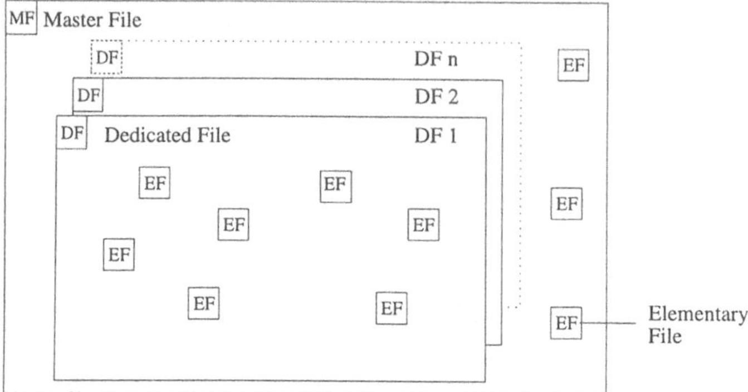

Fig. 7. File organisation

ISO/IEC 7816-4 [11-4] distinguishes two categories of Elementary Files (EFs). While the data contained in a Working Elementary File (WEF) cannot be interpreted by the operating system, the data stored in a Secret File (ISF) must be interpreted by the operating system. A DF may contain more than one WEF but only one ISF. Typical information stored in an ISF are keys which are to be protected from read by the outside world. Smart card operating systems such as STARCOS® may, in addition, support files which are of "mixed" type, i.e. some of their data can be interpreted while the remaining data cannot be interpreted by the operating system.

4.2.2 File structures

The structure of an EF depends on its use. The following four types are defined in ISO/IEC 7816-4 [11-4]:

- *transparent files* consisting of a sequence of bytes having an amorphous structure;
- *linear fixed files* consisting of records having the same length;
- *linear variable files* consisting of records of variable length;
- *cyclic files* having records of fixed length which are organised in a ring structure, the oldest entry will be overwritten by the entry to be stored.

The four basic structures can be extended by other file structures. Examples are database applications (database files), electronic purses (compute files) or public key applications (internal public key files).

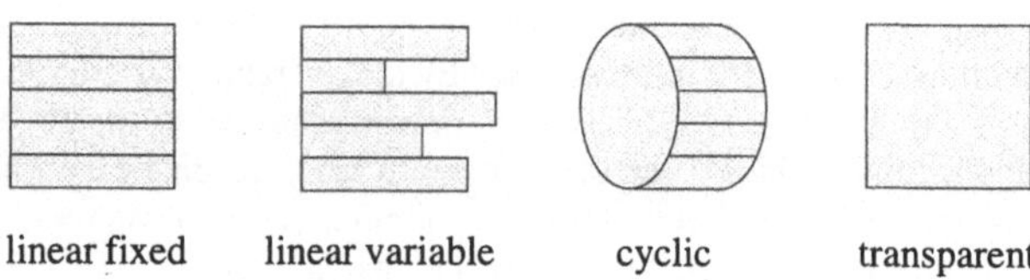

Fig. 8. Data structures of elementary files

The definition of the category, the type and the access possibilities (read, write, update, delete,...) with their respective access conditions (never, always, PIN, special authentication,...) are stored in the file header of each file.

4.2.3 Commands

The functionality of an operating system is not only reflected in the number of available commands but also in their complexity. This grows with the need for more security for an application. The commands defined in ISO/IEC 7816-4 [11-4] are the *basic* command set and can be grouped in the following functional classes:

- file selection;
- read data;
- modify and delete data;
- generate data;
- compare data;
- authenticate using cryptographic functions.

Due to the complexity of today's applications approximately 60% of the commands of an operating system are "private-use-commands" and not defined in [11-4]. Examples are commands for mutual authentication and cryptographically secured counter functions representing chained basic functions, so-called macros.

4.3 Security

Operating systems quite often administer applications with high security requirements. The security of a smart card is a combination of the security of the chip (hardware) and the operating system (software). To obtain optimum security the programming of an operating system requires extensive knowledge of the properties of the hardware in particular with respect to electrical characteristics, detectors, interrupts, timing and other features which may influence the security. From a software point of view one can distinguish between functional security and security against manipulations.

Functional security can be guaranteed by:

- transport protocol;
- command interpreter;
- file organisation, file structure, data objects;
- functions;
- layer separation;
- error detection and correction functions.

Security against manipulation can be obtained by implementing:

- secure messaging;
- identification and authentication;
- state machines;
- object protection;

- digital signature;
- proof retention;
- random numbers generators.

4.4 Card life cycle

The operating system contributes considerably to the guaranteed card life cycle of a smart card. The card life cycle – from production to deactivation of the card (see figure 9) – is divided into the following five phases according to ISO 10202-1 [12]:

- Phase 1 – chip and card manufacturing
  - development of the operating system and transport to chip manufacturer;
  - implementation of the operating system, usually as a ROM mask;
  - chip production and transport to card manufacturer.
- Phase 2 – card preparation
  - initialisation and prepersonalisation of the card, i.e. loading constants and system-related data;
  - dispatching the cards to the issuers.
- Phase 3 – application preparation
  - assignment, personalisation and activation of one or several applications.
- Phase 4 – application phase
  - using global card functions and application access;
  - using management functions (administration) of applications (lock, release).
- Phase 5 – termination of use
  - delete keys and the complete application.

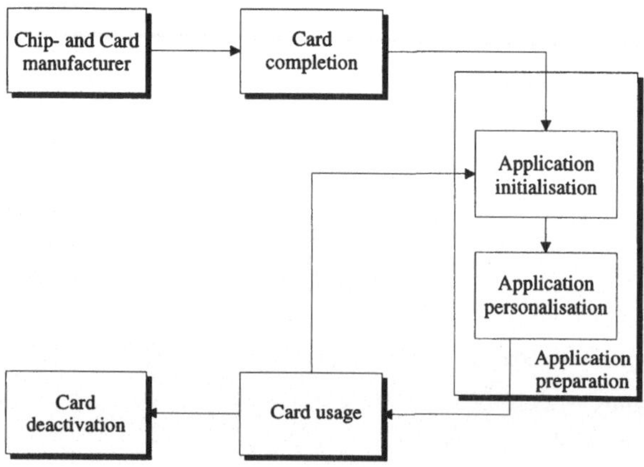

Fig. 9. Card life cycle

5 Applications

The main security functions of a smart card are the identification of the user, the authentication of the card by the system and, depending on the application, the authentication of the system by the card. This involves the secure storing of secret information as well as the secure execution of cryptographic algorithms.

The identification of a user is usually done by means of a Personal Identification Number (PIN). The PIN is verified by the microcomputer of the card which compares in its RAM the PIN presented with the PIN stored. If the comparison is negative, the CPU will refuse to work. The chip also keeps track of the number of consecutive wrong PIN entries. If this number reaches a pre-set (card specific) threshold (usually three), the card blocks itself against any further use. The change of the PIN by the user, which is a standard feature of smart cards, may be subjected to the exclusion of trivial values such as "0000", "123456", "4711", "0815" or the date of birth of the user (if known) which could be stored in the card.

At the beginning of any session is authentication, the "corroboration that an entity is the one claimed" (ISO/IEC 9798-1). Entity authentication is achieved by using a "challenge-response method" employing a cryptographic algorithm such as the DES or a public key function. To authenticate the card the system challenges the card by sending a random number. The card uses this and its own card specific key as input to its cryptographic algorithm. The output of the calculation, the response, is transmitted to the system. This compares the value received with the one calculated itself. If the two match the card is considered to be genuine. Analogously the card can check the authenticity of the system.

5.1 GSM

The Global System for Mobile communications (GSM) is a digital, cellular radio system with more than 170 networks on air in about 100 countries. Access to all these networks is controlled by smart cards, the so-called Subscriber Identity Modules (SIMs), in form of ID-1 or Plug-in cards. GSM is the first world-wide application based on smart cards and has given a huge impetus to their standardisation and use.

Subject only to their size and roaming agreements between the network operators all SIMs work in all mobile phones in all networks. The general properties of the interface between the SIM and the mobile phone are laid down in GSM 11.11 [5] while GSM 11.12 [6] was the first specification for a 3V card interface. The latest specification is GSM 11.14 [7] the so-called SIM Application Toolkit. This provides the operators with a toolbox to create their own applications on the card. Information containing commands and data may be downloaded over the air into the SIM, transparently to the mobile phone. After interpreting the information, the SIM will react as requested by, for instance, updating a file or creating a new file or even a new application. The SIM may also become pro-active and request the mobile phone to act on its behalf. For more information on GSM and SIMs the reader is referred to [16].

5.2 Company ID-cards

Using a smart card as a company ID-card and, at the same time, for access control to the company computer network can solve quite a few security and handling problems arising from passwords and access control in general.

The communication between the card and the system can be protected by message authentication codes to secure the update of sensitive data in the card. The combination of a smart (company) card with PIN has several advantages over Password and User-ID. Access to the network requires the possession of the card *and* the knowledge of the PIN. Knowing just the PIN is not sufficient. If the holder "lends" the card to someone else this will be temporary. Without the card the holder may not be able to leave or enter the premises or parts thereof. Passwords can be passed on indiscriminately; company cards will not. A proper log-off can be enforced by making the entry to the (computer) room depend on the card and by checking the presence of the card, creating security entries if the card is forcefully removed.

To enhance the security, visual information can be engraved into the card body for example by means of a laser beam. This information could consist of the usual data such as date of issue and card number as well as typical data of an ID-card such as the photograph and the signature of the card holder.

5.3 Banking Cards

The first nation-wide smart card application was a banking card in France. The field-test using memory chips took place in Lyon in 1983. Shortly afterwards microcontrollers from Motorola with EPROM as non-volatile memory were introduced at national level. The transition from magnetic stripe to microcontroller with on-board memory had the following advantages from a security point of view:

- administration, storage and validity check of the PIN by and in the chip;

- authentication and enciphering of data using a cryptographic algorithm.

Smart cards supporting electronic purse applications were introduced on a national level in Austria in 1995 and in Germany in 1996. All these cards, which were issued as a replacement for the eurocheque cards with magnetic stripe, were hybrid cards with both a magnetic stripe and a microcomputer. Apart from the usual Point of Sale (POS) functions of the magnetic stripe the chip supports an electronic purse. Since the credit limit is also administered by the chip POS transactions can now be handled off-line. As in most electronic purse systems (e.g. Proton, VISA Cash, STARCOIN) the loading of the purse is done on-line while purchasing can be done off-line. With the availability of on-board hardware units for special arithmetic the symmetric cryptoalgorithms such as DES are expected to be replaced with asymmetric (public key) algorithms in the coming few years. This will significantly improve both the key management and the cryptographic protection.

5.4 Other Applications

Not only banking cards will benefit from the introduction of digital signatures and authentication mechanisms using public key based cryptosystems. The recent development of such systems and the improvement of the added hardware units now allow the computation of a digital signature employing a key of 512 to 1024 bits in much less than 1 second. This is of particular importance for the security of information and transactions for applications in world-wide network.

Contactless cards are supposed to be ideal for applications where the cards of a large number of people may have to be handled in a very short time. A typical example is a "city card" with an electronic purse for the combined use of (public) transport and public amenities such as libraries, swimming pools and museums. An example of a hybrid card in both meanings of this term is the Lufthansa chip card. Its magnetic stripe (and embossing) serve as a credit card, its contactless microcontroller chip is used for electronic ticketing and boarding, and the memory chip with contacts as a subscriber card for the public German telephone system.

Another large application are the so-called health cards which may contain a database of the medical history, allergies and other relevant information such as address and insurance number of the card holder. Access and update rights depend very much on the data and may be divided between the patient and the respective medical doctors. Access is defined on the level of data objects and not on files. A surgeon, for instance, may thus only access data relevant for the diagnosis and treatment of cancer.

6 Standardisation

Though chip cards are quite standardised objects this does not mean that there is nothing left to do. Conformance testing and common test methods are one subject. A standard, solely dedicated to test methods, is ISO/IEC 10373 (1993). This is one of the many topics within the scope of SC 17, a subcommittee of the Joint Technical Committee 1 (Information technology) of ISO and IEC. SC 17 is also responsible for the International Standard ISO/IEC 7816 and other basic International Standards for identification cards. One of its Working Groups (WG) develops a multipart standard for contactless cards ISO/IEC 10536. The first three parts of this standard correspond to the respective parts of ISO/IEC 7816. The work on proximity contactless cards (ISO/IEC 14443) has just started.

Another international standardisation body dealing with smart cards is ISO/TC 68 "Banking and related financial services". It edits two multipart standards: *Messages between the integrated circuit card and the card accepting device* (ISO 9992) and *Security architectures of financial transaction systems using integrated circuit cards* (ISO 10202).

Within Europe chip cards are standardised by CEN, the Comité Européen de Normalisation, and by ETSI, the European Telecommunications Standards Institute. The Technical Committee TC 224 of CEN has 13 working groups which specify items from physical characteristics (WG 1) to airline applications (WG 14). Its WG 9 was closely related to the former ETSI SubTechnical Committee STC TE9 which produced a standard on application independent card requirements for telecommunications [4]. This specified for the first time commands which went beyond the basic erase, read, write and update functions. This is now to a large extent superseded by part 4 and the draft versions of parts 7 and 8 of ISO/IEC 7816. WG 10 of TC 224 develops standards for electronic purse systems. ETSI produced several smart card specifications for specific applications. The most important ones are probably those written for GSM. Others are the DECT Authentication Module (DAM) specification for the Digital Enhanced Cordless Telecommunications and its test specification. ETSI has now formed a new Technical Committee to write a generic core specification for telecommunication cards.

More information on the committees mentioned can be obtained from the national standards institute of the relevant country.

Apart from the standards produced and developed by regional and international standards organisations there exist quite a few "industry standards". The most important one of these is the so-called EMV-specification for payment systems which has been written jointly by

EUROPAY, MasterCard and VISA [3]. The three parts define the IC Cards, the card terminal and the card application.

7 Outlook

The advancement of the semiconductor technology in the last years had a dramatic effect on the memory provided by smart card chips. While 5 years ago chips providing 8 kByte of ROM, 128 Byte of RAM and 3 kByte of EEPROM were the flagships of all manufacturers, it can be assumed that in a few years time semiconductor technology will have a characteristic distance below 0.6 μm and high end microcomputers will have in excess of 64 kByte of ROM and 32 kByte of EEPROM. The values given in table 1 as maximum will be part of the typical range within the next year. This and the progress of the tools will have a strong influence on the development of operating systems and applications. It will become common to write most of the ROM code not only in C but also in the form of modular blocks. This will immensely improve the portability of the software from one microprocessor to another, even if they have "incompatible" CPUs, as well as the design of true multiapplication operating systems, which require in excess of 10 kByte of ROM without the coding of the application.

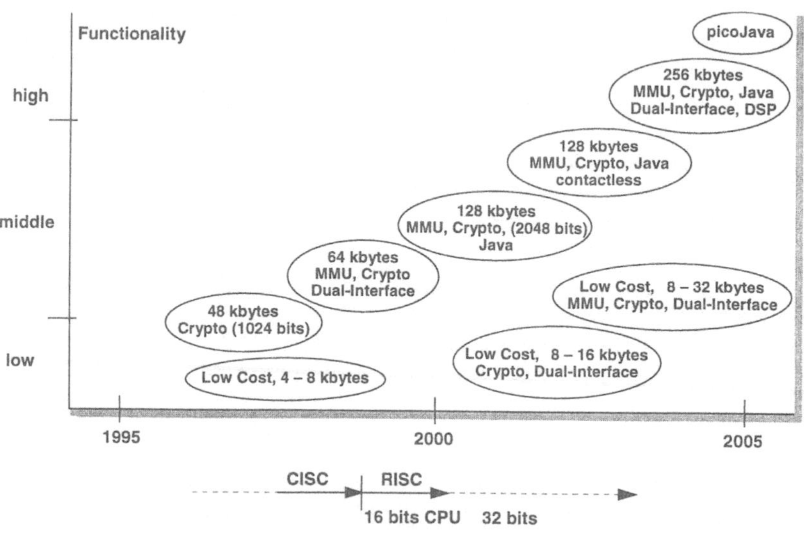

Fig. 10. Trends of smart card microcontroller

As the applications (to be) supported by a smart card become more and more complex, today's CPUs with their CISC (Complex Instruction Set Computer) architecture and their 8 bit controllers, which are by comparison with other areas quite outdated, need to be improved. Several chip manufacturers have started extending the cores of their microcomputers by introducing additional instructions and ways of addressing memory as well as higher clock rates (e.g. 10 MHz). This upgrading has the advantage that existing software is supported and can easily be ported. There is, however, a limit to the improvement of the performance. This can only be achieved on a major scale by replacing the core by a RISC-CPU (Reduced Instruction Set Computer). This architecture will, for instance, allow internal frequencies of 35 MHz and higher. It will also further the introduction of high level languages. Several manufacturers have already announced their intention to introduce this architecture which has been widely used in other areas for several years (figure 10).

Larger microcomputers with an improved performance will increase the potential for multiapplication cards and the requirement to download applications after the card has been issued to the user. When loading a new application into a card several issues have to be considered very closely. Is the new application independent of the hardware (CPU and memory) of the chip and the implementation of the operating system? How does it interact with the resident application(s)? Can security problems definitely be excluded? For these reasons the loading of another application with *executable code* needs to be evaluated by the software engineers who are responsible for the operating system and the resident applications. To ease this process and to achieve the goal of downloading applications into issued cards Memory Management Units (MMU) and Interpreter Languages will be used. Memory Management Units support the operating system in controlling the memory (firewall function) and allow the use of the CPU by the downloaded application with executable code. Existing interpreter solutions are mostly proprietary and slow down the execution of applications and programs.

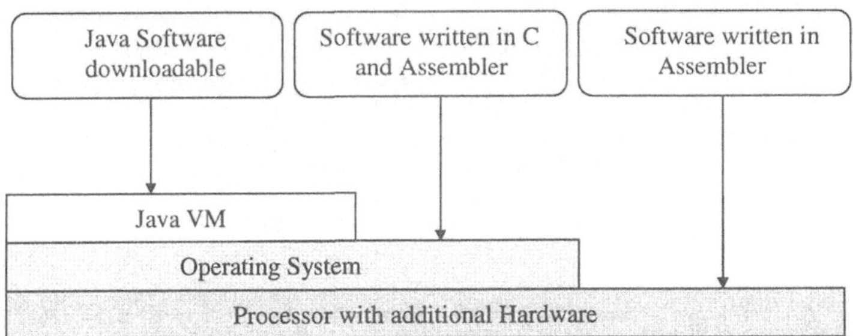

Fig. 11. Software designer's view

In 1995 SUN Microsystems presented Java, an interpreter language for a large variety of microprocessor platforms. It was originally intended for the linkage of set-top-boxes, copiers and other electronic consumer goods with a microprocessor. Since Java was provided free of charge on the Internet and was running on such a variety of hardware platforms it immediately became the language of the Internet. Java has completely realised the concept of object orientation and provides security mechanisms which make it suitable also for smart cards. The compiler produces a byte code which is interpreted by a virtual machine, the so-called Java Virtual Machine (JVM). The functionality of this machine is independent of the processor and can, by means of the Java API (Application Program Interface), access the operating system of the smart card. It can thus also make use of hardware dependent functions such as some security, utility and I/O routines (see figure 11). The standardisation of the Java byte code, the JVM and the JC (Java Card) API for smart card is progressed with great determination by SUN and several smart card manufacturers. As with all interpreter solutions Java has, however, an adverse effect on the performance of the microprocessor.

Several chip manufacturers have announced the design of new hardware to support the JVM. These products, which will greatly improve the performance, are expected to be available within the next couple of years [17]. Complete solutions, i.e. the JVM in hardware (pico Java) for smart cards, will probably not be available in this millennium.

All present Java implementations on smart cards make use of a two-part Java virtual machine (JVM) consisting of an Off-Card JVM in a PC and an On-Card JVM on the card itself. The Off-card JVM does the pre-processing of the "JAVA byte-code" and, in particular, the class files. This optimisation is needed for the processing of the byte code by the CPU of the smart card, since current smart card chips are not powerful enough for the efficient execution of the standard JAVA byte code. For this reason a standardisation of the downloading interface between the Off-card JVM and the On-card JVM is important to avoid the development of proprietary solutions.

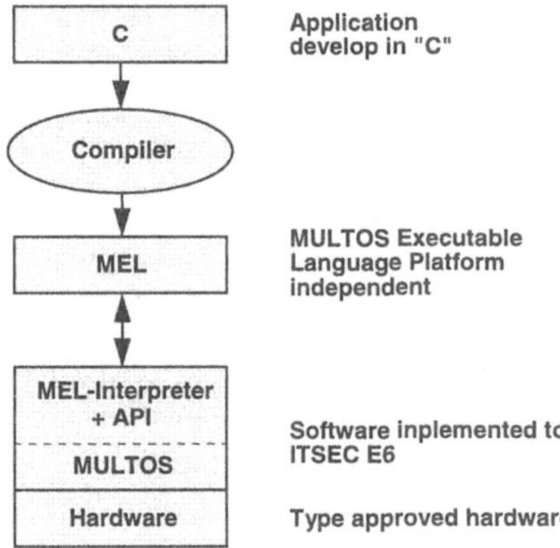

Fig. 12. Application design with MULTOS

The other new interpreter operating system is MULTOS which is supported by MasterCard and MONDEX. Similarly to Java, the core of the operating system is an interpreter which allows the application to be developed independently of the underlying hardware. Applications can be written in the high level language C and are then translated with the help of a C-compiler into the interpreter language MEL (MULTOS Executable Language). MEL is thus a specified interface for downloading software. The MEL code is interpreted by the smart card interpreter using the API (see figure 12). The security of MULTOS is expected to satisfy ITSEC E6, the highest level of the ITSEC certification. The first application to run on MULTOS is probably the electronic purse application from MONDEX. Similarly to Java, applications requiring the execution of complex or time critical (security) functions will need a special API to provide an acceptable execution time for the application. As is the case with all interpreter languages, the widespread use of this language will depend on the development of hardware specific for MULTOS. This new hardware could be an extension of the CPU with respect to the virtual machine or an additional processor dedicated to the specific interpreter.

Smart cards are also security devices and will take over more and more security related functions of the systems they are used in. The introduction of technologies of 0.6 µm and below will not only allow the production of chips providing more memory, it will also provide an hitherto unknown protection against potential future attacks. Extra hardware for public key cryptography (be it RSA with a key length of 2084 bit or elliptic curve systems) will provide the means for secure (off-line) authentication and digital signatures. Digital Signal Processors (DSP) on-board a smart card microprocessor will pave the way for the usage of biometric systems for user authentication and may make the PIN obsolete.

The smart card of the future will be a PC in pocket size with sensors for biometric features and a human interface.

Glossary

ATR	Answer To Reset
CEN	Comité Européen de Normalisation
CPU	Central Processing Unit
DES	Data Encryption Standard
DF	Dedicated File
EF	Elementary File
ETSI	European Telecommunications Standards Institute
EEPROM	Electrically Erasable Programmable ROM
EPROM	Electrically Programmable ROM
FRAM	Ferro electrical RAM
GSM	Global System for Mobile communications
IC	Integrated Circuit
ICC	Integrated Circuit(s) Card
IEC	International Electrotechnical Commission
IFD	InterFace Device
ISF	Internal Secret File
ISO	International Oganization for Standardization
JVM	Java Virtual Machine
MF	Master File
OSI	Open System Interconnect(ion)
PIN	Personal Identification Number
POS	Point of Sale
RAM	Random Access Memory
ROM	Read Only Memory
UART	Universal Asynchronous Receiver/Transmitter
WEF	Working Elementary File

References

[1]   CEN Draft ENV 1375-2: Identification card systems - Intersector integrated circuit(s) card additional formats - Part - 2: ID-00 Card size and physical characteristics.

[2]   Data Encryption Standard (DES). Federal Information Processing Standards Publication 46, National Bureau of Standards 1977.

[3]   EMV '96, Specification for Payment Systems, EUROPAY, MasterCard and VISA, version 3.0, June 1996.

[4]   EN 726-3:1994, Identification card systems - Telecommunication(s) integrated circuit(s) cards and terminals - Part 3: Application independent card requirements.

[5]   GSM 11.11 (ETS 300 608), Digital cellular telecommunications system (Phase 2); Specification of the Subscriber Identity Module - Mobile Equipment (SIM-ME) interface.
GSM 11.11 (ETS 300 977), Digital cellular telecommunications system (Phase 2+); Specification of the Subscriber Identity Module - Mobile Equipment (SIM - ME) interface.

[6]   GSM 11.12 (ETS 300 641), Digital cellular telecommunications system (Phase 2); Specification of the 3 Volt Subscriber Identity Module - Mobile Equipment (SIM - ME) interface.

[7]   GSM 11.14, Digital cellular telecommunications system (Phase 2+); Specification of the SIM Application Toolkit for the Subscriber Identity Module - Mobile Equipment (SIM - ME) interface.

[8]   ISO/IEC 7810 (2nd edition): 1995, Information technology - Identification cards - Physical characteristics.

[9]   ISO/IEC 7811 (2nd edition): 1995, Information technology - Identification cards - Recording technique.

[10]   ISO/IEC 7813 (3rd edition): 1990, Information technology - Identification cards - Financial transaction cards.

[11]   ISO/IEC 7816, Information technology - Identification cards - Integrated circuit(s) cards with contacts.
Part 1: 1987, Physical characteristics. (Under review).
Part 2: 1988, Dimensions and location of the contacts. (Under review).
Part 3: 1989, Electronic signals and transmission protocols. (Under review).
Part 4: 1995, Interindustry commands for interchange.
Part 5: 1994, Numbering system and registration procedure for application identifiers.
Part 6: 1994, Interindustry data elements.
Part 7 (Draft): Interindustry commands for Structured Card Query Language (SCQL).
Part 8 (Draft): Security related interindustry commands.
Part 9 (Draft): Enhanced interindustry commands.
Part 11 (Draft): Security architecture.

[12]   ISO 10202-1: 1991: Banking, securities and other financial services - Financial transaction cards: Security architecture of financial transaction systems using integrated circuit cards - Part 1: Card life cycle.

[13]   M. Paterson, Secure Single Chip Microcomputer Manufacture, in: D. Chaum (ed.), Smart Card 2000, North Holland 1991, 29-37.

[14]   STARCOS S2.1, Reference Manual, 10/96, Giesecke & Devrient GmbH, Munich.

[15]   J. Svigals, Smart Cards. The ultimate personal computer. Mac Millan Publ. 1985.

[16]   K. Vedder, GSM: Security, Services and the SIM, in: Springer Lecture Notes in Computer Science 1997.

[17]   P. Wayner, Sun Gambles on Java Chips, in: Byte Nov. 1996, 79-88.

[18]   F. Weikmann, Chipkarten - Entwicklungsstand und weitere Perspektiven, in: PIK 1/93, 28-34.

Neue Wege für Smartcard-Anwendungen

Stephan Ondrusch

Siemens Semiconductors
Application Engineering
steve.ondrusch@hl.siemens.de

Zusammenfassung

Das seit Jahren in der Chipkartenbranche vorherrschende Schlagwort von der Multiapplikationskarte beschreibt bis zum heutigen Tag ein Phantasieprodukt. In einem sehr beschränkten Umfang ist es zwar schon gelungen Multifunktionalität zu erreichen, aber eine Karte, die mehrere Applikationen unter einen Hut bringt, existiert de facto nicht. Dafür gibt es eine Reihe von Gründen, von denen viele nichttechnischer Natur sind und deswegen hier nicht diskutiert werden sollen. Aber auch vom technischen Standpunkt her gesehen überstiegen die Ansprüche eines echten Multiapplikations-Betriebssystems bisher die Möglichkeiten der Industrie erheblich. Dank des Fortschritts im Bereich der Halbleitertechnologie ist es jedoch heute möglich, Microcontroller herzustellen, die genügend Ressourcen bieten, um das lang beschworene Ziel technisch erreichbar zu machen. Die ersten Vertreter dieser Controllergeneration werden von Siemens Halbleiter gegenwärtig am Markt eingeführt und sollen in diesem Vortrag näher dargestellt werden.

1 Einführung / Problemstellung

Seit ihren Anfangszeiten waren Smartcard-Anwendungen geschlossene Systeme, die nur den Einsatz für jeweils einen einzigen Zweck gestatteten. Eine Electronic-Purse-Karte kann nicht in einem Handy benutzt werden, die Handy-Karte verschafft keinen Zugang zu einem PC und mit der Webbrowser-Karte kommt man in einer Telefonzelle nicht weit. Mit meiner elektronischen Geldbörse kann ich noch nicht einmal in einem Laden bezahlen, wenn er ein anderes Börsensystem verwendet.

Aus technischer Sicht gibt es dafür im Wesentlichen nur einen Grund: Smartcard-Controller sind einfache 8-bit-Mikroprozessorsysteme, die es nicht gestatten, die Daten einer Anwendung vor dem Zugriff einer anderen zu schützen, wenn diese aktiv auf dem Prozessor Programmcode ausführen darf. Maschinencode, der auf der CPU abläuft, findet einen linearen Adreßraum vor und kann jeden Speicher der Karte beliebig lesen oder schreiben. Folglich konnte es bislang eine friedliche Koexistenz zweier Anwendungen auf einer Karte nur dann geben, wenn sich die verschiedenen Service-Provider und der Hersteller des Betriebssystems vor der Einführung der Karte auf eine gemeinsame Spezifikation einigten und sich gegenseitig vertrauten. Dies führte konsequenterweise zu „Multiapplikationskarten" mit sehr eng begrenzten Möglichkeiten, z.B. GSM-Karten, die auch in Telefonzellen funktionieren.

2 Lösungsansatz

Vor etwa zwei Jahren wurde von verschiedenen Herstellern jedoch der Versuch unternommen, dieses Problem mit Hilfe von Software zu lösen. Ein in das Betriebssystem integrierter Interpreter ermöglichte es, daß anwendungsspezifischer Programmcode in die Karte nachgeladen und kontrolliert ausgeführt werden kann. Vor der Ausführung jedes Anwendungsbefehls wird dabei geprüft, ob die referenzierten Daten innerhalb der für die Anwendung zulässigen Bereiche liegen. Werden diese Bereichsgrenzen verletzt, wird der Zugriff nicht ausgeführt und die Anwendung typischerweise mit einer Fehlermeldung beendet.

Bei diesen Interpretern handelt es sich um proprietäre Skriptsprachen, die von den Entwicklern der Chipkartenbetriebssysteme definiert wurden und die zueinander nicht kompatibel waren. Der Sprachumfang eines derartigen Interpreters lehnt sich stark an die Möglichkeiten eines 8-bit-Prozessorbefehlssatzes an und reicht aus, um die wesentlichen Aufgaben einer klassischen Chipkartenanwendung zu erledigen: Manipulation der Chipkartendaten abhängig von den aktuellen Zugriffsbedingungen. Aufwendigere oder zeitkritische Operationen können jedoch nicht sinnvoll in den interpretierten Sprachen codiert werden. Aus diesem Grund bietet das Betriebssystem der Anwendung neben dem Interpreter eine Reihe von Bibliotheksfunktionen, die in nativem Maschinencode programmiert sind.

3 Technische Einschränkungen

Die Anforderungen einer derartigen Interpreterlösung an den Smartcard-Chip sind jedoch erheblich:

- Der Codespeicher (ROM) muß zusätzlich zum Betriebssystem den Interpreter aufnehmen, dessen Umfang natürlich stark von der Mächtigkeit der verwendeten Skriptsprache abhängt. Hinzu kommt ein Satz von Bibliotheksfunktionen, der für eine leistungsfähige Unterstützung der Applikationen möglichst viele Funktionen enthalten soll. Dies umfaßt z.B. kryptographische Algorithmen zum Hashen, Signieren und Verschlüsseln von Daten oder verschiedene Protokolle für die Kommunikation mit der Außenwelt.

- Der nichtflüchtige Datenspeicher (EEPROM) muß nun die Userdaten mehrerer Anwendungen speichern. Hinzu kommt daß er außerdem noch den jeweiligen applikationsspezifischen Interpretercode halten muß. Die Größe dieses Codes hängt hier wesentlich von der Codeeffizienz des verwendeten Interpreters ab. Ein leistungsfähiger Interpreter kann den benötigten Speicher im EEPROM stark reduzieren, belegt aber dafür mehr Speicher im ROM. Da jedoch die Siliziumfläche einer EEPROM-Zelle nach der Faustformel etwa viermal so groß ist wie die einer ROM-Zelle, versucht man konsequenterweise das Interpretersystem zugunsten eines geringeren EEPROM-Speicherbedarfs zu optimieren.

- Vor allen anderen Speichern verlangen Interpretersysteme aber nach größeren flüchtigen Speichern (RAM), die den Applikationen (und dem Betriebssystem) als Arbeitsspeicher dienen. Der Speicherbedarf wächst hier insbesondere dann schnell an, wenn das Betriebssystem ein Umschalten zwischen den Anwendungen gestattet, so daß der Kontext mehrerer Applikationen gleichzeitig im RAM gespeichert werden muß.

- Außerdem ist es wichtig, daß die CPU eine möglichst hohe Rechenleistung bietet, damit der Leistungsverlust duch das interpretierende Ausführen des Codes für den Anwender nicht spürbar wird. Diese Anforderung versucht man im Wesentlichen durch das Verwenden von Bibliotheksfunktionen zu entschärfen.

- Und schließlich stellt sich die Notwendigkeit bei einer offenen Multiappliaktionskarte, das Nachladen von Applikationen sicher und kontrolliert durchführen zu können. Dafür bieten sich asymmetrische Kryptoverfahren an, die dieses Problem mit elektronischen Signaturen und Zertifikaten lösen. Damit diese aufwendigen Algorithmen mit einem vertretbaren Zeitaufwand berechnet werden können, benötigt die CPU jedoch einen Coprozessor, der auf modulare Arithmetik spezialisiert ist.

Diese Anforderungen konnten von den auf dem Markt verfügbaren Chipkarten-ICs nur in Maßen erfüllt werden. Der Hauptgrund dafür ist in erster Linie die Forderung, daß ein Chipkarten-IC aufgrund der mechanischen Belastbarkeit nach den ISO-Standards nicht größer als 25 mm² sein darf - wobei die Anforderungen des Marktes inzwischen noch deutlich strenger sind [ISO87, ISO93/1]. Konsequenterweise lagen die maximalen Speichergrößen von Smartcardcontrollern bei 24 KByte ROM, 16 KByte EEPROM und 800 Byte RAM. Diese größten Speicher waren allerdings nicht alle zusammen auf einem Chip verfügbar, sondern nur einzeln auf spezialisierten Derivaten zugunsten eines bestimmten Speichertyps. Insbesondere war keiner dieser Extremalwerte bei einem Controller mit integriertem Arithmetik-Coprozessor möglich.

Ein weiterer limitierender Faktor ist außerdem die Energie, die das Anwendungsumfeld dem Chip zur Verfügung stellen kann. Die meisten Chipkartengeräte bieten nur eine sehr begrenzte Stromversorgung - bei portablen Geräten liegt diese in der Größenordnung von wenigen Milliampere. Das bedeutet, daß Performancesteigerungen für den Chip kaum mit klassischen Lösungen wie Taktvervielfachern oder schnellen Bustimings (die starke Treiber voraussetzen) möglich sind.

4 Eine neue IC-Generation

Diese Situation wird sich jetzt jedoch grundlegend ändern. Siemens Halbleiter führt mit der „Triple-E-Familie" eine neue Serie von Chipkarten-ICs ein, die durch die Verwendung modernster Halbleitertechnologie die Leistung aller bisherigen Smartcard-Controller bei weitem übertrifft. Triple-E steht hier für „Enhanced Memory, Enhanced Performance, Enhanced Security", was man am ersten Vertreter dieser Familie deutlich sehen kann. Der SLE 66CX160S bietet

- 32 KByte ROM,

- 16 KByte EEPROM und

- 1280 Byte RAM.

Zusätzlich verfügt er über eine Reihe von Peripheriemodulen, von denen einige erstmals auf einem Chipkarten-IC verfügbar sind:

- Coprozessor für modulare Arithmetik, der für Operanden bis zu 1100 bit Länge optimiert wurde (Advanced Crypto Engine)

- Hardware-Generator für Zufallszahlen

- 16-bit Timer

- CRC-Modul zur schnellen Berechnung von Prüfsummen.

Die Integration dieser Module zusammen mit den großen Speichern wurde in erster Linie möglich durch den Übergang zu 0,6 µm CMOS-Technologie. Das Design einer CPU, die die oben beschriebenen Anforderungen unter den strengen Chipkarten-Randbedingungen erfüllt, machte es jedoch zudem notwendig, neue konzeptionelle Wege zu gehen.

Die Vorgängerfamilie der Triple-E-Bausteine basierte auf einer reinen 8-bit-Architektur, dem Intel-8051-Standard. Die zwei wesentlichen Vorzüge dieser Architektur sind eine sehr hohe Codeeffizienz und eine sehr gute Rechenleistung bei einem sehr kleinen Rechnerkern. Diese Vorteile wollte man natürlich beibehalten und so entschloß man sich, die Architektur nur an den Stellen zu erweitern, an denen Schwächen offensichtlich geworden waren.

So verlangen Chipkarten-Betriebssysteme beispielsweise nach leistungsfähigen 16-bit-Zeigermanipulationen und -Adressierungsarten, ohne daß es notwendig wäre, deswegen gleich die gesamte CPU auf 16-bit-Architektur hochzurüsten. Eine Neuorganisation des internen Bustimings ermöglichte es, die Rechenleistung der CPU ohne zusätzlichen Hardwareaufwand auf das Sechsfache der Standard-8051-Architektur zu beschleunigen. Außerdem wurde eine weitere Erhöhung der CPU-Leistung durch die Implementierung von DMA-ähnlichen Blocktransferbefehlen erreicht, die den Datentransport zwischen den einzelnen Chipmodulen erheblich beschleunigen.

Zur Unterstützung der zunehmenden Programmierung in Hochsprachen wurden neue, stackorientierte Adressierungsarten vorgesehen. Außerdem erhöhen eine Reihe von kombinierten Befehlen die Codeeffizienz sowohl für Hochsprachencompiler als auch für Assemblerprogrammierung.

Damit stellt die Architektur des SLE 66CX160S den idealen Kompromiß für die gegenwärtigen Anforderungen von Chipkartenbetriebssystemen dar. Eine klare Absage an überdimensionierte ausgewachsene 16-bit-Prozessoren und erst recht an 32-bit-RISC-Maschinen, die ebenfalls schon für den Smartcard-Einsatz getestet werden. Hauptargument für derart performante Prozessoren sind üblicherweise biometrische Verfahren und asymmetrische Kryptoalgorithmen.

5 Neue Peripheriemodule

Diese Kryptoalgorithmen werden jedoch - nach wie vor - am effizientesten von dedizierter Hardware berechnet, die von einer optimierten CPU kontrolliert wird. Die Advanced Crypto Engine des SLE 66CX160S ist eine erweiterte Version des bisherigen Siemens Crypto-Coprozessors CCP. Sie bietet eine leistungsfähige Architektur zur Programmierung von asymmetrischen Kryptoverfahren, die auf modularer Arithmetik beruhen. Ein Satz von fünf Registern mit 1100 bit Länge kann mit einem Befehlssatz bearbeitet werden, der alle notwendigen Operationen wie Addition, modulare/nichtmodulare Multiplikation, Reduktion oder Schiebeoperationen zur Verfügung stellt.

Der Modulus der Hardware-Modulooperationen kann dabei maximal 1100 bit lang sein, so daß sich optimal Verfahren bis 1100 bit Länge realisieren lassen, der RSA-Algorithmus aber auch bis zur doppelten Registerlänge (mit Hilfe von Chinesischem Restsatz oder kleinen öffentlichen Exponenten). Die dabei erreichten Rechenzeiten sind der derzeit optimale Kompromiß zwischen Performance, Chipfläche und Stromverbrauch. Bei 5 MHz benötigt der Prozessor für eine volle 512-bit RSA-Operation 110 ms, für 1024 bit sind es 880 ms. Dementsprechend läßt sich eine 1024-bit RSA-Signatur mit Hilfe des Chinesischen Restsatzes in 250 ms berechnen. Die Erzeugung eines 1024-bit RSA-Schlüssels dauert keine Minute.

Ein Hauptcharakteristikum des hier verwendeten Coprozessors ist, daß er darauf optimiert ist, die modularen Berechnungen mit einem Minimum an Elementaroperationen auszuführen [On97]. Folglich ist die arithmetische Grundperformance des Prozessors bereits sehr hoch und es kann darauf verzichtet werden, ihn mit Hilfe eines Frequenzmultiplizierers zu beschleunigen. Durch die daraus resultiererende extrem niedrige Stromaufnahme von wenigen Milliampere kann die Advanced Crypto Engine problemlos in allen Arten von Chipkartenanwendungen eingesetzt werden.

Für das Erzeugen von Schlüsseln und den Einsatz in diversen kryptographischen Protokollen bietet der SLE 66CX160S als erster Chipkartenprozessor einen Hardware-Generator für Zufallszahlen an. Die technische Herausforderung besteht hierbei vor allen Dingen darin, im gesamten Arbeitsbereich des Chips Zufallswerte mit einwandfreien statistischen Eigenschaften zu liefern. Die hier gewählte Lösung besteht aus einer analogen Rauschquelle, die eine digitale Nachbearbeitungsstufe speist. Als stabile Quelle wurde eine Anordnung von gekoppelten spannungsgesteuerten Oszillatoren gefunden, die nicht von externen Parametern wie Temperatur oder elektrischer Spannung beeinflußt wird. Dies wird durch Zertifikate der in Deutschland wesentlichen Institutionen garantiert. Darüberhinaus besteht aber auch für jeden Anwender die Möglichkeit, die Quelle mit den üblichen Testsuites für Zufallszahlen selbst zu prüfen (z.B. [Ri95]).

Eine weitere Peripheriekomponente, die bisher auf Chipkartencontrollern noch nicht verfügbar war, ist ein Hardwaremodul zur Berechnung von Prüfsummen. Für den Einsatz im Chipkartenbereich bietet sich hier der für das Kommunikationsprotokoll T=1 [ISO97] standardisierte Cyclic Redundancy Check an, der mit der 16-bit Frame Checking Sequence des HDLC-Protokolls identisch ist [ISO93/2]. Die effiziente Berechnung dieser Prüfsummen in Hardware gestattet es nun, diese nicht nur zu Absicherung übertragener Protokollblöcke zu verwenden, sondern auch alle auf dem Chip gespeicherten Daten regelmäßig auf ihre Integrität zu überprüfen. Damit erreicht man nicht nur eine erhöhte Zuverlässigkeit des Bausteins im Feld sondern auch höhere Sicherheit gegenüber physischen Manipulationen der Speicher.

6 Neue Sicherheitsfeatures

Diese Erhöhung der Sicherheit war - wie bereits erwähnt - neben der Erhöhung der Speichergrößen und der Rechenleistung der dritte Hauptaspekt bei der Entwicklung der Triple-E-Familie. Als weitere Neuerung auf diesem Gebiet verschlüsselt der SLE 66CX160S sämtliche chipinternen Daten auf den Bussen und in den Speichern per Hardware. Diese Verschlüsselung läuft ohne Zeitverzögerung und völlig transparent für das Betriebssystem ab. Damit bietet der Chip einem Angreifer eine deutlich reduzierte Angriffsfläche sowohl gegen physisches Auslesen als auch Modifizieren der Daten.

Zum Schutz der Daten im RAM der CPU bietet der SLE 66CX160S darüber hinaus einen schnellen aktiven Mechanismus, der das RAM löscht, sobald die Sicherheitssensoren des Chips detektieren, daß die Umgebungsbedingungen der Bereich der Spezifikation verlassen (z.B. Versorgungsspannung oder Taktfrequenz).

Nicht übersehen werden darf außerdem die Tatsache, daß allein die Verwendung immer kleinerer Strukturbreiten im Halbleiterbereich neben der Betriebssicherheit auch die Angriffssicherheit der Chips zunehmend vergrößert. Je kleiner die verwendeten Strukturen sind, desto teurer wird es, einen derartigen Chip zu analysieren und desto schwieriger wird es, das nötige Know-how zu beschaffen. So gestattet es die hier verwendete 0,6 µm-Technologie, die aufgeführten Komponenten auf einem Chip unterzubringen, der kleiner als 20 mm² ist.

7 Ausblick

Damit stellt dieser Baustein eine ideale Plattform dar für die Implementierung von interpreterbasierten Multiapplikations-Betriebssystemen. Dennoch wird die Entwicklung natürlich nicht an dieser Stelle Halt machen. Bereits vor einigen Monaten wurde ein Betriebssystem für Smartcards vorgestellt, dessen Interpretersprache eng an Java™-Bytecodes angelehnt war. Vor allem wegen der oben dargestellten Ressourcenprobleme mußten am Funktionsumfang einige Abstriche gemacht werden, aber die Marschrichtung war klar: Durch die Verwendung einer gemeinsamen Interpretersprache für Personal Computer und Smartcards würden sich die beiden Welten einander erheblich annähern. Dies ist auch das Ziel der Java™Card API, einer für Smartcards geeigneten Version von Java™ [Sun97].

Die Revision 2.0 dieser Spezifikation wurde Ende 1997 fertiggestellt und stellt erhebliche Ansprüche an das Chipkarten-Zielsystem. Obwohl diese Anforderungen vom SLE 66CX160S komplett abgedeckt werden, ist es das Ziel von Siemens Halbleiter, dieses ehrgeizige Projekt mit einer noch leistungsfähigeren Plattform zu unterstützen.

Zu diesem Zweck wurde im Juli 1997 ein Vertrag mit SUN Microsystems geschlossen, dessen Gegenstand die Entwicklung eines Chips ist, der Java™-Bytecode direkt ausführt. Offensichtlich ist, daß diese Hardwarelösung gegenüber einer Softwarevariante erhebliche Geschwindigkeitsvorteile hat. Außerdem entfällt die Notwendigkeit, die Interpretersoftware im ROM unterzubringen, es bleibt also mehr Platz für zusätzliche Funktionen. Ein sehr wichtiger Aspekt für Chipkartenanwendungen ist aber auch die Zertifizierung des Betriebssystems durch herstellerunabhängige Institutionen. Dies ist zur Gewährleistung der Sicherheit einer Applikation notwendig und - abhängig von der angestrebten Sicherheitsstufe - sehr zeit- und geldaufwendig. Eine derartige Zertifizierung kann durch die Verwendung einer einmal zertifizierten Java™-Hardwarelösung deutlich erleichtert werden.

Ein weiteres Leistungsmerkmal dieses Java™-Chips wird sich ebenfalls sehr günstig auf die Verarbeitungsgeschwindigkeit und die Codeeffizienz auswirken: eine Memory Management Unit (MMU). Diese MMU wird es gestatten, die Speicherbereiche der verschiedenen Applikationen per Hardware voneinander zu trennen. Damit entfällt die Notwendigkeit für das Betriebssystem, das Einhalten der Segmentgrenzen in Software zu überwachen. Neben einer höheren Rechengeschwindigkeit erreicht man so natürlich auch eine höhere Sicherheit der Applikationen auf der Karte.

Diese für Multiapplikations-Betriebssysteme optimierte Familie von Chipkarten-Controllern wird 1999 erneut Standards auf dem Markt setzen und damit der echten multifunktionalen Karte endgültig den Weg ebnen.

Literatur

[ISO87]   ISO 7816-1, Identification cards - Integrated circuit(s) cards with contacts - Part 1: Physical characteristics, 1987

[ISO93/1]   ISO/IEC 10373, Identification cards - Test methods, 1993

[ISO93/2]   ISO/IEC 3309, Information technology - Telecommunications and information exchange between systems - High-level data link control (HDLC) procedures - Frame structure, 1993

[On97]   Ondrusch, Stephan: The Siemens Implementation: Still The Most Efficient Way To Do Crypto On A Smart Card? In: Public Key Solutions 1997, Conference Proceedings

[Sun97]   Sun Microsystems Inc.: Java Card 2.0 Language Subset and Virtual Machine Specification, Revision 1.0 Final, 1997

[Ri95]   http://www.io.com/~ritter/NEWS2/TESTSBBS.HTM

SmartXA
Verifizierbare Sicherheit
für Multiapplikations Smart Cards

Thomas Wille

Philips Semiconductors
Productgroup Identification & Automotive
Thomas.Wille@hamburg.sc.philips.com

Zusammenfassung

Der Trend zu Multi-Applikationskarten für nachladbare Anwenderprogrammen verschiedener Serviceanbieter stellt völlig neue Sicherheitsanforderungen an Smart Card Controller und an die auf ihnen implementierte Software. Für die Multi-Applikations-/Multi-Provider-Karten werden heute Lösungen auf Basis von Interpreter-Konzepten wie z.B. JAVA oder MULTOS vorgeschlagen. Diese Interpreter sind jedoch noch längst nicht stabil spezifiziert. Zudem kann ein Interpreter nicht mit einer nativen Applikation auf einer Hardware Plattform koexistieren, ohne daß für die spezielle Konfiguration der Karte erhebliche Sicherheitsprobleme verifizierbar gelöst werden müssen. Mit der neuen 16 bit Smart Card Controller Plattform - SmartXA - für Multi-Applikations-/Multi-Provider-Karten können die gravierenden Sicherheitsprobleme über einen Hardware-FIREWALL Schutz zwischen den geladenen Karten-Applikationen gelöst werden. Zudem bietet die 16-bit Architektur einen ca. 30-fach höhere Rechenleistung gegenüber 8051 Controllern. Auf dem ersten Derivat dieser Familie, dem SXA08W01, wird der Co-Prozessor FameX für schnelle Public-Key Verschlüsselung zusammen mit 32K ROM, 8K EEPROM und 1,5K RAM implementiert. Die SmartXA-Familie bildet damit die optimale Plattform für zukünftige Smart Cards.

1 Multi-Applikations-Karten

Smart Cards im engeren Sinne das heißt Smart Cards mit Micro-Controllern sind in den letzten Jahren besonders im Bereich Bankkarten, GSM und Karten für Pay-TV als neues Tool für den elektronischen Zahlungsverkehr erfolgreich eingeführt worden und haben so ihre erste Bewährungsprobe bestanden. Obwohl die erwähnten Märkte für Smart Cards sich noch überwiegend in der Etablierungsphase befinden, wird bereits über weitergehende Anwendungen von Smart Cards nachgedacht und neue Konzepte werden dafür entwickelt. Danach sollen Smart Cards nicht nur für eine spezifische Anwendung genutzt werden, sondern mehrere Anwendungen sollen parallel auf einer Smart Card untergebracht werden und unabhängig voneinander operieren können. Solche Smart Cards sind wenigstens zum Teil bereits heute realisiert, wenn z.B. ein Systemanbieter etwa eine Kundenkarte mit einer Telefonkartenanwendung kombiniert hat. Diese Smart Cards könnte man als *geschlossenen* Multi-Applikationskarten oder *Single-Provider* Karten bezeichnen.

Das Anwendungsfeld von Smart Cards ließe sich jedoch erheblich erweitern, gelänge es, ein Konzept für *offene* Multi-Applikationskarten oder *Multi-Provider* Karten zu entwickeln. Diese Karten sollen per Prinzip mehrere Applikationen verschiedener Service Anbieter unabhängig voneinander zur Verfügung stellen können. Zudem sollte auch ein Nachladen von kompletten Applikationen durch andere Service Provider frei möglich sein.

2 Interpreter Smart Cards

Ein derartiges Konzept wirft naturgemäß erhebliche und vielfältige Sicherheitsfragen auf, da es bei allen Smart Card Anwendungen im Grunde immer um Transfer von Geld geht. Wie ist beispielsweise sichergestellt, daß eine Applikation nicht in die Datenbereiche einer anderen hineinschreiben kann? Oder wie kann das Lesen über Applikationsgrenzen hinweg verhindert werden? Wie sollte das Nachladen von Applikationen organisiert sein? Viele weitere Fragen müßten hier noch gestellt werden. Andererseits ist auch klar, daß eine erfolgreiche Lösung all dieser Sicherheitsfragen die Sicherheit von Smart Cards insgesamt, also auch für Single Provider Karten deutlich erhöhen würde. Es lohnt sich also, für diesen Problemkreis nach Lösungen zu suchen.

Ein erster Versuch zur Realisierung von Multi-Provider Karten wird mit der Entwicklung von Smart Card Betriebssystemen auf JAVA-Basis betrieben. Ohne in die Detaildiskussion einzusteigen, stellt sich bei JAVA Cards immer die Frage nach der verifizierbar sicheren Abschottung der Applikationen und ihrer Datenbereiche untereinander. Dieses Problem ist nicht zufriedenstellend allgemein lösbar. Außerdem sind JAVA Cards über Interpreter realisiert, die zwangsläufig Ausführungszeiten für Befehle und damit die Programmlaufzeit verlängern.

Aus dieser kurzen Diskussion wird bereits klar, eine Lösung für Multi-Applikations/Multi-Provider Smart Cards kann nur mit einer geeigneten Unterstützung durch die Hardware Plattform für die gebotenen Sicherheitsebene designed werden.

3 Die SmartXA-Familie

Mit der Smart*XA*-Familie stellt Philips seine neue Hardware Plattform für Multi-Applikations/Multi-Provider Smart Cards vor. Folgende typische Hauptmerkmale machen die der Smart*XA*-Familie zum Wegbereiter dieser neuen Smart Card Generation:

- die Smart*XA* Architektur ist die erste 16-bit Architektur für Smart Cards

- der Smart*XA* ist zu einem 'echten' Computer konfiguriert. Der User-Modus ist vom Kernel- oder System-Modus vollständig isoliert. Damit ist per Hardwaredesign eine sogenannte FIREWALL auf dem Chip implementiert

- der Smart*XA* verhält sich kompatibel zum 8051, und 8051-Software kann 1:1 auf SmartXA-Code übersetzt werden, dann aber mit deutlich erhöhter Rechenleistung

- die SmartXA Architektur ist hochsprachenorientiert, so daß typische Hochsprachenkonstrukte direkt auf Assemblerniveau unterstützt werden. Dadurch wird die schnelle Entwicklungszeit von Karten-Betriebssystemen garantiert.

- der SmartXA stellt eine Rechenleistung bereit, die im Vergleich zum 8051 Controller etwa um den Faktor 30 höher liegt.

- der SmartXA ist eine echte Multi-Tasking Maschine. Mehrere Tasks können parallel, aber im Daten- und Codebereich völlig getrennt voneinander, sicher bearbeitet werden, so daß auch ein extrem schneller Kontextwechsel für die FIREWALL-Funktion möglich ist.

- der SmartXA liefert durch seinen umfangreichen und mächtigen Befehlssatz eine hervorragende Code-Dichte, so daß bei gleicher ROM-Größe mehr Code gespeichert werden kann oder ein kleineres ROM, also kleinere IC verwendet werden können.

Besonders wichtig für die Multi-Applikations/Multi-Provider Smart Card sind die Multi-Tasking-Fähigkeit zusammen mit der ON-CHIP FIREWALL. Dies erlaubt einen Betriebssystemsoftware als Kernel zu schreiben, der die Kontrolle über alle Hardware-Resourcen der Smart Card hat, aber zudem noch überschaubar in Funktionalität und Größe ist.

Dadurch kann die Verifizierbarkeit und Zertifizierbarkeit der Kernel Software für den System-Modus sichergestellt werden. Das Sicherheitsfundament der Smart Card wird somit durch dem System-Modus Kernel gebildet und alle User-Modus Programme können auf diese gemeinsame Sicherheitsfunktionen zugreifen, während aber alle User-Modus Programme gegeneinander vollständig isoliert sind (siehe Abb. 1).

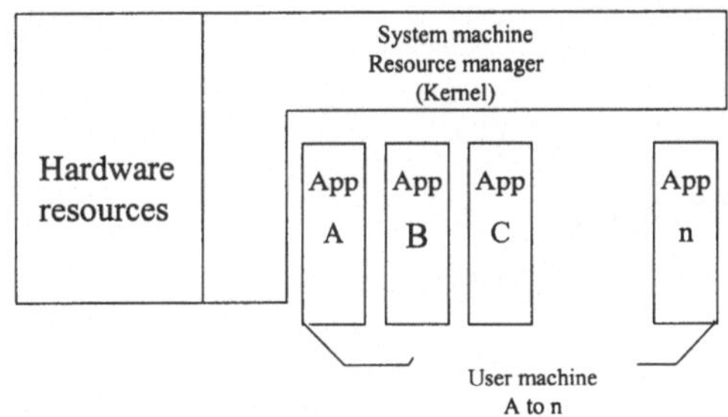

Abb. 1: Offene Multi-Applikations/Multi-Provider Smart Card

4 SmartXA mit FameX

Aus dem Blockdiagram des SmartXA (Abb. 2) erkennt man die typischen Funktionsblöcke eines Smart Card Controllers, wie EEPROM, RAM, ROM, Power-On-Reset, Interrupt und Clockgenerator. Zusätzlich zu diesen Grundfunktionen steht neben Timer und UART auch der derzeit leistungsfähigste Crypto-Coprozessor Fame$X$ für asymmetrische Verschlüsselung wie z.B. RSA zur Verfügung. Eine Vielzahl von weiteren Sicherheitsfunktion sind dem Stand der Technik entsprechend ebenfalls in die SmartXA-Familie integriert.

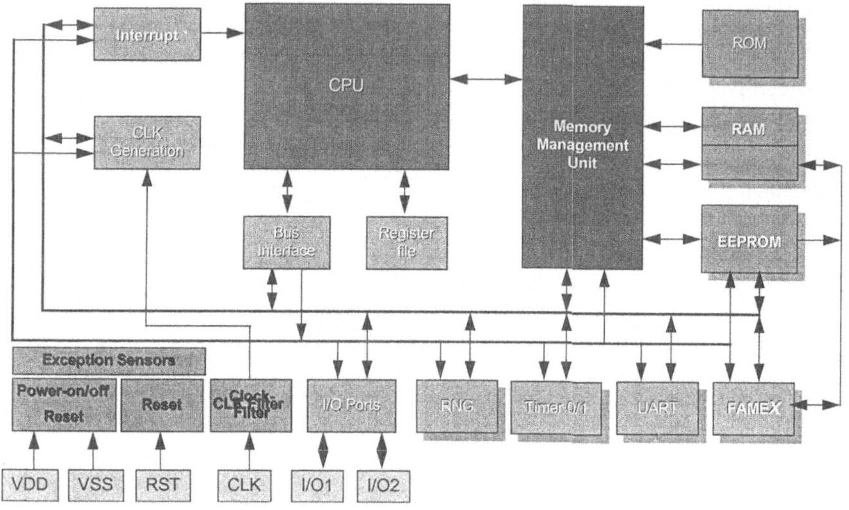

Abb. 2: Blockdiagramm des SmartXA

5 Zusammenfassung

Abschließend kann gesagt werden, daß Philips mit der SmartXA-Familie die erste 16-bit-Plattform für Smart Cards der nächsten Generation bereitstellt. Die SmartXA-Familie garantiert Code-Kompatibilität zur 8051-Architektur. Die ON-CHIP-FIREWALL im Kombination mit der Multi-Tasking-Fähigkeit, der hohen Rechenleistung und der Hochsprachenorientierung ermöglichen mit dem SmartXA das sichere, schnelles Design einer Multi-Applikations/Multi-Provider Smart Card. Zudem wird die Zertifizierung der Applikations-Software erheblich vereinfacht, wenn nicht gar erst durch die Partitionierung in System- und User-Modus-Software per Prinzip ermöglicht. Die Implementierung eines JAVA - Betriebssystems ist auf dem SmartXA effizient möglich, ohne daß durch eine dedizierte JAVA-Hardware die Flexibilität für zukünftige Weiterentwicklungen verbaut wird.

Patientenkarten
Betroffenenorientierte Technik-
bewertung und Gestaltung

Claus Stark

Technische Universität Darmstadt
Institut für Soziologie
cs@computer.org

Zusammenfassung

Technik wird oft von nur wenigen Akteursgruppen gestaltet, die Betroffenen und späteren Nutzer werden selten in den Gestaltungsprozeß integriert. Am Beispiel der Gestaltung der Patientenkarte, einer Chipkarte mit medizinische und gesundheitliche Daten eines Bürgers, soll diese traditionelle Rollenteilung zwischen Technikgestalter und Technikbetroffenen kritisch diskutiert werden. Welche Akteure gestalten die Patientenkarte? Welche Rolle spielt der Patient in diesem Gestaltungsprozeß? Welche Anforderungen und Interessen könnte er bei der Gestaltung der Patientenkarte einbringen? Dieser Beitrag möchte dazu anregen darüber nachzudenken, ob die Gestaltung von Chipkartenapplikationen für das Gesundheitswesen (oder für andere gesellschaftliche Kernbereiche) nicht eine gesamtgesellschaftliche Aufgabe ist, die von möglichst allen relevanten gesellschaftlichen Gruppen aktiv unterstützt werden sollte. Dieses kann erreicht werden, indem Vertreter dieser Gruppen adäquat an den Gestaltungsprozessen beteilt würden. Partizipative Technikgestaltung bietet dazu ein reichhaltiges Methodenspektrum an, um aus passiv Betroffene aktive Beteiligte zu machen.

1 Mit verdeckten Karten spielen: Das Problem selektiver Technikgestaltung

Chipkartenanwendungen werden im allgemeinen gemeinsam von Chipkartenexperten und von Experten aus dem speziellen Anwendungsfeld spezifiziert und realisiert. Die Beteiligung von Anwendungsexperten soll sicherstellen, daß "die richtigen Probleme mit adäquaten Mitteln gelöst werden". Diesem umfassenden Anspruch werden die Technikgestalter aber oft nicht gerecht, denn die Auswahl der Akteure aus dem Anwendungsfeld ist nicht selten selektiv - und wirkt damit ausgrenzend: Relevante Akteure, die wesentliche Beiträge zur Gestaltung leisten könnten (wie beispielsweise die potentiellen Anwender), werden oft an der Technikgestaltung nicht explizit beteiligt, bzw. aktiv daran gehindert. Die Entwicklung findet oft genug mit nur wenigen Akteursgruppen hinter verschlossenen Türen statt - man möchte sich nicht in die Karten schauen lassen. Das kann dazu führen, daß die resultierenden Lösungen eher den berufsspezifischen Interessen der Gestalter als den Bedürfnissen der Betroffenen dienen - was nicht selten zu gesellschaftlichen Konflikten führt.

Partizipative Technik*gestaltung (*als eine spezielle Form *betroffenenorientierter Technikbewertung und -gestaltung*) kann den o.g. Gesellschaftskonflikten aktiv begegnen, indem bereits in der Planungsphase eines (Karten-)Projektes Betroffene und andere relevante Akteure identifiziert und eingeladen werden, ihre spezifischen Bedürfnisse und Interessen an die zu gestaltende Technik zu formulieren - und schließlich auch adäquat in den Gestaltungsprozeß einzubringen. So ist es theoretisch wie praktisch möglich, innovative Technik zu entwickeln, die frühzeitig von breiten gesellschaftlichen Gruppen akzeptiert, genutzt und gar gefördert wird - indem nämlich die Technik und die konkreten Projektziele für sie verständlich gemacht, und idealerweise Gestaltungsanforderungen dieser Grruppen explizit berücksichtigt würden. Dissens zwischen den Interessengruppen würde sehr früh diskutierbar gemacht, Konsens führte zu schnellen Gestaltungsentscheidungen - was in aller Interesse sein sollte.

An einem Beispiel aus dem Anwendungsfeld Gesundheitswesen, der Gestaltung der "Patientenkarte", soll diese traditionelle Rollentrennung zwischen Technikgestalter (u.a. Ärztekammern, Krankenkassen und Apothekerverbände) und Betroffenen bzw. Technikanwender (hier: Bürger und Patienten) kritisch diskutiert werden. Bürger und Patienten sollen diese Patientenkarte auf vielfältige Weise anwenden und Nutzen daraus ziehen. Patientenvertreter sind bisher aber nicht an der Gestaltung der Patientenkarte beteiligt worden, haben bisher keine Anforderungen an die Patientenkarte "aus der Sicht des Patienten" entwickeln können.

Leider können daher in diesem Papier keine konkreten Ergebnisse betroffenenorientierter Bewertung und -gestaltung von Patientenkarten vorgelegt werden. Es soll somit als Anregung für Technikgestalter dienen darüber nachzudenken, ob ihre geplante Kartenanwendung nicht auch eine gesamtgesellschaftliche Bedeutung hat - und ob es in diesem Falle nicht sinnvoll wäre, darüber frühzeitig einen möglichst breit angelegten Gestaltungsdiskurs mit allen relevanten Gruppen zu führen (Technikfolgenabschätzung aus verschiedenen Perspektiven, gesamtgesellschaftliche Technikbewertung, diskursive Technikgestaltung). Alle Karten sollten dabei offengelegt werden! Gerade für innovative Technologien, bei deren Gestaltung nur wenig Erfahrungswissen vorliegt, wäre das eine wichtige vertrauensbildende Maßnahme. Die hier entwickelten Überlegungen gelten für viele gesellschaftliche Anwendungsfelder. Neben dem Gesundheitswesen wären beispielsweise zu nennen: Verkehr, Hochschule, Öffentliche Verwaltung und Handel.

2 Chipkarten im Gesundheitswesen

Seit der Einführung der Krankenversichertenkarte (KVK) im Jahre 1995 sind einige Projekte zur Patientenkarte in Deutschland initiiert worden (Übersicht "Kartenprojekte im deutschen Gesundheitswesen" in [DuD97]: DuD-Forum). Diese Karten sollen, im Gegensatz zur staatlich verordneten KVK, nicht nur administrative Daten enthalten, sondern auch medizinische und gesundheitliche Daten tragen: Diagnosen, Therapien, Unverträglichkeiten und allerlei Pässe (z.B. Impfpaß, Notfallausweis, Organspendepaß, Röntgenpaß) ließen sich in wenigen Bytes speichern. Neben Karten für spezielle Zielgruppen wie Diabetiker, an Krebs Erkrankte und andere chronisch kranke Menschen (u.a. DiabCard, OnkoCard, DefiCard) soll es in Zukunft auch Karten für die Allgemeinheit geben:

- Medikamentenkarten sollen helfen, Fehlverschreibungen von Pharmaka und Unverträglichkeiten zu vermeiden (z.B. ACard der Bundesvereinigung dt. Apothekerverbände).

- Krankenkassenkarten sollen helfen, einen gesunden Lebensstil zu pflegen (z.B. VitalCard der AOK).

- Die Industrie experimentierte mit dem elektronischen Mutterpaß (z.B. HealthCard der Firmen Bayer und IBM).

- Die Ärzte erproben seit einiger Zeit in Neuwied ihr Konzept der Patientenkarte (Medizinische Patientenkarte u.a. der Kassenärztlichen Bundesvereinigung).

Es gibt sehr viele Ideen, Chipkarten und Computernetze im Gesundheitswesen einzusetzen. Interessant ist die Frage nach der Rolle des Patienten in diesem Umfeld: "Where is the Patient in Patient-Centered User-Oriented Design in Health Care Systems Development?" - diese Frage wurde auf der Participatory Design Conference 1996 in Cambridge wissenschaftlich diskutiert [CPSR96]. Das Fazit war: Es gibt ihn dort (noch) nicht. Dabei könnte er dort einiges beitragen.

3 Patienten und Patientenkarten

In allen deutschen Projekten zur Patientenkarte spielte die Information und die Beteiligung der Bürger und Patienten bisher eine untergeordnete Rolle. Beispielsweise wurden lokale und überregionale Patientengruppen selten von den Projektverantwortlichen eingeladen, damit diese sich mit ihnen und ihren Kartenkonzepten auseinandersetzen könnten. Die direkte Beteiligung von sachverständigen Bürgern aus Selbsthilfegruppen steht auch heute nicht zur Debatte. Erfahrungen aus der Arbeit mit den Chipkartengestaltern im Gesundheitswesen belegen die These, daß der Patient bei der Gestaltung der Patientenkarte nicht erwünscht ist. Dabei ist Angst vor dem Patienten fehl am Platze: Patientengruppen beispielsweise stehen der Patientenkarte sehr offen gegenüber und würden sich gerne aktiver mit ihr beschäftigen.

3.1 Akteure im Gesundheitswesen

Wer sind die aktiven Gestaltungsakteure, die dem abstrakten Gegenstand "Patientenkarte" erst die konkrete Form geben? Wer verfügt im Gesundheitswesen über die "Gestaltungsmacht"? Dazu soll ein Blick auf die im Gesundheitswesen aktiven Gruppen gerichtet werden [Wanek94]. In diesen Gruppierungen können diejenigen identifiziert werden, die aktiv gestalten - und die vielen eher passiven Gruppen.

Es gibt eine Vielzahl von Akteuren im Gesundheitswesen: Ärzte, Pflegepersonal, Patienten, Forscher, Industrie, Politiker, Krankenkassenfunktionäre, Juristen, Ökonomen, Versicherte, Kranke und Gesunde und (Medizin-) Informatiker sind hier zu finden. Diese Gruppen unterteilen sich noch weiter (beispielsweise in Allgemein- und Fachärzte), und es gibt noch weitere Gruppen. Eine übliche Gruppierung ist die nach "Professional" und "Patient":

- Die "Professionals" sind durch ihren Beruf sehr leicht zu identifizieren und zu erreichen. Sie haben sich zumeist organisiert, um ihre berufsspezifischen Interessen im Gesundheitswesen gezielt zu vertreten. Beispielsweise sind die Ärzte durch Kassenärztliche Vereinigungen und Ärztekammern vertreten, die Kassen durch ihre Spitzenverbände. Der Organisationgrad und der berufspolitische Einfluß der verschiedenen Berufsverbände variiert jedoch sehr stark.

- Die Gruppe der "Patienten" hingegen ist sehr schillernd, und sie existiert offenbar nur als Phantom: Jeder ist Patient, und gleichzeitig ist niemand Patient "von berufs wegen". Patienten gelten per Definition als krank und somit als nicht belastbar. Obwohl von allen Akteursgruppen im Gesundheitswesen die der Patienten mit Abstand die größte ist, hat sie doch nur eine sehr schwache Stellung. So ist es heute üblich, daß kleine aktive Professionals-Gruppen weitreichende Entscheidungen "im Sinne des Patienten" treffen - und diese auch gegen explizite Patienteninteressen durchsetzen können.

3.2 Interessen bei der Kartengestaltung

Aber warum sollten Patienten bei der Kartengestaltung mitwirken, wo doch Technikentwicklung "ingenieursgemäß" stets nach objektiven Kriterien verläuft? Kommt da nicht immer nur die beste Lösung für ein erkanntes Problem zu Zuge? Dem ist nicht immer so: Informations- und Kommunikationstechnik ist in höchstem Maße gestaltungsfähig, Sie läßt sich hervorragend für konkrete Vorgaben (z.B. berufspolitische Interessen) sehr zielgenau entwickeln. Technik wird immer in einen sozialen Kontext eingebettet, und kann daher nicht ohne diesen betrachtet werden: Allein die Frage, welche Probleme im Gesundheitswesen mit welchen Mitteln eigentlich vorrangig gelöst werden sollten, ist - je nach Interessenlage des betrachtenden Akteurs - höchst unterschiedlich zu beantworten. Zunächst soll daher ein Blick auf die Ziele aktueller Kartenprojekte geworfen werden - können Patienten mit ihnen zufrieden sein? Danach soll diskutiert werden, welche Gestaltungsoptionen dem Patienten offenstehen könnten.

3.2.1 Projektziele ...

In vielen Projekten wird als Hauptziel das "Wohl des Patienten" angegeben, dennoch steht dieses doch eher abstrakte Ziel hinter sehr konkreten Zielen wie Kostenbegrenzung und Kundenbindung zurück. Zwei Beispiele sollen dieses illustrieren:

- **Beispiel ACard**: Dieses Kartenprojekt wurde 1995 von der ABDA, dem Bundesverband der Apothekerverbände, initiiert. Neben dem Ziel, den Arzt bei der Verschreibung von Medikamenten via Karte zu unterstützen, um so rechtzeitig gefährliche Wechselwirkungen von Pharmaka oder Medikamentenunverträglichkeiten beim Patienten zu erkennen standen von Anfang an Ziele wie Kundenbindung und die Positionierung der Apotheken gegen die Konkurrenz der Drogeriemärkte an vorderer Stelle. Ärzte und Drogeriemärkte sollten auf der Karte nicht speichern können - so würde Sie aber schwerlich dem Anspruch einer "allgemeinen" Medikamentenkarte gerecht. Die Karte wurde anscheinend unter dem Fokus eigener berufspolitischer Zwecke entworfen.

- **Beispiel AOK VitalCard**: Im Raum Leipzig sollten 1995 alle 500 000 AOK-Mitglieder eine Chipkarte erhalten, die eine Vielzahl an medizinischen Daten speichern sollte. Mit ihr sollte es leichter sein, ein gesundes Leben zu führen (z.B. durch Monitoring entsprechender Laborparameter wie den Cholesterinspiegel). Der Landesbeauftragte für den Datenschutz des Landes Sachsen rief daraufhin zum Boykott dieser Chipkarte auf, denn die Ausgabe war mit ihm und der Leipziger Ärzteschaft nicht abgesprochen gewesen - und die waren vom Projekt gar nicht begeistert, vermuteten Sie doch Kontrolle ärztlicher Tätigkeit. Die Karte sollte offenbar als "Marketinginstrument" Verwendung finden im Wettbewerb zwischen den Krankenkassen.

Beide Karten sollten dem Bürger und Patienten nützlich Helfer sein, und doch spielten Patienteninteressen keine explizite Rolle bei deren Gestaltung. Patientenferne Interessen formen diese Technologie ebenfalls kräftig mit. Diese Interessen können durchaus legitim sein. Es besteht aber die Gefahr, daß hier Patientenkarten entwickelt und durchgesetzt werden, die für Patienten nur wenig Nutzen bringen oder sogar konträr zu ihren originären Interessen stehen.

3.2.2 ... Konflikte ...

Zu den beiden o.g. Patientenkarten konnten bisher keine praktischen Erfahrungen gesammelt werden, da sie beide bisher nicht eingeführt wurden. Daher spielten sie in der gesellschaftlichen Diskussion bisher auch keine Rolle. Immerhin rief aber der Sächsische Landesbeauftragte für den Datenschutz zum Boykott der "VitalCard" auf, was deutlich auf latent vorhandenes Konfliktpotential hinweist. Welcher Art die gesellschaftlichen Konflikte sein könnten, wurde aber schon bei der Einführung der - datenschutzrechtlich bisher eher unbedenklichen - Krankenversichertenkarte deutlich: Sie wurde 1994 und 1995 von massiven Protesten einiger gesellschaftlicher Gruppen begleitet [GesLad94]. Der Bundesverband der PatientInnenstellen und das Institut für Informationsökologie riefen gemeinsam zum Boykott der KVK auf und forderten ein Moratorium der Einführung von Karten im Gesundheits- und Sozialwesen. Sie befürchteten, daß hiermit ein wirksames Werkzeug für Rationierung und Selektion bereitgestellt würde, was die Stellung des Patienten gegenüber den anderen Akteuren noch weiter verschlechtern würde. Zunächst solle über die Ziele eines computergestützten Gesundheitswesens Klarheit und Einigung geschaffen werden, bevor über technische Maßnahmen zur Erreichung dieser Ziele diskutiert werden könnte. Spitzenverbände der Zahnärzte beschäftigten sich 1994 ebenfalls sehr kritisch mit der Krankenversichertenkarte und boykottierten diese teilweise . Innerhalb der Ärzteschaft finden z.T. sehr themenspezifische Auseinandersetzungen um die Computerisierung statt - ein Beispiel ist die ärzteschaft-interne ICD-10-Diskussion von 1996. Die Datenschutzbeauftragten des Bundes und der Länder haben bereits einige sehr konkrete Anforderungen an Chipkartenanwendungen im Gesundheitswesen formuliert - die wiederum von der Bundesärztekammer sehr kritisch analysiert wurden. Die Projektierung von Patientenkarten seitens der Krankenkassen wurde bisher von der Ärzteschaft stets sehr kritisch verfolgt. Mit der zunehmenden Computerisierung im Gesundheitswesen wird in der Gesellschaft auch vermehrt über Ziele und Design von EDV- und Kartenanwendungen diskutiert werden. Beispielsweise ist zu erwarten, daß sich zur Erprobung und Einführung von Health Professional Cards - hier werden Arbeitnehmerinteressen direkt tangiert - Arbeitnehmervertretungen verstärkt zu Wort melden werden.

Der bundesdeutsche Arbeitskreis "Karten im Gesundheitswesen" beschäftigt sich seit 1995 mit der Planung, Erprobung und Einführung von Kartensystemen im Gesundheitswesen (u.a. Patientenkarten und Health Professional Cards). In diesem Arbeitskreis sind ausgewählte Vertreter relevanter Institutionen aus Ärzteschaft, Krankenkassen, Politik, Industrie, Datenschutz, Datensicherheit - allerdings bisher keine seitens der Patientenverbände und der Arbeitnehmerschaft - vertreten. Dieser Arbeitskreis könnte somit prinzipiell die wichtige integrierende Funktion des Interessenausgleichs zwischen den gesellschaftlichen Gruppierungen leisten, würde der Kreis für weitere wichtige - auch kritische - Akteure geöffnet.

3.2.3 ... und Gestaltungsoptionen

Aber was könnten Patienten mitgestalten? Was sind ihre Interessen? Genau das wäre im Diskurs mit ihnen zu klären - und dieser wurde bis heute nicht geführt. Es sind Möglichkeiten auf sehr vielen Ebenen denkbar: Organisatorische und juristische Rahmenbedingungen stehen genauso auf dem Prüfstand wie technische Details des Kartenbetriebssystems und der Zugriffsprogramme. Datenschutz als Persönlichkeitsschutz vor unberechtigter Verdatung, Datensicherheit (Security), persönliche Handlungsautonomie und medizinische Behandlungssicherheit geben mögliche Gestaltungskoordinaten für die Gestaltung einer Patientenkarte vor. Einige konstruierte Beispiele sollen illustrieren, in welche Richtung die Diskussion laufen könnte - vielleicht wären den Patienten aber auch ganz andere Aspekte wichtig.

- **Datensätze definieren:** Patientengruppen wissen im allgemeinen sehr gut, welche Daten für ihre Klientel relevant sind. Sie könnten mitberaten, wie die Datensätze auf der Karte aussehen sollten, damit Ihnen die Karte eine wirkliche Hilfe im Alltag wäre.

- **Wer darf schreiben, lesen, löschen?** Soll der Patient am heimischen PC die Karte lesen können oder nicht? Soll er ggfs. eigene Daten darauf ablegen können - z.B. seine Blutdruckwerte, die er zuhause regelmäßig kontrolliert, um Sie dann mit dem Arzt diskutieren zu können? Oder soll nur der Arzt lesen und schreiben dürfen? Die Urheberschaft der Daten könnte mit digitalen Signaturen eindeutig zugeordnet werden. Es wäre mit einer entsprechenden Sicherungsinfrastruktur sogar denkbar, daß der Patient seine Daten am heimischen PC selber verwaltet: Er hätte signierte Datenbausteine (wie beispielsweise eine Impfbescheinigung) zur Verfügung, die er - je nach Bedarf - in die Karte einspielen oder auch löschen könnte. Das würde ihm ein hohes Maß an persönlicher Datenautonomie gewähren.

- **Zugriffsschutz:** Wie werden die Daten auf der Karte vor fremden Blicken geschützt? Kann der Patient für die einzelnen Datenbereiche ein eigenes Paßwort verwenden, um den Zugriff zu steuern - oder haben alle Ärzte automatisch Zugriff auf alle Daten? Kann der Patient gespeicherte Daten vor einem Arzt geheimhalten - oder sieht der Arzt, daß da etwas vor ihm verborgen wird? Wie kann der Patient eine evtl. sehr komplexe Zugriffskontrolle sicher und intuitiv benutzen? Kann der Patient mitbestimmen, welche Schlüssel auf der Karte verwendet werden? Soll der Patient mit der Karte gar Teile "seiner" Datenbestände beim Arzt und im Krankenhaus "auf- und zuschließen" können? Dadurch wäre der Besitzer der Karte in der Tat - jedenfalls etwas mehr als in den bisher vorgelegten Konzepten einer Patientenkarte - "Herr seiner Daten".

- **Name oder Pseudonym ?** Auf der Karte könnte die vollständige Identität des Kartenbesitzers gespeichert sein. Oder er könnte ein Pseudonym wählen [Pomm96], um so ggfs. der Gefahr der persönlichen Diskriminierung vorzubeugen. Das kürzlich verabschiedete Signaturgesetz sieht die Möglichkeit der Pseudnymisierung explizit vor.

- **Mitentscheiden, was gespeichert wird:** Patienten könnten fordern, daß nur Daten mit ihrer Zustimmung auf der Karte gespeichert werden dürften. Daten seien dann auch wieder zu löschen, wenn Patienten es wünschten (der Datenschutz hat entsprechende Forderungen bereits formuliert). Die Ärzteschaft würde diese Forderung aber wohl nicht unterstützen, denn mit unvollständigen Daten wäre die Karte für Sie nicht vertrauenswürdig. Hier gilt es, zwischen informationeller Selbstbestimmung des Patienten und ärztlichem Berufsethos abzuwägen - und eine gemeinsame Lösung zu finden.

- **Recht auf Nichtvorlage:** Das Recht auf Nichtvorlage der Karte beim Arzt könnte festgeschrieben werden, auch wenn das der Ärzteschaft und den Krankenkassen nicht gefällt. Es könnte beispielsweise für Patienten wichtig sein, eine unvoreingenommene zweite Meinung einzuholen - mit Zwang zur Kartenvorlage wäre das nur schwerlich durchzusetzen.

- **Qualifizierung:** Schließlich gilt es, geeignete Maßnahmen für den qualifizierten Umgang mit der Patientenkarte zu definieren und umzusetzen. Vielleicht könnte es eine Art "Karten-Führerschein" geben, mit dem der Kartenbesitzer sogar Schreibrechte auf der Karte erhielte? Es wäre auch denkbar, daß die Patienten von ihren Ärzten und Pflegekräften entsprechende Qualifizierung erwarten.

- **Option oder Pflicht?** Wird die Benutzung der Patientenkarte völlig freiwillig bleiben oder de facto zur Pflicht werden? Entsprechende Beobachtungen sind mittel- bis langfristig gesellschaftlich zu diskutieren, um rechtzeitig gegensteuern zu können. Es ist beispielsweise rechtzeitig offen und breit zu diskutieren, wie die technische und organisatorische Fortentwicklung der obligatorisch zu nutzenden Krankenversichertenkarte aussehen könnte.

In einem Forschungsprojekt der Technischen Hochschule Darmstadt soll die Rolle des Patienten bei der Gestaltung der Patientenkarte genauer untersucht werden [StaSchm 95, 96, 97]: Wer ist "der Patient" im Gesundheitswesen und in den Kartenprojekten? Was sind seine Anforderungen an eine Patientenkarte? Gibt es Gestaltungsoptionen für ihn - und Möglichkeiten seiner direkten Partizipation? Welche Probleme ergeben sich aus dem Partizipationsansatz? Am Beispiel eines laufenden Kartenprojektes, dem Modellversuch Patientenkarte Koblenz/Neuwied [SemBre97], sollen diese Fragen in der Praxis untersucht werden. Dazu traf sich 1995 und 1996 regelmäßig eine Bürgergruppe aus Vertretern einiger Patientengruppen1 und diskutierten u.a. auch mit Vertretern des Kartenprojektes. Das Experiment wurde 1997 - leider ohne partizipative Ansätze im Gestaltungsprozeß der Patientenkarte praktisch erproben zu können - beendet.

3.3 Umfrage unter Patientengruppen

Von Februar bis März 1997 führte eine Wissenschaftlergruppe der Technischen Hochschule Darmstadt und der Universität Koblenz-Landau in Koblenz und Neuwied eine Umfrage unter lokalen Patientengruppen durch. Ziel war es, das Wissen und die Meinung von Patientengruppen zur Patientenkarte und zum dort laufenden Modellversuch möglichst vollständig zu erheben. Dieser Modellversuch startete im Sommer 1995, d.h. er lief bis zum Zeitpunkt der Umfrage über 1-1/2 Jahre. An der telefonischen Befragung, die eine wissenschaftliche Hilfskraft mit Hilfe eines Interviewleitfadens durchgeführte, nahmen schließlich 33 der insgesamt 52 angeschriebenen Gruppen statt. Jede Gruppe benannte einen Gesprächspartner. Die Umfrageergebnisse werden im Detail in [SMHS97] diskutiert.

Als zusammenfassendes Ergebnis der Umfrage läßt sich festhalten:

- Fast alle befragten Gruppen wußten nur sehr wenig über das Konzept "Patientenkarte" im

---

[1] Im Laufe unserer Untersuchungen zur Rolle des Patienten in der Technikgestaltung beschränkten wir uns auf einen Teilbereich der Akteursgruppe "Patient": Wir betrachteten die in Vereine und Selbsthilfegruppen organisierten, zumeist chronischen Patienten. Diese sind stark von sehr speziellen Aspekten von Krankheit und Gesundheit betroffen, sie sind mit den Strukturen, Akteuren und Entscheidungsprozessen im Gesundheitswesen vertraut , sie sind für das Thema " Patientenkarte" leicht zu interessieren - und sie sind einfach zu erreichen.

allgemeinen, bzw. vom konkreten Modellversuch vor Ort.

- Die Gruppen hatten großes Interesse, mehr über die Patientenkarte zu erfahren.

- Die Gruppen schätzten im allgemeinen die Chancen der Patientenkarte größer ein als deren Risiken.

- Die Bedeutung von Partizipation wird von den befragten Patientengruppen als wichtig eingeschätzt.

Weniger als 30% der Befragten kannten das Konzept der Patientenkarte, weniger als 10% waren über den lokalen Modellversuch informiert. Nicht ein Befragter kannte eine Person aus seinem Umfeld, die eine Patientenkarte besaß. Das Informationsdefizit bei den Patientengruppen ist (u.a.) erklärbar durch die nicht-optimale (d.h. nicht auf den Patienten ausgerichtete) Informationspolitik der Projektinitiatoren.

Die Gruppen hatten sehr großes Interesse, mehr über das Konzept "Patientenkarte" zu erfahren: Mehr als 80% der Befragten wünschten Informationsmaterial und würden Veröffentlichungen in ihren Mitgliederpublikationen begrüßen, ca. 70% zeigten Interesse an Vorträgen und Diskussionsrunden. Entsprechende Konzepte könnten von den Projektverantwortlichen - in Zusammenarbeit mit den örtlichen Bildungs- und Selbsthilfeträgern wie Volkshochschule oder Gesundheitsamt - leicht umgesetzt werden.

Die Befragten standen der Technologie "Patientenkarte" sehr interessiert und offen gegenüber: Sie wurden für die Allgemeinheit mit ca. 60% als nützlich eingestuft, für ihre eigene Klientel lag diese positive Einschätzung mit 45% etwas niedriger. Ganz negative Einschätzungen waren selten. Als Chance sahen Sie u.a. die Verfügbarkeit in Notsituationen und den Einsatz als Gedächtnishilfe an. Aber auch die Kontrolle über den Arzt war für einen Befragten ein wichtiger Aspekt. Risiken hingegen wurden hauptsächlich bezüglich des verletzlichen Datenschutzes vermutet, der "gläserne Patient" wurde durchaus als reales Risiko empfunden. Diese offene Grundhaltung scheint ideal zu sein für die unvoreingenommene Beschäftigung mit der Patientenkarte.

Über 60% der Befragten waren der Meinung, daß explizite Patienteninteressen bei der Entwicklung der Patientenkarte adäquat berücksichtigt werden sollten. Über 70% hielten die direkte Beteiligung von Patientenvertretern an Kartenprojekten für realistisch. Aber die Mehrheit der Befragten sah sich außerstande anzugeben, wie Partizipation konkret aussehen könnte. Hier sind vorrangig die Projektinitiatoren - in Zusammenarbeit mit entsprechenden lokalen oder überregionalen Einrichtungen wie Volkshochschulen, Gesundheitsämtern und politischen Bildungszentren - gefordert, Ihnen adäquate Angebote zu unterbreiten. Die methodischen Möglichkeiten sind hierzu sehr vielfältig.

4   Karten aufdecken erwünscht! Chancen partizipativer Kartengestaltung wahrnehmen

Unter der Prämisse, daß Chipkartenanwendungen in den diversen gesellschaftlichen Feldern zu wichtig sein werden, um als Marketing-Gag oder als berufspolitisches Instrument einzelner Akteure mißbraucht zu werden, ist der offene gesellschaftliche Diskurs um deren bürgernahen Gestaltung zu führen. Der Gestaltungsdiskurs sollte daher nicht nur mit einzelnen Experten eines Anwendungsfeldes, sondern auch mit Betroffenengruppen, interessierten Einzelpersonen und mit repräsentativ ausgewählten Bürgern (à la Schöffenpflicht) geführt werden. Es sollte mit offenen Karten gespielt werden, um aktiv Vertrauen zu schaffen!

Am Beispiel "Patientenkarte", deren Gestaltung wegen der hohen gesellschaftspolitischen Relevanz als "gesamtgesellschaftliches Projekt" begriffen werden kann, wurde aufgezeigt, daß es sinnvoll sein kann, patientenorientierte Technologien unter Beteiligung von Patienten bzw. Patientengruppen zu entwickeln. Dieses wurde bisher noch nicht realisiert, so daß keine Erfahrungen vorgelegt werden können.

Ist Partizipation realistischerweise auch umzusetzen? Es gibt theoretische wie praktische Vorwände, Partizipation als "nicht machbar" abzutun und die Betroffenen auf ihre Plätze zu verweisen. Viele Beispiele zeigen aber, daß "Participatory Design" praktisch umsetzbar ist: In Dänemark wurde mit repräsentativ ausgewählten Bürgern eine mehrtägige Konsensuskonferenz zur "Bürgerkarte" (Anwendungsfeld: Öffentliche Verwaltung) durchgeführt, die zu einem Bürgergutachten für das dänische Parlament geführt hat [Stripp96]. In Deutschland wurden eine Reihe von Zukunftswerkstätten und Planungszellen zu den verschiedensten Themen mit Betroffenen und mit repräsentativ ausgewählten Bürgern und Experten erfolgreich durchgeführt. In den USA wird über die Umsetzung basisdemokratischer Prinzipien in der Technikgestaltung theoretisch wie praktisch diskutiert [CSPR96].

Die Projektinitiatoren sollten bei der Entwicklung zukunftsfähiger Kartenkonzepte auf die Sachkompetenz und den Elan der Betroffenen und anderer Interessierter setzen und adäquate Beteiligungsmodelle anbieten. Es reicht einfach nicht aus, allgemeine Informationsveranstaltungen für die Bürger anzubieten, durch die sich keiner angesprochen fühlt. Methodisch gibt es sehr viele bessere Ansätze, aus denen aber gezielt gruppen-, themen- und zielspezifisch ausgewählt werden muß. Durch adäquate betroffenenorientierte Technikbewertung und -gestaltung ließe sich das für innovative Technikkonzepte notwendige Vertrauen in der Bevölkerung leicht und dauerhaft gewinnen. Die Technikgestalter haben aber bisher die Chance, mit ehrlicher Offenheit der negativen Mythenbildung entgegenzuwirken, anscheinend noch nicht verstanden. Es bedeutet natürlich auch ein Risiko, den Betroffenen reale Gestaltungsoptionen zu eröffnen: Die eigenen Konzepte wären dem kritischen und öffentlichen Diskurs auszusetzen - und diese könnten sich schließlich als unerwünscht oder unpraktikabel erweisen. Aber dieses überschaubare Risiko sollte unbedingt eingegangen werden, denn ihr stehen die großen Chancen gesellschaftlich breit akzeptierter Kartenanwendungen gegenüber.

Literatur

[CPSR96]   Computer Professionals for Social Responsibility: PDC '96 - Proceedings of the Participatory Design Conference, MIT, Cambridge/USA, 1996.

[DuD97]   DuD: Themenheft "Gesundheitswesen" der Zeitschrift Datenschutz und Datensicherheit (DuD), Wiesbaden, Nr. 10, 1997.

[GesLad94]   Gesundheitsladen Köln: Die Krankenversichertenkarte gefährdet unsere Gesundheit!, Broschüre des Bundesverbandes der PatientInnenstellen, Köln, 1994.

[Möhr96]   Möhring, Michael: Modellversuche zur Einführung von Patientenchipkarten, in: FIfF Kommunikation (Zeitschrift des Forum InformatikerInnen für Frieden und gesellschaftliche Verantwortung e.V.), Bonn, Nr. 1, 1996, S. 19 f.

[Pomm96]   Pommerening, Klaus: Chipkarten und Pseudonyme, in: FIfF Kommunikation (Zeitschrift des Forum InformatikerInnen für Frieden und gesellschaftliche Verantwortung e.V.), Bonn, Nr. 1, 1996, S. 9 ff.

[SemBre97]   Sembritzki, J.; Brenner,G.: Introduction of a Patient Data Card into outpatient medical care and the pharmacy sector (CARDLINK test site Koblenz), Proceedings Health Cards 1997, Amsterdam/NL, 1997, S. 292ff.

[StaSchm95]   Stark, Claus; Schmiede, Rudi: Patients assess the Patient's Card - A Participatory Project to Technology Assessment in Medical Informatics, in: Proceedings of the 14th International Congress on Cybernetics, Namur/B, 1995, S. 808.

[StaSchm96]   Stark, Claus; Schmiede, Rudi: Ist Bürgerbeteiligung bei der Gestaltung einer Patientenkarte wünschenswert und machbar?, in: Datow, M. et al (Hrsg.): Die Chipkarte im Alltag (MultiCard 96), Berlin, 1996, S. 235 ff.

[StaSchm97]   Stark, Claus; Schmiede, Rudi: Keine Angst vor dem Patienten! Plädoyer für eine offene Informationskultur in Kartenprojekten im Gesundheitswesen, in: Fluhr M. (Hrsg.): Die Chipkarte - Eine Welt der Möglichkeiten (OmniCard 97), Berlin, 1997, S. 19 ff.

[Stark97]   Stark, Claus: Patientenorientierte Gestaltung der Patientenchipkarte, in: Datenschutz und Datensicherheit, Wiesbaden, Nr.10, 1997.

[SMHS97]   Stark, C.; Möhring, M.; Herrmann, H.; Schmiede, R.: Patients assess the Patient's Card - Results of the Survey among Patients Groups concerning the Pilot Project Patient's Card Koblenz/Neuwied, Proceedings Health Cards 1997, Amsterdam/NL, S. 238ff.

[Stripp96]   Stripp, Steffen: Die Bürgerkarte - Ein kleiner Helfer zum Schutz der Privatsphäre? , in: FIfF Kommunikation (Zeitschrift des Forum InformatikerInnen für Frieden und gesellschaftliche Verantwortung e.V.), Bonn, Nr. 1, 1996, S. 12 f.

[Wanek94]   Wanek, Volker: Machtverteilung im Gesundheitswesen, Frankfurt, 1994.

Chipkarten als Sicherheitswerkzeug in einem regionalen klinischen Tumorregister

Peter Pharow[1] · Bernd Blobel[1] · Petra Wohlmacher[2]

[1]Otto-von-Guericke-Universität Magdeburg
Institut für Biometrie und Medizinische Informatik
Peter.Pharow@Medizin.Uni-Magdeburg.DE
Bernd.Blobel@MRZ.Uni-Magdeburg.DE

[2]Universität Klagenfurt
Institute für Informatik – Systemsicherheit
petra@ifi.uni-klu.ac.at

Zusammenfassung

Die Implementierung einer verteilten elektronischen Krankenakte über die Organisations-grenzen einer Leistungsstelle hinaus erfordert eine vertrauenswürdige Kommunikation und Kooperation, die durch rechtliche, organisatorische und technische Bedingungen und Maßnahmen zu gewährleisten sind. Dies gilt um so mehr für onkologische Dokumentationssysteme mit ihren besonderen medizinischen, sozialen und ethischen Implikationen. Alle Dimensionen der Datensicherheit und des Datenschutzes sind dabei zu berücksichtigen, wobei kryptographische Techniken das Mittel der Wahl zur Realisierung der Sicherheitsanforderungen sind. Im regionalen klinischen Tumorregister Magdeburg/Sachsen-Anhalt wurden Chipkarten als Sicherheitswerkzeuge und eine entsprechende Sicherheitsinfrastruktur pilotierend eingeführt. Im vorliegenden Beitrag werden Bedrohungen, Sicherheitsanforderungen und Sicherheitslösungen in verteilten Informations- und Kommunikationssystemen diskutiert. Die Architektur der verwendeten Chipkarten und insbesondere die organisierte und implementierte TTP-Struktur werden vorgestellt. Aufgrund der Verallgemeinerung der Ergebnisse sind die Lösungen sowohl für das Gesundheitswesen als auch für andere Bereiche von Interesse.

1 Einleitung

Der zunehmende Einsatz der Informations- und Kommunikationstechnik (IK-Technik) stellen eine große Herausforderung für Gesetzgeber, Entwickler, Administratoren, Verantwortliche und Nutzer dar. Die IK-Technik ermöglicht enorme Steigerungen der Effizienz und Qualität der Prozesse, erschließt neue Anwendungsgebiete, birgt jedoch zugleich neue bzw. größere Risiken für die Betroffenen. Das gilt insbesondere für verteilte Informations- und Kommunikationssysteme, die die Abteilungsgrenzen, Unternehmensgrenzen und sogar Landesgrenzen überschreiten und zudem umfassende, oftmals sehr vertrauliche Informationen behandeln. Auch im Gesundheitswesen soll durch verzahnte Strukturen entsprechend dem *Shared-Care*-Paradigma [ElKö93] und deren adäquate informationelle Unterstützung [Blob95] eine Zunahme der Qualität und Effizienz der Versorgungsprozesse erreicht und der alle Industrieländer ergreifenden Krise des Gesundheitswesens begegnet werden. Verteilte Informations- und Kommunikationssysteme im öffentlichen Umfeld sind besonderen Bedrohungen hinsichtlich der Sicherheit der Daten sowie des Schutzes der Betroffenen ausgesetzt. Bedrohungen, Erfordernisse und Lösungen für die Datensicherheit werden in den folgenden Abschnitten am Beispiel einer regionalen elektronischen Krankenakte vorgestellt. Die Generalisierung der Betrachtungen erlaubt jedoch auch eine Transformation in andere Anwendungsbereiche. Die vorgestellten Aktivitäten sind in Projekte eingebunden, die von der Europäischen Kommission im Rahmen des *Telematics Applications Programme* gefördert werden.

2 Datensicherheitstypen und -domänen

Die Datensicherheit in verteilten (medizinischen) Informationssystemen betrifft die Sicherheit der Kommunikation aber auch die Sicherheit der Anwendungen, die miteinander kommunizieren und ggf. sogar kooperieren (Abb. 1). Grundlage jeglicher vertrauenswürdiger Kommunikation und Kooperation ist die Authentizität *(Authentication)* der beteiligten Partner *(Principals)*, wie Nutzer, Anwendungen oder Systeme, sowie die Integrität *(Integrity)*, Verfügbarkeit *(Availability)*, Vertraulichkeit *(Confidentiality)* und Verbindlichkeit *(Accountability)* der Daten bspw. die Nichtabstreitbarkeit *(Non-repudiation)* der Herkunft *(Origin)*, des Inhalts *(Content)* und des Empfangs *(Receipt)* der Daten. Da verschiedene *Principals* mit mehreren *Principals* kommunizieren und kooperieren können und die *Security Services* auf die Authentizität des *Principals* bezogen sind, kann die Kommunikationssicherheit mit den entsprechenden Services global verwaltet werden. Neben diesen *Security Services* oder Dimensionen der Kommunikationssicherheit ist die Anwendungssicherheit im Sinne der Zugriffsbeschränkung berechtigter *Principals* auf die Funktionen und Daten der Anwendungen, die Authentizität der Urheber bzw. Nutzer der Informationen sowie die Verbindlichkeit der Prozesse und Daten, die gespeichert und verarbeitet werden, zu gewährleisten.

Die Summe der rechtlichen Grundlagen, Verordnungen, Richtlinien, Beschränkungen, Befugnisse, Sanktionen, organisatorischen und technischen Bedingungen für die Vorbereitung, Implementierung und Nutzung von Informationssystemen wird als *Policy* bezeichnet. Um komplexe Informationssysteme beim Design, bei der Implementierung und beim Management hinsichtlich des Bedrohungsmodells und der Sicherheitsspezifikation handhaben zu können, gruppiert man Komponenten und Services nach gemeinsamen Gesichtspunkten wie dem rechtlichen Rahmen, dem Anwendungsgebiet, der Organisationsstruktur, Funktionalitäten, Verantwortlichkeiten, der technischen Basis etc. in sogenannte Domänen. Solche Sicherheitsdomänen repräsentieren Gruppen gegenseitigen Vertrauens, für die ein bestimmtes Risikoniveau und entsprechende Sicherheitsmaßnahmen definiert werden können. So können verteilte, einrichtungsübergreifende Informationssysteme gruppiert werden nach

- der *Security Policy Domain,*

- der *Security Environment Domain*, die die *Message Protection Domain* und die *Identity Domain* einschließt, und

- der *Security Technology Domain.*

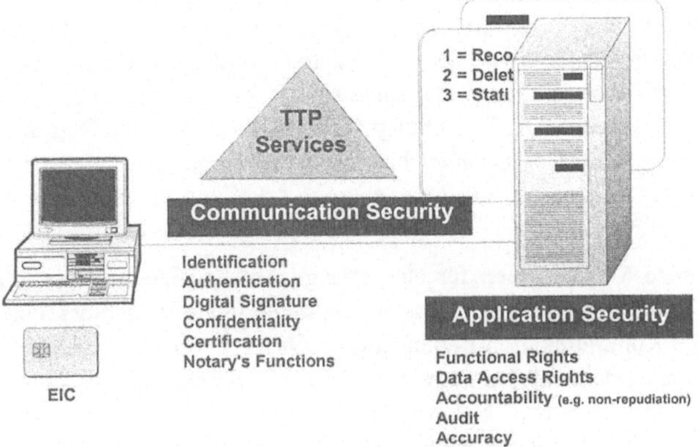

Abb. 1: Typen der Datensicherheit

Für weitere Details siehe auch [Blob97, BBM+96, SEIS96].

Die Anwendungssicherheit, die Legitimation im Sinne der Autorisierung, Verbindlichkeit und Verantwortlichkeit basiert auf einem entsprechenden rechtlichen und ethischen Rahmen, einer gemeinsamen *Security Policy* und *Codes of Conduct*. Die Zugriffskontrolle und das Management von Rechten und Privilegien ist hierbei lokal zu verwalten [Blo96a, Blob97, BlHo97, OMG_95].

3   Das regionale klinische Tumorregister Magdeburg/ Sachsen-Anhalt

Am Universitätsklinikum Magdeburg wurde in den vergangenen Jahren ein klinisches Tumorregister eingerichtet, welches von Beginn an als ein regionales elektronisches Informations- und Kommunikationssystem zur Unterstützung der onkologischen Versorgung in der Region Magdeburg/Sachsen-Anhalt konzipiert und implementiert wurde [Blo96a, Blob97]. Es betreut ein Einzugsgebiet mit etwa 1,2 Millionen Einwohner und umfaßt gegenwärtig faßt 60 Kliniken sowie die onkologische Nachsorgeleitstelle der KV Sachsen-Anhalt. Eine zunehmende Zahl von Einrichtungen arbeitet mit dem Register on-line. Im Jahre 1998 sollen alle Einrichtungen on-line mit dem Register kommunizieren und kooperieren. Das klinische Register dient der Verbesserung der Effizienz und Qualität der Versorgung von Tumorpatienten, der Unterstützung der Forschung sowie der Aus-, Weiter- und Fortbildung. Die Registeranwendung mit dem Namen „Gießener TumorDokumentationsSystem GTDS" [GTDS97] basiert auf der ORACLE-Entwicklungsumgebung. Ein derartiges offenes und verteiltes Dokumentationssystem, welches schrittweise zu einer verteilten interoperablen elektronischen Krankenakte für die Onkologie entwickelt wird, stellt besondere Anforderungen an den Datenschutz und die Datensicherheit, da es institutionen- und folglich domänenübergreifend ist. Das gilt sowohl für die *Policy*, die zwischen den beteiligten Partnern (z.B. Krankenhäuser, Ambulanzen, niedergelassene Ärzte, Reha-Einrichtungen) differiert, als auch für die Sicherheitsumgebung und die eingesetzten Sicherheitstechnologien. Solche Architekturen sind auf der Basis statischer und dynamischer Rechteverwaltung (*Negotiation/Commitment*) hinsichtlich ihrer *Policy* (*Policy Bridging*), ihrer Technologien einschließlich zugrundeliegender Systemdienste, ihrer Komponenten sowie der verwendeten Protokolle und Services offen zu gestalten und zu harmonisieren, d.h. neue Domänen sind zu bilden. Dies setzt eine strikte Orientierung auf Standards voraus. Die Bedrohungs- und Risikoanalyse, die Definition der *High Level Policy* sowie der *Codes of Conduct* basieren auf Ergebnissen des SEISMED- bzw. ISHTAR-Projektes der EU, in das die Magdeburger Abteilung eingebunden ist [SEIS96, ISHT97].

Ab 1993 wurden erste Mechanismen für eine strenge Authentifizierung der über Analog- bzw. später ISDN-Verbindungen angeschlossenen Systeme sowie Verschlüsselungsverfahren für eine vertrauliche Kommunikation implementiert. Dabei wurden für analoge Verbindungen ein *Modem Access Control System* sowie für ISDN-Verbindungen zwischen dem sicheren Server und „sicheren" externen Netzwerken spezielle LAN-Boxen bzw. zwischen dem Server und „sicheren" *Stand-alone*-PCs vergleichbare *Plug-in*-Karten eingesetzt. Das klinische Tumorregister war die erste Routineanwendung, die fortschrittliche Sicherheitsmechanismen für eine strenge Authentifizierung von *Principals* und für Integrität und Vertraulichkeit der Daten benutzte. Jedoch waren die starken Services und Mechanismen auf Systeme bezogen und lediglich über schwache Verfahren wie Paßwörter mit dem Nutzer verknüpft.

4 Security Services und Security Mechanisms

Security Services basieren auf *High Level Policies*, *Codes of Conduct* sowie organisatorischen und technischen Maßnahmen. Die Mechanismen, die die *Security Services* realisieren, beruhen im allgemeinen auf kryptographischen Methoden, auf die hier aber nicht im Detail eingegangen werden soll. Tabelle 1 zeigt die Beziehungen zwischen *Security Services* und *Security Mechanisms*.

Tabelle 1: Beziehungen zwischen Security Services und Security Mechanisms

Security Services	Security Mechanism(s)
Authentication (Peer and Data Origin)	Digital Signature, Authentication Protocols
Authorisation and Access Control	Digital Signature, Access Control Lists
Integrity	Check Values, Digital Signature
Confidentiality	Encryption
Availability	Backup Concepts, Accident Management, Key Recovery, Key Escrowing
Accountability	AuditTrails, Logs, Receipts
Non-repudiation (Origin and Receipt)	Check Values, Digital Signature

Die Sicherstellung der Verfügbarkeit der Informationen unter allen Bedingungen ist ein Erfordernis, welches auch im Falle der Vorhaltung von Daten in verschlüsselter Form zutrifft. Neben den üblichen Backup- und Havarie-Konzepten ist folglich auch bei Verlust oder Zerstörung des Verschlüsselungsschlüssels eine Entschlüsselung zu garantieren. Dazu existieren prinzipiell zwei Verfahren: die sichere Hinterlegung eines weiteren Exemplars des Schlüssels (*Key Escrowing*) oder die Generierung des Schlüssels aus dem Verschlüsselungsschlüssel (*Key Recovery*) ggf. unter Zuhilfenahme eines oder mehrerer Teilschlüssel, wobei in beiden

Verfahren die Zugriffsprozeduren entscheidend für die Akzeptanz dieser Mechanismen sind. Nach unserem Selbstverständnis dürfen diese Werkzeuge ausschliesslich den Eigentümern bzw. Verantwortlichen für die Daten, d.h. dem entsprechenden medizinischen Personal (in der Medizinischen Fachabteilungen bzw. am Arztarbeitsplatz), zugänglich sein. Es ist zu beachten, daß im Gegensatz dazu in manchen Ländern der algorithmen-orientierte terminus technicus „Key Escrowing" synonym als Möglichkeit zur Durchsetzung rechtlicher und staatlicher Erfordernissen verwendet wird.

Ziel des TrustHealth-1-Projektes[1] [THT197] war die Realisierung vertrauenswürdiger Informationssysteme auf der Basis einer Reihe von Spezifikationen für *Security Services* und *Interfaces* sowie einer entsprechenden Infrastruktur von *Trusted Third Party* (TTP) *Services*. Die *Security Services* werden hierbei durch Krypto-Chipkarten für Berufstätige im Gesundheitswesen, sogenannte *Health Professional Cards* (HPC), unterstützt [Wohl97].

4 1 Health Professional Cards

Unter Einbeziehung der Anforderungen aus der Telematik-Initiative des Landes Sachsen-Anhalt für die pilotierende Schaffung einer verteilten medizinischen Krankenakte in der Onkologie wurde im Mai 1997 die HPC nach der Spezifikation von TrustHealth 1 eingeführt. Die Karte wird in diesem Zusammenhang als Sicherheitswerkzeug betrachtet und für die Speicherung sowohl von geheimen Schlüsseln für die Authentifizierung, die Digitale Signatur und die Verschlüsselung als auch von anderen sicher zu verwahrenden Daten wie z.B. Links zu den Zertifikaten mit den öffentlichen Schlüsseln benutzt. Die HPC kann auf Basis verschiedener, bereits existierender Kartenbetriebssysteme implementiert werden, ebenso kann sie in verschiedene Konfigurationen integriert werden. Da Ergebnisse von Trust Health 1 in die Standardisierung eingeflossen sind, folgt die Karte neben den existierenden Standards von ISO 7816 auch bereits den in Vorbereitung befindlichen dieser Reihe. Die HPC unterstützt sowohl asymmetrische als auch symmetrische kryptographische Algorithmen.

Die Version der HPC, die die Anforderungen der Magdeburger Anwendung in der Zukunft zu erfüllen hat, wird letztendlich vier verschiedene geheime Schlüssel tragen. Einmal wird wie heute bereits neben der Verschlüsselung die Verwendung für Signaturen von der Verwendung für die Authentifizierung getrennt. Zum anderen wird ein weiterer vierter Schlüssel als Gruppenschlüssel (*Class Key*) für Gruppen von Professionals eingesetzt, der eine sichere Kommunikation, aber auch einen autorisierten Zugriff auf die gespeicherten Daten auf Patientenkarten oder in Archiven im Notfall oder bei Ausfall des eigentlich Berechtigten ermöglicht.

Optional kann die Karte einen symmetrischen Schlüssel enthalten, der für das Zusammenwirken mit Patientenkarten genutzt wird. Auch Daten für andere Anwendungen, z.B. deren Verschlüsselungsschlüssel, oder Zugriffsrechte für den Schutz des Systems oder für den Zugriff auf Dateien können auf der Karte abgelegt werden. Die letztere, der Anwendungssicher-

---

[1] Das TrustHealth-1-Projekt ist eng koordiniert mit dem „Deutschen Modellversuch - Health Professional Cards" (HPC), bei dem HPC für eine starke Authentifizierung sowie TTP Communication Security Services eingesetzt werden [AHPC97]. Die vorgestellte Lösung ist Teil dieser Vorhaben.

heit zuzuordnende Funktionalität wird jedoch der Anwendungsphilosophie entsprechend in Magdeburg nicht realisiert.

4.2 Trusted Third Party Services

Eine *Trusted Third Party* (TTP) darf formell keinesfalls mit den Interessen eines der kommunizierenden Partner verbunden sein, sondern muß unabhängig von ihnen allen handeln. Nur so ist es möglich, daß die beteiligten Partner Vertrauen in die Infrastruktur und in die angebotenen *Security Services* haben. Diese Dienste müssen durch die TTP in einer sicheren und vertrauenswürdigen Art und Weise realisiert werden, wobei die TTP durchaus als ein Oberbegriff für alle, in einer definierten Weise einen *Security Service* anbietenden unabhängigen Organisationen zu verstehen ist.

Um die Hauptaufgabe, das Anbieten der *Security Services,* realisieren zu können, ist einmal ein gewisser Grad an technischen und organisatorischen Maßnahmen notwendig. Dazu gehören ein sicheres IT-System sowie ein sicheres Kommunikationssystem. Von gleicher Bedeutung ist andererseits aber auch die rechtliche Stellung der TTP innerhalb ihrer Domäne. Im folgenden sollen diese Komponenten speziell für die hier betrachtete Infrastruktur näher erläutert werden. Dabei liegt das Hauptaugenmerk auf der praktischen Umsetzung der vorliegenden Ergebnisse. Auch hierzu finden sich detailliertere Ausführungen über die Funktionsverteilung innerhalb der Institutionen einer TTP und deren Zusammenwirken in einer Infrastruktur in [THT197].

Über Schnittstellen (Abb. 2) werden die notwendigen Informationen aus der uns bekannten „realen" Welt (Meldestellen, Prüfungsämter) an die Institutionen der TTP übermittelt. Mit diesen Informationen werden authentische Zusammenhänge zwischen einem Individuum und seinem öffentlichen Schlüssel (*Public Key Certificate*) sowie zwischen einem Individuum und seinem Beruf (*Professional Certificate*) hergestellt, die für die zu erbringenden *Security Services* benötigt werden. Basis für alle diese Dienste ist die Vergabe eines eindeutigen Namens (*Unique Name* UN, *Distinguished Name* DN) für alle Zertifikate. Für diese Aufgabe wird eine Instanz zur Namensvergabe (*Naming Authority* NA) benötigt. Unter Einbeziehung von Registrierungsinstanzen (*Registration Authority* RA bzw. *Professional Registration Authority* PRA) für die Bestätigung von Identität bzw. Beruf werden von den Zertifizierungsinstanzen (*Certification Authority* CA bzw. *Professional Certification Authority* PCA) Zertifikate ausgestellt und anschließend in öffentlichen Verzeichnissen (*Directory Service* DS) gespeichert und verwaltet (Änderungsdienst, Sperrlisten).

Die Rolle der hier aufgeführten Autoritäten (*Authorities*) wurden für die Anwendung Tumorregister beispielsweise von der Universität Magdeburg und der Landesärztekammer Sachsen-Anhalt übernommen, die hier die vertrauenswürdigen Instanzen in der realen, in Zukunft aber auch in der elektronischen Welt darstellen. Sie können dann z.B. sowohl als Autorität zur Vergabe eindeutiger Namen innerhalb der elektronischen Welt und zur Registrierung identitäts- und berufsbezogener Daten als auch als Autorität zur Erstellung von Zertifikaten agieren. In der ersten Phase des Modellversuchs wurde die Ärztekammer des Landes Sachsen-Anhalt als Autorität zur Bestätigung beruflicher Qualifikationen (*Professional Authentication Authority* PAA) eingesetzt. Die Ärzte und das administrative Personal stellten beim Tumorzentrum auf einem Papierformular einen Antrag auf ihre HPC. Die Anträge wurden von dort an die GMD weitergeleitet, die alle Autoritäten in der elektronischen Welt übernahm. Die personalisierten Chipkarten wurden von der GMD an das Tumorregister ausgehändigt, das die HPCs dann an alle Antragsteller persönlich aushändigte.

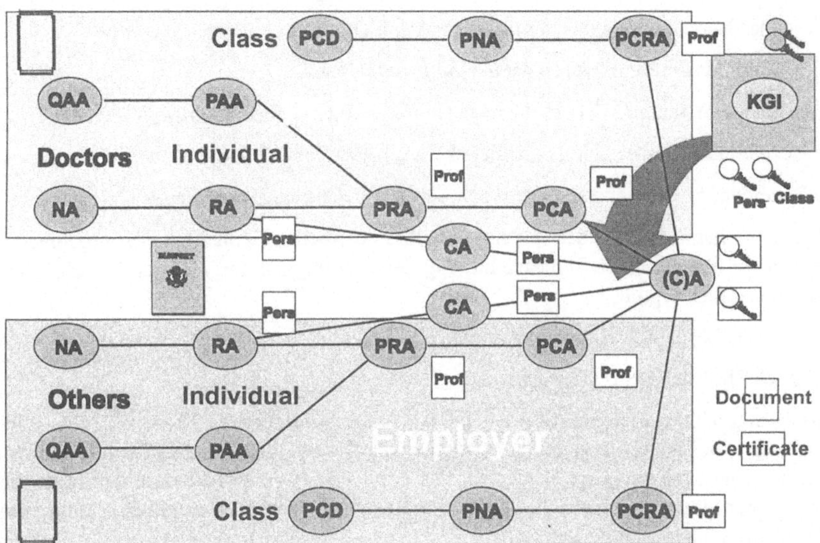

Abb. 2: TTP-Struktur im Klinischen Tumorregister Magdeburg/Sachsen-Anhalt

4.3 Directory

Um den Inhalt der erstellten PSE bzw. der Zertifikate in einem DS zu plazieren, ist eine weitere Schnittstelle erforderlich. Für die hier beschriebene Infrastruktur wurde das *Lightweight Directory Access Protocol* (LDAP) verwendet [UMIT96]. Die technische Seite der X.500-Lösung basiert auf dem Client-Server-Prinzip. Der Nutzer wird durch einen *Directory User Agent* (DUA), einer Komponente der Anwendungssoftware am Arbeitsplatz des Nutzers, unterstützt, welche den Zugriff auf den eigentlichen *Directory Service* realisiert. Der Dienst wird auf dem Server durch einen *Directory System Agent* (DSA) bereitgestellt. Und da die Nutzer diesen Dienst unabhängig von etwaigen Änderungen im Netzwerk (neuer Server, neue Verbindungen) nutzen können müssen, ist eine separate Ausführung der genannten Komponenten unabdingbar.

4.4 Naming

Die enge Verbindung zwischen der Vergabe des Namens und dessen Benutzung sowohl in den *Public Key*-Zertifikaten als auch den Berufszertifikaten (Attributzertifikaten) erfordert ein weltweit eindeutiges Namensschema. Daraus resultieren Anforderungen bezüglich Organisationsstruktur und funktioneller Struktur, denen sich die als Namensvergabe-Instanz (*Naming Authority* NA) wirkende Institution stellen muß.

Eingebunden werden

- der Name des Kartenbesitzers, hinter dem entweder

    - eine Person mit ihrem Namen (*Common Name* CN) bzw.

    - der Name der Organisation (*Organisation* O) oder

    - Organisationseinheit (*Organisation Unit* OU) steht,

- eine eindeutige Nummer innerhalb eines Numerierungsschemas (*Subject Number* SN),

- die Bezeichnung des Anbieters dieses Schemas (*Distributor* D) sowie

- die Bezeichnung des Landes, in dem der Anbieter beheimatet ist (*Country* C).

Das Namensschema besteht deswegen aus mehreren Ebenen, wobei die Verknüpfung der Attribute aller Ebenen unter Berücksichtigung der nachfolgend aufgeführten Strukturen die eindeutigen Namen ergibt:

- Fall 1: CN||SN||D||C,

- Fall 2: O||SN||D||C bzw. OU||O||SN||D||C .

In beiden angeführten Fällen wird die Eindeutigkeit des vergebenen Namens bereits durch die Attribute UN, D und C erzielt. Die Attribute CN, O bzw. OU werden hinzu- bzw. davorgesetzt, um eine Lesbarkeit der Namen für den Nutzer bzw. Anwender, also den Menschen zu ermöglichen. Im genannten Fall wurden nur Namen für Personen vergeben. Damit ergab sich die in Abb. 3 dargestellte Struktur.

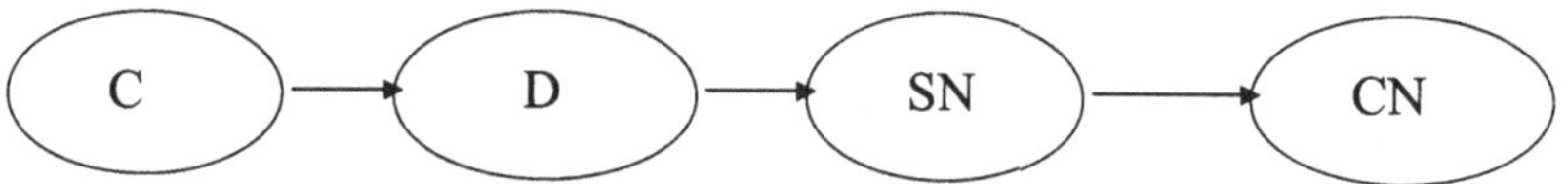

Abb. 3: Struktur der Verknüpfung der Komponenten des Namens

Für die Erzeugung der DN innerhalb des Modells der HPC-Anwendung wurde dieses allgemeingültige und für alle Bereiche anwendbare Schema detailliert und an die praktischen Gegebenheiten angepaßt. Als Anbieter des eindeutigen Numerierungsschlüssels (SN) und damit als NA agierten einmal die Ärztekammer Sachsen-Anhalt (LKS) für alle involvierten Ärzte sowie das Tumorzentrum e.V. (TRM) für die anderen Beteiligten (medizinische Dokumentationsassistenten und Forscher). Daraus ergab sich die in Abb. 4 dargestellte Struktur der Namen oder in der Form der Verknüpfung der einzelnen Attribute angegeben:

- PeterPharow.004.TRM.DE   bzw.

- HansEisenbart.007.LKS.DE

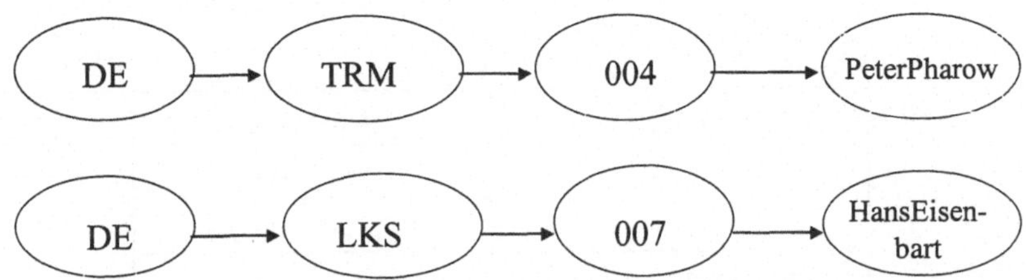

Abb. 4: Beispiele für die Namensvergabe durch TRM bzw. LKS

4.5 Security Toolkit

Für die Schnittstelle zwischen der Infrastruktur der Karte und der eigentlichen Anwendung wurde das Sicherheits-Toolkit SecuDE (*Security Development Environment*) [Secu97] verwendet und unter maßgeblicher Mitarbeit der GMD für die spezifischen Belange der Anwendung Tumorregister angepaßt. SecuDE stellt neben einer Implementierungsunterstützung verschiedene *Security Services* wie die Authentifikation des Benutzers, die Integrität der Daten, die Nichtabstreitbarkeit von Herkunft und Empfänger sowie die Vertraulichkeit der Daten zur Verfügung. Dies wird auf der Basis von symmetrischen (wie DES, Triple DES oder IDEA) und asymmetrischen (wie RSA oder DSS) kryptographischen Algorithmen realisiert. Außerdem sind verschiedene bekannte *Hash*-Funktionen (wie MD5, SHA, Sqmodn oder RIPEMD160) implementiert. Über ein *Personal Security Environment* (PSE) sind die angegebenen Funktionalitäten ansprechbar und benutzbar.

SecuDE ist in der Lage, die sicherheitsrelevanten Funktionen von *Cross*-Zertifizierung und Zertifikats-Sperrlisten zu behandeln und auf X.509 *Public-Key*-Zertifikaten aufzusetzen. Außerdem werden Funktionen und Bibliotheken angeboten, die für die Arbeit einer Zertifizierungsinstanz sowie für das Zusammenspiel zwischen der Instanz und den von ihr zertifizierten Nutzern erforderlich sind [Hühn97].

5 Die Magdeburger Version der HPC

Die für die Anwendung Tumorregister verwendete und am Markt verfügbare Chipkarte ist vom Typ STARCOS PK 1.0 [STAR94]. Die STARCOS PK basiert auf dem Single-chip Philips 83C852, dessen technischen Details aus Tabelle 2 und Speicherpartitionierung aus Abb. 5 ersichtlich sind. Da diese Karte nicht alle Forderungen erfüllen konnte, ist die Umsetzung auf die neue Version STARCOS PK 2.0 im ersten Halbjahr 1998 geplant [Meis97].

Die eingesetzte Chipkarte wurde durch den Hersteller bereits vorpersonalisiert geliefert. Das bedeutete, daß die Dateistruktur für die Karte bereits definiert und generiert war und sich auf der Karte befand. Dabei wurde für jede der Dateien eine bestimmte Größe an Speicherplatz benötigt, der durch den Inhalt der jeweiligen Datei festgelegt und durch das verwendete Sicherheits-Toolkit SecuDE bestimmt worden war.

Tabelle 2: Technische Details der Philips 83C852

Storage:	256 byte RAM
	6 kbyte ROM
	2 kbyte EEPROM
EEPROM-characteristics:	100.000 write-/erase cycles
Addressing	8 bit
Frequency:	1 to 6 MHz
Voltage:	5V
Contacts:	corr. to ISO/IEC 7816-2

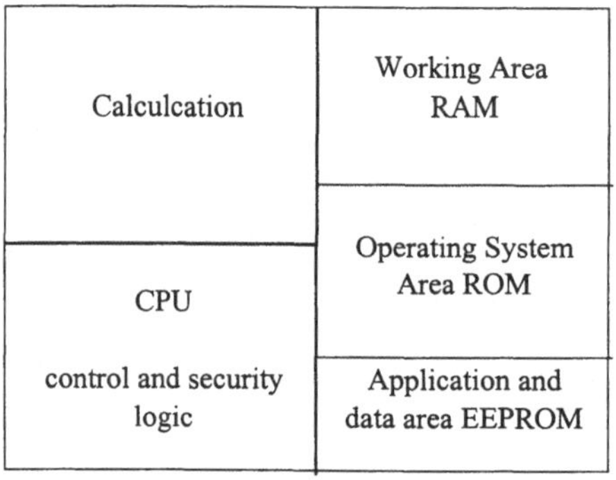

Calculcation	Working Area RAM
CPU control and security logic	Operating System Area ROM
	Application and data area EEPROM

Abb. 5: Speicherpartitionierung

In diesem vorbereiteten Zustand wurden die Karten an die GMD übergeben. Dort erfolgte die eigentliche Personalisierung. Dafür wurden die *Distinguished Names* (DN) benötigt, die vorher durch die Landesärztekammer bzw. das Tumorzentrum definiert wurden. Ein *Application Identifier* wurde im DF_AID gespeichert und das PSE-Objekt (*Personal Security Environment*) erstellt. Wegen des begrenzten Speicher der bisherigen STARCOS PK-Karte war es nicht möglich, die komplette PSE auf der Karte abzulegen. Eine Konfigurationsdatei enthält nun alle die PSE-Objekte, die auf der Karte gespeichert sind. Alle anderen Objekte sind in einer sogenannten SC-PSE-Erweiterung in einem Verzeichnis auf der Festplatte abgelegt. Diese SC-PSE-Erweiterung wird automatisch während der Erzeugung einer PSE aufgebaut. Die PIN für das Öffnen der PSE-Erweiterung ist aus Sicherheitsgründen auf der Karte gespeichert. Sie verbindet den Teil der PSE, der sich auf der Karte befindet, mit der auf der Platte abgelegten SC-PSE-Erweiterung.

Dieser Prozeß der Authentifizierung kann entweder zyklisch oder aber immer dann wiederholt werden, bevor eine sicherheitsrelevante Operation (wie z.B. die Erzeugung einer Digitalen Signatur) durchgeführt wird.

Abb. 6 zeigt die implementierte Dateistruktur der Karte. Die Datei EFɪᴄᴄsɴ enthält eine eindeutige Seriennummer des ICC in der Form eines *Tag-Length-Value*-Datenobjekts (TLV). Die Datei EFᴅɪʀ existiert nur, wenn die Karte die indirekte Anwendungsauswahl unterstützt. Die Datei EFᴄʜɴ schließlich enthält den Namen des Karteninhabers, ebenfalls als TLV-Datenobjekt [Stru97].

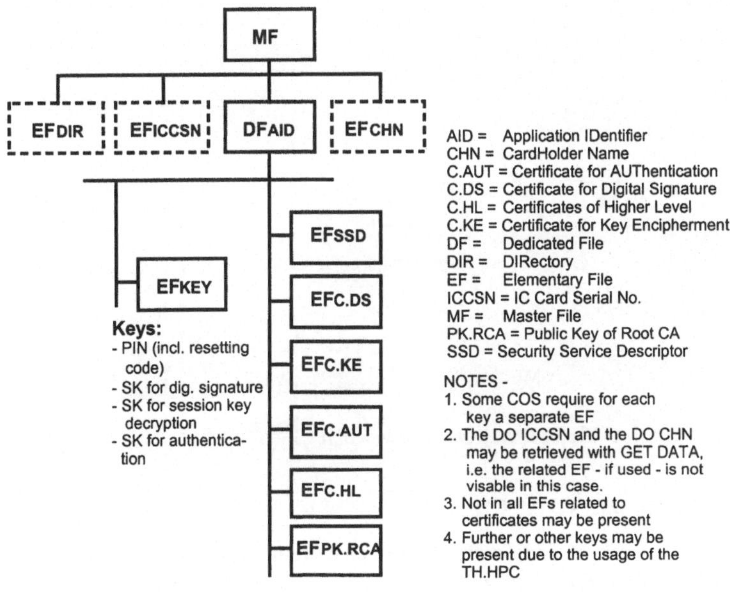

Abb. 6: Die Dateistruktur der HPC

6 Die Magdeburger Implementierung

Um auf die Karte lesend zugreifen zu können, wird ein Multifunktionales Kartenterminal [AMKT96] verwendet, welches mit einer Tastatur nach ISO/IEC 9564 und einem LCD-Display ausgestattet ist.

Speziell zugeschnitten auf die Anforderungen der medizinischen Anwendung und orientiert an den aktuell verfügbaren Chipkarten, mußten anfangs kleine Abstriche an der Durchsetzung der eigentlichen Planung und Vorbereitung des Einsatzes der Karten gemacht werden. Zwei verschiedene Schlüsselpaare wurden durch SecuDE erzeugt: ein 512 Bit (in Zukunft 1024 Bit) langes RSA-Schlüsselpaar für die Erzeugung bzw. Prüfung der Digitalen Signatur, und ein 512 Bit (in Zukunft 1024 Bit) langes RSA-Schlüsselpaar für Ver- und Entschlüsselung. Die Zertifikate, in denen neben weiteren Informationen die korrespondierenden öffentlichen Schlüssel enthalten sind, entsprechen dem Standard X.509 V1. SecuDE ist in der Lage, zwei verschiedene X.509 V1 Zertifikate zu verwalten, die für mehrere Zwecke benutzt und durch ein innerhalb von SecuDE geführtes *Naming* unterschieden werden können. Zertifikate nach dem Standard X.509 V3 sind geplant und werden für die nächste Phase der Implementierung eingesetzt, zusammen mit der nächsten Generation von Chipkarten, die dann den innerhalb von TrustHealth 1 definierten Spezifikationen entsprechen werden.

Abb. 7 gibt einen Überblick über die Prozesse zwischen Client und Server.

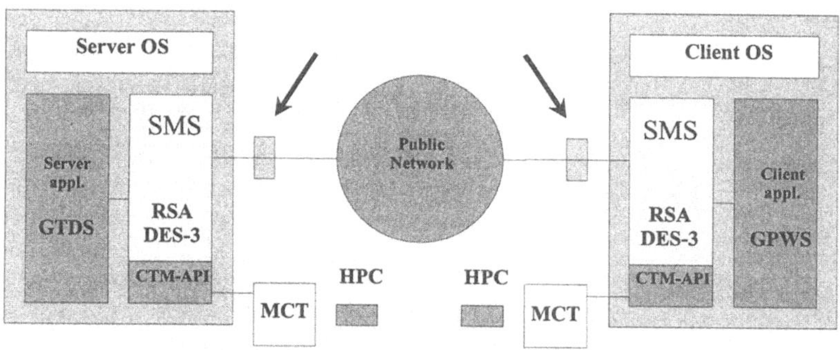

Abb. 7: Die Prozesse zwischen Client und Server

7 Schlußbemerkungen

Durch die EU wurden in den letzten Jahren eine Reihe von Projekten wie TrustHealth, ISHTAR, EUROMED-ETS und andere initiiert, die sich mit Sicherheitsfragen befassen [BlPh97]. Die dabei vor allem von TrustHealth spezifizierten Lösungen für *Security Services*, Schnittstellen und eine Sicherheitsinfrastruktur können genutzt werden, um sichere Anwendungen zu entwickeln - und das nicht nur im Gesundheitswesen. Die Möglichkeiten der HPC und der zugehörigen, von der TTP angebotenen *Security Mechanisms* und *Security Services* wurden herangezogen, um ein verteiltes und interoperables Informationssystem wie die elektronische Krankenakte (EHR) zu implementieren. Weitere Schritte von Forschung und Entwicklung sowie erste Implementationsstufen sind heute darauf gerichtet, die Lösungen und die auf HPC und TTP basierende Sicherheitsinfrastruktur von TrustHealth auch in *Middleware-Architectures* [Blob97, BlHo97] zu integrieren. Die auf dem Prinzip des *Shared Care* beruhenden Informationssysteme im Gesundheits- und Sozialwesen erfordern bereits heute derartige Lösungen, die auf internationalen oder zumindest europäischen Standards basieren.

8 Danksagung

Die Autoren danken der Europäischen Kommission sowie dem Ministerium für Bildung und Wissenschaft des Bundeslandes Sachsen-Anhalt für ihre Unterstützung. Außerdem danken sie den Partnern innerhalb des Projektes TrustHealth und dabei insbesondere der GMD

Darmstadt für die sehr gute Zusammenarbeit bei der Planung, Vorbereitung und Einführung der Chipkartenlösung für die medizinische Anwendung Tumorregister Magdeburg.

Literatur

[AMKT96]   Arbeitsgemeinschaft (1996): AG „Karten im Gesundheitswesen", GMD-Forschungszentrum Informationstechnik GmbH: Multifunktionale Karten-Terminals (MKT) für das Gesundheitswesen und andere Anwendungsgebiete. Spezifikation Version 1.0.

[AHPC97]   Arbeitskreis (1996): AG „Karten im Gesundheitswesen", AK „Health Professional Cards": Deutscher Modellversuch „HPC". Version Dezember 1997.

[BBM+96]   Blobel, B., Bleumer, G., Müller, A., Flikkenschild, E., and Ottes, F. (1996): Current Security Issues Faced by Health Care Establishments. *Deliverable of the HC1028 Telematics Project ISHTAR*, October 1996.

[Blob95]   Blobel, B. (1995): GSG 93 und GNG 95 - Umstrukturierung der Krankenhaussysteme. *klinikarzt* 10/24, S. 491-499.

[Blo96a]   Blobel, B. (1996a): Clinical Record Systems in Oncology. Experiences and Developments on Cancer Registries in Eastern Germany, in *Preproceedings of the International Workshop "Personal Information - Security, Engineering and Ethics"* pp 37-54, Cambridge, 21-22 June, 1996, also published in *Personal Medical Information - Security, Engineering, and Ethics* (edr. R. Anderson), pp 39-56. Spinger, Berlin 1997.

[Blob97]   Blobel, B. (1997): Security Requirements and Solutions in Distributed Electronic Health Records, in *Information Security in Research and Business* (edrs. L.Yngström and J.Carlsen), pp 377-390. Chapman & Hall, London.

[BlHo97]   Blobel, B. and Holena, M. (1997): Security Threats and Solutions in Distributed, Interoperable Health Information Systems Using Middleware, in *New Technologies in Hospital Information Systems* (edrs. J. Dudeck, B. Blobel, W. Lordieck and T. Bürkle). IOS Press, Amsterdam.

[BlPh97]   Blobel, B., Pharow, P. (1997): EUROMED; ISHTAR, HANSA und MED-SEC, in: *Datenschutz und Datensicherheit (DuD)*, 10/97, S. 598-599.

[ElKö93]   Ellsässer, K.-H., und Köhler, C.O. (1993): Shared Care - Konzept einer verteilten Pflege - Kurz- und langfristige Perspektiven in Europa. *Informatik, Biometrie und Epidemiologie in Medizin und Biologie* 24, H.4, S. 188-198.

[GTDS97]   GTDS (1997): Gießener Tumordokumentationssystem (GTDS), *Programmdokumentation 1.0.* Entwicklung des Instituts für Medizinische Informatik der Universität Gießen.

[Hühn97]   Hühnlein, D. (1997): On the development of a security toolkit for open networks - New security features in SECUDE. *DuD-Fachbeiträge: Verläßliche IT-Systeme*, S. 113-118

[ISHT97]   The ISHTAR Consortium (edr.) (1997): Implementing Secure Healthcare Telematics Applications in Europe. http://euromed.iccs.ntua.gr/

[Meis97]   Meister, G. (1997): Classification and Structure of Certificates, in *Health Cards 97*, S. 216-222, IOS Press, Amsterdam

[OMG_95]   OMG (1995): The CORBA Security Specification. *OMG Doc.*No. 95-12-01.

[Secu97]   GMD (1997): SecuDE 5.1 - Hyperlink Documentation, http://www.darmstadt. gmd.de/secude/doc/

[SEIS96]   The SEISMED Consortium (edr.) (1996) Data Security for Health Care, Volume I - III. IOS Press, Amsterdam.

[STAR94]   Giesecke & Devrient (1994): STARCOS-Beschreibung: Referenz-Handbuch STARCOS PK, 1.0. Ausgabe Aug. 94.

[Stru97]   Struif, B. (1997): Security Functions of Health Professional Cards, in: *Health Cards 97*, S. 211-215, IOS Press, Amsterdam.

[THT197]   The TRUSTHEALTH Consortium (edr.) (1997): Trustworthy Health Telematics 1 project: Project Deliverables, 1996-1997, http://www.ehto.be/projects/ trusthealth

[UMIT96]   University of Michigan Information Technology Division: LDAP servers, client library and sample text based UNIX clients, copyright (c) 1992-1996 Regents of the University of Michigan. ftp://terminator.rs.itd.umich.edu/ldap/ ldap-3.3.tar.Z, http://www.umich.edu/rsug/ldap/

[Wohl97]   Wohlmacher, P. (1997): Die Sicherheitsdienstleistungen von Health Professional Cards. In: BSI (Hrsg.): *Mit Sicherheit in die Informationsgesellschaft*, Tagungsband 5. Deutscher IT-Sicherheitskongreß des BSI 1997, SecuMedia Verlag, Ingelheim, 1997, S. 261-277.

Individualisierte Kommunikationsumgebung mit kontaktlosen Chipkarten

Hans-Günter Lindner[1] · Igor Jaceniak[2]

[1]humanIT Human Information Technologies GmbH, GMD TechnoPark
lindner@humanIT.de

[1,2]GMD – Forschungszentrum Informationstechnik GmbH
Igor.Janceniak@gmd.de

Zusammenfassung

Überall nutzbare und personalisierte Informationsversorgung (UPI), gewinnt aufgrund der zunehmenden Informationsüberflutung an Bedeutung. Die Möglichkeit einfach und sicher Informationen zu erhalten, die an die Situation, den Kontext und die individuellen Bedarfe angepaßt sind, ist technisch zwar schon heute realisierbar aber aufgrund der Komplexität des Zusammenspiels zwischen Mensch und Maschine und der technischen Komponenten noch nicht im kommerziellen Einsatz.

Ein Versuch, den Nutzen von UPI-Systemen in einem abgegrenzten Umfeld zu kommunizieren, ist der Einsatz einer individualisierten Kommunikationsumgebung für Konferenzen und Messen. Die Besucher sind hierbei auf zielgerichtete und persönliche Informationsversorgung in einer veränderlichen Umgebung angewiesen. Die Software für diese Umgebung wurde bereits auf der UM97 getestet und durch eine Teilnehmerbefragung evaluiert. Der Einsatz kontaktloser Chipkarten ersetzt dabei nicht nur die manuelle Eingabe einer persönlichen Identifikation, sondern eröffnet auch die Chance erforderliche Sicherheitsbedingungen zu erfüllen.

Im Rahmen des Projektes „aKISS für UPC" wird die am GMD – Forschungszentrum Informations technik GmbH, Abteilung Mensch-Maschine Kommunikation entwickelte Software auf der OMNI-CARD98 in Berlin eingesetzt. Das System verschafft den Teilnehmern Zugang zum E-Mail und der Versorgung mit Informationen, die für die Konferenzteilnahme relevant sind.

1 Überall nutzbare und personalisierte Informationsversorgung - UPI

1.1 Begriffsdefinition

Überall nutzbare und personalisierte Informationsversorgung bzw. Ubiquitous Personalized Information (UPI) ist die Vision der schnellen und einfachen Lieferung persönlich relevanter Informationsdienste an jedem Ort, unabhängig von der Funktionsweise der jeweiligen Endgeräte. UPI basiert auf den Konzepten benutzeradaptiver Systeme [Kobs93, UM97] und des

Ubiquitous Computing [Weis91], das mittlerweile synonym zu „mobile computing" verwendet wird.

Das Ziel von UPI ist die Kombination von Terminal- und Personenmobilität. Terminalmobilität impliziert eine fest definierte Distanz zwischen Nutzer und Endgerät wie z.B. einem Mobiltelefon. Personenmobilität hingegen beinhaltet die Identifizierung, beispielsweise durch eine Chipkarte, an einem feststehenden Terminal, das dann die nutzerbezogenen Daten liefert. Die häufig diskutierten gegensätzlichen Alternativen der Chipkarte als zentralem Element gesicherten Datenzugriffs an unterschiedlichen Terminals und des multifunktionalen persönlichen Endgerätes verschmelzen damit zu einer Einheit. Die Nutzung unterschiedlicher Endgeräte wird dabei so alltäglich wie das Verwenden von Papier und Bleistift. Die nutzende Person tritt in den Vordergrund und nicht wie bisher das Endgerät. Das bloße „Nehmen" eines Computer macht ihn zum eigenen - mit allen Daten, Privilegien und Leistungen des Nutzers. Das „Weglegen" macht den Computer dann wieder zu einem beliebigen Gegenstand wie z.B. ein Block Papier [Walk94].

UPI-Systeme sind sozio-technische Systeme. Sie erfordern schon bei der Konzeption die intensive Auseinandersetzung mit den Nutzern, die Berücksichtigung des persönlichen Umfeldes und die dafür notwendige Privatsphäre. Weitere Systembestandteile sind neben Netzwerk und Endgeräten die notwendige Energieversorgung. Der wesentliche Erfolgsfaktor für derartige Systeme ist die effektive Interaktion. Dies erfordert eine intensive Analyse des gegenseitigen Bewußtseins und der Interaktion zwischen Mensch und Maschine während der Nutzung.

1.2 GMD - Prototypen und Vorteile

Im Rahmen der Forschungsarbeiten zur Mensch-Maschine Kommunikation am GMD – Forschungszentrum Informationstechnik GmbH, Sankt Augustin, wurden mehrere Prototypsysteme zur benutzeradaptiven und ubiquitären Informationsversorgung entwickelt. Der einfache und sichere Zugang zu diesen Systemen erfolgt mit Hilfe von Chipkarten, die benutzerspezifische Daten und Präferenzen speichern. Ziel ist es, die Einsatzmöglichkeiten individueller Anpassung von Informationsdienstleistungen an die Nutzer ubiquitärer Systeme zu untersuchen und den Nutzen für den praktischen Einsatz zu bestimmen.

Für jeden Benutzer, der mit seiner Chipkarte einen entsprechend ausgestatteten Computer oder ein Terminal aufsucht, wird automatisch eine individuelle Informationsumgebung bereitgestellt. Die auf einer Chipkarte befindlichen Informationen werden für Anpassungen der Benutzungsoberfläche und Funktionalitäten verwendet wie z.B. Landessprache, Schriftgröße, Kontrast, verwendete Ein-/ Ausgabegeräte, Internet-Konfiguration etc. Im weiteren sind bevorzugte und am Standort sinnvolle Anwendungsprogramme in aufgabenspezifischer Konfiguration mit den aktuellen nutzerbezogenen Daten sofort verfügbar.

Als Vorteile für den Benutzer ergeben sich:

1. bekannte Bedienung und Benutzungsoberfläche durch benutzeradaptive Anpassung,
2. vereinfachte Identifikation durch Chipkarten,
3. Wahrung der Privatsphäre durch Sicherheitsmechanismen,
4. konsistente Datenhaltung durch die überall verfügbaren, eigenen Daten,
5. fehlerfreieres und schnelleres Arbeiten durch automatische Konfiguration.

Zur Evaluation wurden mehrere Versionen der Personalisierungssoftware entwickelt, die in den Projekten TEDIS, AVANTI und auf der Konferenz User Modeling 1997 - UM97 zum

Einsatz kam. Unterschiedliche Personengruppen beurteilten das System auf Messen, Tagungen und Seminaren durchweg positiv.

Im Projekt TEDIS wird Behinderten und älteren Menschen die Rechnernutzung ohne fremde Hilfe erleichtert. Das System paßt sich selbsttätig den Behinderungen an und bietet den Zugriff auf den persönlichen Arbeitskontext, die geeignete Darstellung in Kombination mit unterschiedlichen Zugangsmöglichkeiten wie z.B. Kopfmaus oder die persönliche Spracherkennung. In den meisten Fällen wird durch die automatische Identifizierung und Adaption eine Nutzung von Computern überhaupt erst ermöglicht. [PiGH97]

Das System für AVANTI liefert Nutzern im Rahmen eines regionalen Informationssystems individuelle Informationen. AVANTI erlaubt einem Nutzerkreis mit verschiedenen Vorkenntnissen und Interessen, z.B. Touristen, Einwohnern, Reisekaufleuten, aber auch bestimmten Behindertengruppen wie Blinden und Rollstuhlfahrern eine adäquate multimediale Informationsversorgung über eine Region. Für die Individualisierung der Information werden Annahmen über Interessen, Kenntnisse und Fähigkeiten modelliert. Da das System Internet-Technologie nutzt, kann es von zu Hause, von öffentlichen Informationsterminals, von Büros oder von mobilen Geräten aus genutzt werden. [FiKN97], [FiKJ97]

Im Rahmen eines Feldexperimentes auf der UM97 in Sardinien [UM97] wurde ein UPI-System für den praktischen Einsatz evaluiert. Die kontaktlose HITAG^{TM} -Technologie von Philips Semiconductors kam hierbei zum Einsatz. Die Energieversorgung der Karten und der Datenaustausch mit dem Terminal basieren auf einer Radiofrequenz-Technologie (RFID). Die Karten müssen demnach nicht in ein Kartenterminal eingeführt werden und können beispielsweise in der Hosentasche oder Geldbörse verbleiben. Die Chips können aber auch in beliebige Gegenstände des täglichen Gebrauchs wie Armbanduhren integriert werden. Die Distanz zwischen Karte und Terminal kann eine Reichweite von bis zu 1 m betragen.

Die Konferenzteilnehmer bewerteten die Bedienung als einfach, effizient und klar. Sie erachteten die Nutzung kontaktloser Chipkarten im technischen und sozialen Bereich als unkritisch. Für eine erfolgreiche Nutzung nannten die Teilnehmer als wichtigste Voraussetzungen die Transparenz hinsichtlich Sicherheit und Privatsphäre, die Zuverlässigkeit der Software und die überall vorhandene Verfügbarkeit der Endgeräte. [Jace97]

Auf der Grundlage dieser theoretischen, praktischen und empirischen Ergebnisse ist ein neues System für „Überall nutzbare und personalisierte Informationsverarbeitung" implementiert worden, das im Januar 1998 auf der Omnicard 98 vorgestellt wird. [LiWo97]

2 Das adaptive Kernsystem

Die überall vorhandene Nutzungsmöglichkeit von Informationsdiensten durch verfügbare Endgeräte und die sichere Datenübertragung mit Hilfe Chipkarten ist nur ein erster Schritt, um UPI Wirklichkeit werden zu lassen. Entscheidend für die Nutzbarkeit und Akzeptanz von UPI ist die Bereitstellung individuell relevanter Daten und Dienste. Dies erfordert vielfältige Anpassungsleistungen an Personen, Sicherheit, Interaktionsverhalten, Ort, Zeit, Energieversorgung, Netzwerk und die verfügbaren Endgeräte. Diese Komponenten sind häufig derart miteinander kombiniert, daß hoch-nichtlineares Verhalten in Verbindung mit diskreten und kontinuierlichen Werten berücksichtigt werden muß.

Die Darstellung des Mensch-Maschine-Systems als modellgestütztes, adaptives Regelsystem erlaubt einerseits die Reduzierung der Komplexität für die Konzeption und Entwicklung andererseits die Darstellung der dynamischen Zusammenhänge, unabhängig vom gewählten Einsatzbereich.

Die modulare Darstellung in Abbildung 1 zeigt den Aufbau eines UPI-Systems; die einzelnen Funktionen können sich jedoch in realen Anwendungen aus Gründen der Praktikabilität überschneiden. Das System besteht aus Anwendungssystem (AS), Nutzer, HyperStructure Adaptor (HSA), Chipkartensystem und verteilten Daten- und Modellbanken sowie deren Verwaltungsmodule.

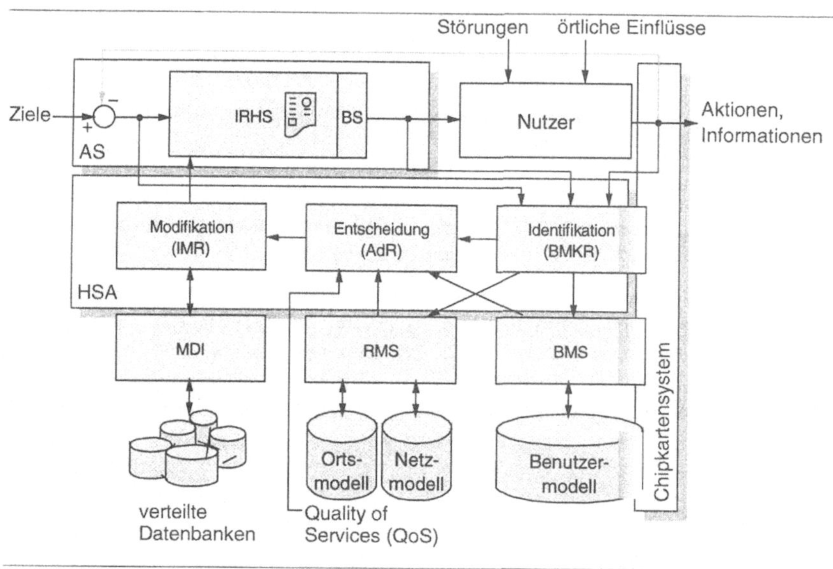

Abb. 1: Das adaptive Mensch-Maschine-System für UPI

Die Vorgaben, die sich aus den Anforderungen der Nutzer und dem Kontext ergeben, sind die Ziele und damit die Eingabe des Gesamtsystems. Die Zielwerte werden dann mit den tatsächlich erreichten Zielen des Benutzers verglichen. Die ermittelte Abweichung wird einem InformationsRegler für HypermediaStrukturen (IRHS) zugeführt. Der IRHS bestimmt daraufhin die Stellgröße, die über die Benutzungsschnittstelle (BS) gezielt auf den Benutzer einwirkt. Die Aktionen des Benutzers und die erzielten Informationen bilden den Ausgang des Regelkreissystems.

Der adaptive Kern besteht aus HSA, Multimedia Database Interface (MDI) sowie den Verwaltungskomponenten Ressourcen- und BenutzerModellierungsSystem (RMS und BMS). Der HSA umfaßt die drei Stufen Identifikation, Entscheidung und Modifikation:

Die Identifikationsstufe erfaßt das Abweichungssignal sowie die Interaktion mit der Benutzungsschnittstelle und wickelt somit auch die Identifizierung und Authenifizierung in Zusammenarbeit mit dem Chipkartensystem ab.

Die Entscheidungsstufe enthält Adaptionsregeln (AdR), die das dynamische Konstruieren von Hypermedia-Seiten steuern. Die Entscheidungen können über Parameter beeinflußt werden, die die Qualität auf die Dienste (QoS) einwirken.

Die Modifikationsstufe dient der angepaßten Änderung der Inhalte und umfaßt InhaltsModellRegeln (IMR) die z.B. den Detaillierungsgrad beschreiben. Über das MDI greift diese Stufe auf die multimedialen Inhaltsobjekte zu, die in heterogenen, verteilten Datenbanken vorliegen können. Die Modifikation wird schließlich an den IRHS weitergeleitet.

Das BMS, das als Server realisiert ist [KoPo95], kann eine individuelle inhaltliche Anpassung an den Benutzer vornehmen. Es speichert die Benutzermodelle aber auch Protokolldaten in Datenbanken und auf den Chipkarten der Nutzer. Bei der Benutzermodellierung werden Annahmen über das Wissen, die Ziele und die Vorstellungen des Benutzers mit Hilfe von BenutzerModellKonstruktionsRegeln (BMKR) getroffen. Hierzu zieht das BMS Rückschlüsse aus dem Interaktionsverhalten des Benutzers. Das BMS repräsentiert und verwaltet die Annahmen, löst Inkonsistenzen auf und bedient das Anwendungssystem mit aktualisierten Daten. Ebenso verwaltet das BMS das konzeptuelle Wissen des Anwendungsgebietes, die Merkmale und Zuordnungen unterschiedlicher Benutzergruppen sowie das didaktische Wissen.

Das RMS erhält Informationen zu den Ressourcen von Orten und Netzwerken, d.h. Endgeräte, Schnittstellen, Leitungskapazität und legt diese Daten in entsprechenden Datenbanken ab.

Das Chipkartensystem besitzt Schnittstellen zur Benutzungsschnittstelle, den BMKR, dem BMS und dem Benutzermodell. Die Sicherheit der Privatsphäre und des Betriebs sowie die Robustheit werden in Verbindung mit dem Einsatz von Chipkarten dadurch erhöht, daß die Zusammenhänge der Dienste und die Referenzen sowie die Annahmen über den Benutzer auf der Chipkarte gespeichert werden. Das Benutzermodell kann daher verteilt sowohl in der Datenbank auf einem Server, auf einem Client, aber auch auf der Chipkarte vorliegen. Das BMS liefert somit nur dann sinnvolle Annahmen, wenn die im verteilten System gespeicherten Daten zusammengeführt werden und auch zusammenpassen. Der Einsatz von Chipkarten bietet zudem den Vorteil, daß der Benutzer selbst die über ihn getroffenen Annahmen und unabhängig vom Gesamtsystem z.B. auf seinem eigenen Computer oder einem dafür geeigneten Endgerät pflegen kann.

3 Eine individualisierte Kommunikationsumgebung

3.1 Ziel des Projektes

Ziel des Projektes ist der Aufbau eines mobilen Systems, das die Vorteile ubiquitärer Leistungen einem größeren Fachpublikum direkt durch Eigennutzung vermittelt. Der Einsatz einer individualisierten Kommunikationsumgebung soll auf Konferenzen und Messen erfolgen, da dort die Besucher auf zielgerichtete und persönliche Informationsversorgung in einer veränderlichen Umgebung angewiesen sind. Zudem sind die Teilnehmer für eine Diskussion und Evaluation derartiger Systeme im Rahmen der genannten Veranstaltungen offener als an ihrem Arbeitsplatz.

Der Einsatz einer individualisierten Kommunikationsumgebung soll für Themengebiete erfolgen, für die in Zukunft Inhaltspersonalisierung eine wichtige Funktion übernehmen wird. Für das Fachpublikum im Themenfeld Chipkarten wird dieses Thema an Bedeutung gewinnen, da sich der Markt inhaltlich orientierter Dienste kontinuierlich vergrößert und diese Inhalte ohne persönlichen Zuschnitt und Wahrung der Privatsphäre wenig Aussicht auf Akzeptanz haben werden. Der Absatz von Chipkarten erhöht sich aufgrund der zunehmenden

Anwendungsgebiete und Marktsegmente, wenn inhaltlich benutzerangepaßte Dienste hinzugefügt werden. Da sich die Nutzer eine überschaubare Anzahl persönlicher Chipkarten wünschen und die angepaßten Dienste im Rahmen eines „cross-selling" besser miteinander kombiniert werden können, dient das Miteinander von Chipkarten und benutzeradaptiven Systemen als Katalysator für künftige Erfolge am Markt.

Mit dem Einsatz der individuellen Kommunikationsumgebung solle den Nutzern am eigenen Beispiel die Vorteile von UPI vermittelt werden. Die Präsentation der Funktionalität alleine führt noch zu keinem „Aha-Effekt", erst die eigene Erfahrung der einfachen Nutzung von Chipkarten und individuellen Informationsdiensten. Die Vorteile sind:

- bequemer Zugriff auf persönlich relevante und aktuelle Informationen,

- Informationsversorgung auch ohne eigenen tragbaren Computer,

- Gezielte personen- und ortsangepasste Informationsverteilung,

- Handlungsfähigkeit auch während der Abwesenheit vom Arbeitsplatz,

- leichtere Bedienung durch Anpassung an den Nutzer und die Gegebenheiten am Standort,

- kombinierte Leistungen von unterschiedlichen Standorten und Geräten (z.B. auch das Zusammenspiel des eigenen Endgerätes mit einem Terminal).

Zur weiteren Verbreitung ubiquitärer Systeme scheint eine weiterführende Standardisierung verteilter benutzerangepaßter Informationsdienste, z.B. in Abstimmung mit Protokollen wie HBCI, eine sinnvolle Maßnahme um den gewünschten Kundennutzen über unterschiedliche Plattformen hinweg gewährleisten zu können.

3.2 Ziele der Nutzer

Die Nutzer der individualisierten Kommunikationsumgebung sind auf Konferenzen und Messen einem kontinuierlichen Wandel an Diensten und Informationen unterworfen. Je nach neu aufgenommener Information oder aktuellen Änderungen können sich die Ziele und das Wissen des Nutzers verändern. Weiterhin führen wechselnde Orte des Nutzers zu kontinuierlichen Anpassungen, die in Kombination mit der auf ihn einströmenden Informationsflut die individuellen Belastungen erhöhen.

Für den Nutzer werden deshalb folgende Ziele angenommen:

1. **Minimierung von Entfernungen und Zeit**
   Erstrebt werden kurze Wege zu den Konferenz- oder Ausstellungsräumen und auch die direkte Lieferung der persönlich relevanten Information. Der Nutzer will leicht und schnell erreichbar und aktuell informiert sein. Die Kontaktaufnahme zu anderen Personen soll unkompliziert und direkt möglich sein.

2. **Maximierung von Wissen und Handlungsfähigkeit**
   Der Nutzen des Veranstaltungsbesuchs soll maximiert werden. Der Teilnehmer will deshalb nicht nur schnelle Informationslieferung, sondern auch vollständige, d.h. er muß das maximale Wissen über Produkte, Entwicklungen und Ansprechpartner schnell nutzen können. Die Nutzung soll nach dem Veranstaltungsbesuch auch direkt am Arbeitsplatz verfügbar sein. Trotz Abwesenheit vom Arbeitsumfeld will er handlungsfähig bleiben. Die ortsunabhängige Erreichbarkeit des eigenen Unternehmens und die Einbindung in den Arbeitsprozeß mit minimalem Aufwand soll auf eigenen Wunsch möglich sein.

3. Maximierung von Transparenz und Sicherheit
Die Transparenz der Systemfunktionen und die Sicherheit soll derart gestaltet werden, daß eine vertrauensvolle Nutzung gewährleistet werden kann. Die Sicherheit umfaßt Funktionssicherheit und den Schutz der Privatsphäre. Die weitere Nutzung der persönlichen Interessen darf nur mit Zustimmung der Nutzer erfolgen, die jederzeit die Kontrolle über ihr Daten haben (vgl. hierzu die Feldstudie auf der UM97).

3.3 Parameter für die Anpassung

Für die Anpassung der Informationsdienste an die Nutzer und die Orte können eine Viezahl von Parametern genutzt werden. Die Ausprägung der Parameter muß direkt durch Auswertung der Bedienung oder durch Befragung der Nutzer vom System aus erfolgen.

Die individuellen Merkmale der Nutzer sind der wichtigste Bereich. Dieser umfaßt das fachliche Interessenprofil sowie Präferenzen bezüglich aktueller Meldungen und deren Präsentation. Hierbei ist zu klären, was der Nutzer unter Aktualität versteht, d.h. welche Informationen er als aktuell erachtet, wie z.B. kontinuierliche Börsenticker oder Unternehmenskennzahlen der letzten Woche. Ein weiterer Faktor ist die Menge an Information, die verarbeitet werden soll.

Zeitliche Unterschiede bei der Bedienung und Informationsaufnahme können zur Adaption ausgewertet werden. Die Differenzen zwischen dem Lesen einer elektronischen Nachricht und der Antwort, zwischen der Datenbereitstellung und der Nutzung und zwischen der durchschnittlichen Lesezeit aller Nutzer und dem individuellen Lesezeitpunkt ermöglichen Aussagen über Anpassungserfordernisse.

Die **Anzahl der Ereignisse** an gesendeten elektronischen Nachrichten, die Häufigkeit der Kontaktanbahnungen sowie die Menge unterschiedlich genutzter Informationsdienste können ein Kriterium für die Anpassung der Oberfläche und der Funktionalitäten sein.

Die **Einflüsse des Ortes** auf die individuellen Merkmale ergänzen das Spektrum. Kriterien sind hierbei die Verweilzeit am jeweiligen Terminal, die dort mögliche Bewegungsfreiheit, die Umweltgegebenheiten (z.B. Lautstärke, Lichteinfall) und die geographische Lage. Hinweise können für die am schnellsten erreichbaren und nützlichen Orte wie z.B. das nächste nutzbare WC oder der nächstliegend freie Besprechungsraum gegeben werden. Aussagen sind mit Hilfe eines UPI-Systems bezüglich der Wegzeit für einen empfohlenen Rundweg möglich.

4 Das Pilotsystem „aKISS für UPC"

Mit Hilfe von aKISS kann für Konferenzteilnehmer der OMNICARD98 eine individuelle und aktuelle Informationsversorgung gewährleistet werden. Hierzu erhält jeder Teilnehmer eine kontaktlose Chipkarte in Form eines persönlichen Namensschildes. Auf Basis der Benutzeridentifikation und des Benutzermodells stehen dem Teilnehmer dann an den Terminals im Konferenzbereich individuelle Dienste zur Verfügung:

- **Aktuelle Konferenzinformationen:**
  Raum- und Tagungsplan mit örtlich angepaßten Hinweisen, aktuelle Änderungen,

- **Zugriff auf die persönliche elektronische Post:**
  Zustellung der aktuellen E-Mails,

- **Persönliche Konferenzmitteilungen:**
  Nachrichten von Externen, Kurzmitteilungen, Ausrufe mittels E-Mail,

- **Individuell relevante Fachinformationen:**
  Individuelle Bereitstellung von präferierten Informationen aus dem Internet und direkter Zugriff auf Informationen der Konferenzaussteller,

- **Geschäftliche Kontaktaufnahmen:**
  Suche von Konferenzteilnehmern und Terminvereinbarung mittels E-Mail.

Die Leistungen werden den ca. 500 Teilnehmern an fünf Terminals dargeboten, die im gesamten Konferenzbereich an exponierten Stellen bereitgestellt sind. Mit Hilfe seiner kontaktlosen HITAG^{TM} -Chipkarte kann sich der Teilnehmer bequem anmelden. Die Chipkarte liefert neben der Identifikation des Benutzers die Profildaten und Parameter zu den einzelnen Diensten. Die Chipkarte kommuniziert mit einem der fünf Terminals, die ein LAN integriert und als Clients mit dem Server verbunden sind. Jedes Terminal besitzt eine eigene Identifikation, die auf seine Standortparameter verweist.

Beim Einsatz kontaktloser Chipkarten mit RFID-Technolgie ist die Kombination der Faktoren Speicher und Übertragungsfrequenz entscheidend für die Personalisierung. Deshalb ist es angebracht, nur geringe Informationsmengen auf dem Chip abzuspeichern, um eine schnelle Adaption des Systems zu erreichen. Die Personalisierungsleistungen erfolgen deshalb hauptsächlich von einem Server aus. Die Speicherung des Benutzermodells erfolgt für aKISS auf dem Server. Eine verteilte Architektur zur Speicherung des Benutzermodells ist aus Sicherheitsgründen in diesem abgegrenzten Einsatzgebiet nicht erforderlich.

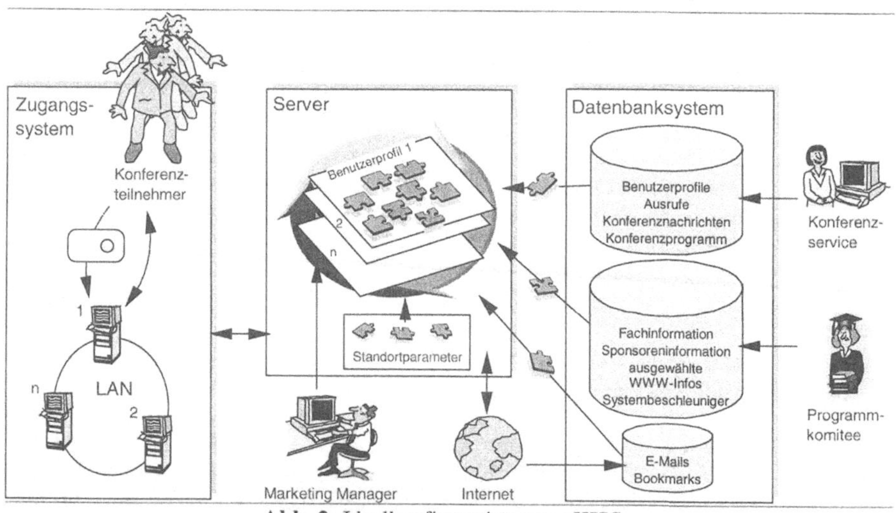

Abb. 2: Idealkonfiguration von aKISS

Der Server verwaltet die Teile der Benutzerprofile und Standortparameter, nimmt Anpassungen vor und ordnet Dienste zu. Die notwendigen Informationen erhält er von der angeschlossenen Datenbank. In dieser sind neben Teilen der Benutzerprofile auch allgemeine und persönliche Konferenzmitteilungen, individuell relevante Fachinformationen, persönliche E-Mails und Bookmarks jedes Teilnehmers gespeichert. Die Zuordnung der Fachinformationen erfolgt über eine Stereotypen-Kennzeichnung. Über den Server erhält der Konferenzteilnehmer auch die Möglichkeit konventionell das Internet und sein E-Mail zu nutzen. Die ideale Konfiguration des UPI-Systems aKISS ist in Abbildung 2 dargestellt.

Aus praktischen Gründen wurde das System für den Einsatz auf der OMNICARD98 in seinem Funktionsumfang teilweise beschränkt. Die Fachinformationen wurden auf die für die Konferenz relevanten URLs beschränkt. Systembeschleuniger wie z.B. Preisausschreiben, die im Rahmen adaptiver E-Commerce eingesetzt werden und die Nutzung deutlich erhöhen, finden keine Anwendung. Für den Einsatz galt das Prinzip „Robustheit vor Funktionsvielfalt".

Erweitert wird aKISS um den Zugang mit kontaktbehafteten Chipkarten, um die Diskussion „kontaktlos oder nicht" nicht in den Vordergrund rücken zu lassen. Nicht die Details des technischen Zugangs, sondern Aspekte der Sicherheit und Adaption sind die Erfolgsfaktoren in der Diskussion um ubiquitäre Systeme.

5 Zukunftsaussichten

Das beschriebene System ist ein erster Ansatz zur praktischen Evaluierung und zur Präsentation einer technologischen Plattform ubiquitärer Leistungen. Die Evaluation und Diskussion auf der OMNICARD98 aber auch auf weiteren Konferenzen wird Licht in die künftige Gestaltung derartiger Systeme liefern. Ein kritischer Erfolgsfaktor hierbei ist die Einbeziehung der Beteiligten in den Gestaltungsprozeß.

Die Vermarktung ubiquitärer Systeme wird aufgrund der Komplexität und der kontroversen Standpunkte der Nutzer noch auf sich warten lassen [Lind97]. Erste Ansätze sind aber derzeit im Rahmen von „cross-selling" im Internet und kombinierten Dienstleistungen auf der Basis von Chipkarten erkennbar.

Entscheidend für den Erfolg überall nutzbarer und personalisierter Informationsversorgung sind sinnvolle Leistungen, die kommerziellen Nutzen erlauben und dabei datenschutzrechtliche Aspekte berücksichtigen. Die Anonymität des Nutzers und die Nichtverfolgbarkeit seiner Handlungen müssen gewährleistet sein. Die gelungene Zusammenführung der Katalysatoren Chipkarten und Benutzeradaption wird dann zum zentralen Element für ubiquitäre Systeme und damit für eine überall nutzbare und personalisierte Informationsversorgung.

Literaturverzeichnis

[FiKJ97]   Fink, J., Kobsa, A., Jaceniak, I.: 'Individualisierung von Benutzerschnittstellen mit Hilfe von Datenchips für Personalisierungsinformation'. In: GMD-Spiegel, Vol. 27, No. 1: S. 16-17.

[FiKN97]   Fink, J., Kobsa, A., Nill, A.: 'Benutzerorientierte Adaptivität und Adaptierbarkeit im Projekt AVANTI', Software-Ergonomie 97, Dresden. Stuttgart: Teubner: 1997, S. 135-144.

[Jace97]   Jaceniak, I.: 'Praktischer Nutzen einiger Adaptionstechniken zur Realisierung persönlicher Arbeitsumgebungen'. In: Schäfer, R., Bauer, M. (Hrsg.), 'ABIS-97', Universität Saarbrücken: 1997. S. 103-108.

[Kobs93]   Kobsa, A.: User Modeling: Recent Work, Prospects and Hazards. In: Schneider-Hufschmidt, Kühne, T., Malinowski, U.: 1993, 'Adaptive User Interfaces: Principles and Practice', Amsterdam: North-Holland. 1993

[KoPo95]   Kobsa, A.; Pohl., W.: 'The user modeling shell system BGP-MS'. In: User Modeling and User-Adapted Interaction, Vol. 4, No. 2: 1995, S. 59-106.

[Lind97]   Lindner, H.-G.: 'Sind benutzeradaptive Anwendungssysteme marktreif?'. In: Schäfer, R., Bauer, M. (Hrsg.), 'ABIS-97', Universität Saarbrücken: 1997, S. 91-96.

[LiWo97]   Lindner, H.-G.; Wohlmacher, P.: Chipkarten und Benutzeradaption als Katalysator für eine überall nutzbare Informationsversorgung. In: Fluhr, M. (Hrsg.): Die Chipkarte auf dem Weg zu Akzeptanz und Nutzung, Konferenzdokumentation OMNICARD 1998, Berlin, 14.-16. Januar 1998, inTIME Berlin 1998, S. 170-183.

[PiGH97]   Pieper, M.; Gappa, H.; Hermsdorf, D.: 'Berufliche Rehabilitation durch computergestützte Telekooperation" - Gebrauchstauglichkeit der internetbasierten Telearbeitsumgebung BSCW und multi-modal Adaptionsoptionen für Behinderte. In: „Software-Ergonomie 97", Workshop 8, Dresden.

[UM97]   UM97: 'User Modeling - Proceedings of the Sixth International Conference UM97', Jameson, A., Paris, C. und Tasso, C. (Hrsg.), 1997, Wien: Springer.

[Walk94]   Walker, J.: 1994, 'Ubiquitous Computation, Global Connectivity, and the End of Privacy', Revision 8 - February 28th, http://www.fourmilab.ch/documents/unicard.doc

[Weis91]   Weiser, M.: 1991, 'The Computer for the 21st Century'. In: Scientific American, Sept. 1991, S. 94-104.

MIFARE®PRO
Kontaktlose Mikrokontrollerkarte
für Dual Interface Betrieb

Wolfgang Haß

Philips Semiconductors
Produktgruppe Identification & Automotive
Wolfgang.Hass@Hamburg.sc.philips.com

Zusammenfassung

Zwei Kartentechnologien, die kontaktbehaftete Mikrokontrollerkarte, die dem Nutzer ein hohes Maß an Sicherheit und Funktionalität bietet, und die kontaktlose Chipkarte mit fest verdrahteter Intelligenz und ihren Vorteilen in der Handhabung haben am Markt ihren Erfolg gefunden. Was liegt näher als diese beiden Technologien zusammen zu fügen und nicht nur die Serviceanbieter sondern auch den Karteninhaber davon profitieren zu lassen und so die Akzeptanz des elektronischen Geldes zu fördern?

1 Einleitung

Es ist immer wieder in jedermanns Munde: die Kombination mehrerer Applikationen auf einer Chipkarte. Besonders deutlich wird dies am Beispiel der elektronischen Geldbörse, für die damit geworben wird, daß sie bei verschiedenen Gelegenheiten wie etwa im Parkhaus, im öffentlichen Nahverkehr oder beim Einkaufen benutzt werden kann. Sie muß in der Lage sein das heutige Bargeld funktional zu ersetzen, d.h., das in ihr gespeicherte Geld muß anwendungsungebunden dem Besitzer zur Verfügung stehen. Voraussetzung für den Erfolg eines solchen Konzeptes ist, daß Märkte mit einem hohen Volumen an Kleinbetragstransaktionen erreicht werden können. Der öffentliche Nahverkehr ist im Hinblick auf die hohe Zahl sich täglich wiederholender Zahlung von Kleinbeträgen eine besonders interessante Anwendung die Akzeptanz des elektronischen Portemonnaies zu erhöhen.

Jedoch hat jede Applikation ihre eigenen spezifischen Anforderungen an die Funktionalität der Smart Card. Geldinstitute vertrauen wegen eines hohen Anspruches auf Sicherheit eher der kontaktbehafteten Smart Card, während Nahverkehrsgesellschaften deutlich die kontaktlose Smart Card wegen ihrer Nutzer- und Wartungsfreundlichkeit bevorzugen.

Folgt man dem Trend der offenen elektronischen Geldbörse und vereint man die Anforderungen an Smart Cards der Geldinstitute mit denen der Nahverkehrsgesellschaften, so wird daraus ein Kartenkonzept mit zwei gleichberechtigten Schnittstellen - die DUAL INTERFACE CARD. -

2 Marktsituation

Bis heute haben sich die Chipkartenanwendungen unabhängig von einander entwickelt. Im Bereich Finance & Banking sind Standards und Sicherheitskonzepte entwickelt worden, die den Einsatz von elektronischen Geldbörsen als Ersatz für das Bargeld gestatten. Das damit verbundene hohe Maß an Sicherheit wird durch den Einsatz von kryptographischen Methoden erreicht, die sich auch bisher im Finanzwesen behauptet haben. So sind die meisten Realisierungen heute auf symmetrische Verfahren und hier zumeist auf DES gestützt. Der aus diesen Lösungen resultierende hohe Rechen- und Kommunikationsaufwand hat wegen bisher fehlender funktionaler Unterstützung auf der Karte, den Einsatz kontaktloser Chipkarten nicht zugelassen.

Es sind zahlreiche Implementationen im Markt zu finden, von denen einige bereits mit einer sehr großen Anzahl Karten eingeführt sind. Die Akzeptanz des elektronischen Geldes hingegen und somit das Transaktionsvolumen liegt jedoch noch weit hinter den Erwartungen zurück. Was der offenen elektronischen Geldbörse fehlt, sind Anwendungen mit großem Transaktionsvolumen und hoher Nutzungshäufigkeit.

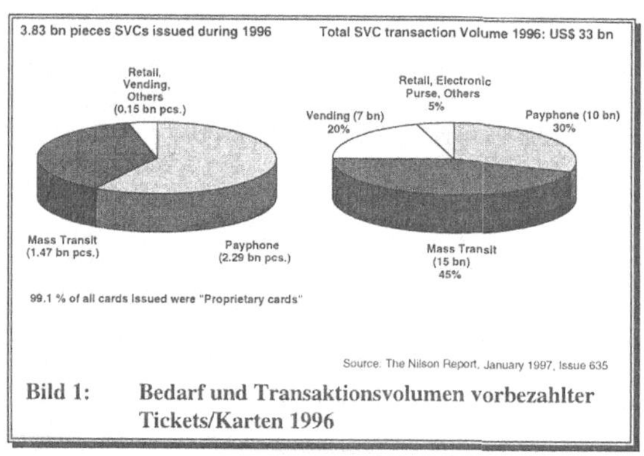

Bild 1:   Bedarf und Transaktionsvolumen vorbezahlter Tickets/Karten 1996

Der Bereich öffentlicher Nahverkehr, in dem die Chipkarten ebenfalls längst Einzug gehalten haben, ist mit einem jährlichen Transaktionsvolumen von 16 bnUS$ ein solcher Markt (vgl. Bild-1). Das Anwendungsgebiet öffentlicher Nahverkehr stellt jedoch an einen elektronischen Fahrschein ganz andere Anforderungen als das Finanzwesen. Es wird hier der Nutzervorteil einer kontaktlosen Chipkarte zum zwingenden Muß, um der hohen Nutzerfrequenz im Zugangsbereich und auch an den Bezahlautomaten gerecht zu werden. Diesem Anspruch hat man mit der auf breiter Basis eingeführten MIFARE® Technologie leicht gerecht werden können. Der Elektronische Fahrschein auf Basis MIFARE® ist mit über 90.000 installierten Terminals und ca. 15 Mio ausgegebenen Karten eine bewiesene und vom Markt bestätigte Technologie mit einem Marktanteil von über 90%. So werden alleine in Seoul, wo die MIFARE® Technologie im Nahverkehr eingesetzt wird, mit über 4 Mio ausgegebenen Karten ca. 85 Mio Transaktionen mit einem Volumen von ca. 35 Mio US$ monatlich durchgeführt. Das sind pro Tag mehr Transaktionen als bisher jährlich in allen E-Purse - Projekten zusammengenommen gezählt werden können.

Soll nun die offene, d.h. anwendungsungebundene, elektronische Geldbörse mit dem elektronischen Fahrschein zusammen gebracht werden und im Nahverkehr (und Parkhäusern, Parkuhren, etc.) zum Einsatz kommen, muß eine technologische Plattform gefunden werden, die die hohen Sicherheitsansprüche einerseits mit dem Nutzervorteil der kontaktlosen Kartentechnik andererseits vereint. Konkret bedeutet dies, daß neben der Bereitstellung beider Schnittstellentechnologien auch die erforderliche Unterstützung zur Bewältigung der Tripple-DES-Operationen und des hohen Kommunikationsaufwandes geschaffen werden muß.

3 MIFARE® PRO - das Bindeglied!

Mit MIFARE®PRO stellt Philips eine neue Produktfamilie vor, die dem genannten Anspruch in vollem Umfang gerecht wird.

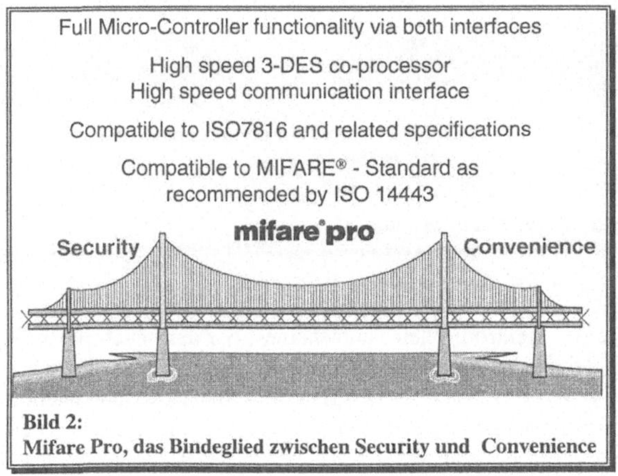

Bild 2:
Mifare Pro, das Bindeglied zwischen Security und Convenience

MIFARE®PRO ist eine Chipkarten IC-Familie, die uneingeschränkte Mikrokontrollerfunktionalität über zwei Schnittstellen zur Verfügung stellt. Die Kontaktschnittstelle entspricht dem bekannten und etablierten Standard ISO 7816 und die kontaktlose Schnittstelle ist 100% kompatibel zum defacto Standard für kontaktlose Chipkarten, der MIFARE® - Plattform, welche zum derzeitigen Stand der Standardisierung für kontaktlose Karten, ISO 14443, voll konform ist.

Die MIFARE® PRO Familie kann an den gleichen Lesermodulen betrieben werden wie die übrigen MIFARE® Produktfamilien -Light, -Standard und -Plus, und ist damit in alle vorhandenen Installationen problemlos integrierbar. Die MIFARE® Schnittstelle zeichnet sich aufgrund der gewählten Übertragungsverfahren durch höchste Sicherheit und Zuverlässigkeit auch in besonders störintensiver Umgebung aus, und ist daher gerade für Anwendungen mit einem gehobenen Sicherheitsanspruch besonders gut geeignet. Der Mikrokontrollerkern basiert auf dem weit verbreiteten 8051-Architektur. Die MIFARE® PRO Familie erlaubt somit die Benutzung ISO7816 konformer Implementierungen an beiden Schnittstellen und ist daher ohne Probleme auch in Kontaktkarten basierenden Anwendungen integrierbar. Erstmalig ist damit die Kombination von kontaktkartenorientierten Anwendungen mit solchen, die eine kontaktlose Kommunikation erfordern, möglich, ohne dabei Kompromisse hinsichtlich Service oder gar der Sicherheit eingehen zu müssen.

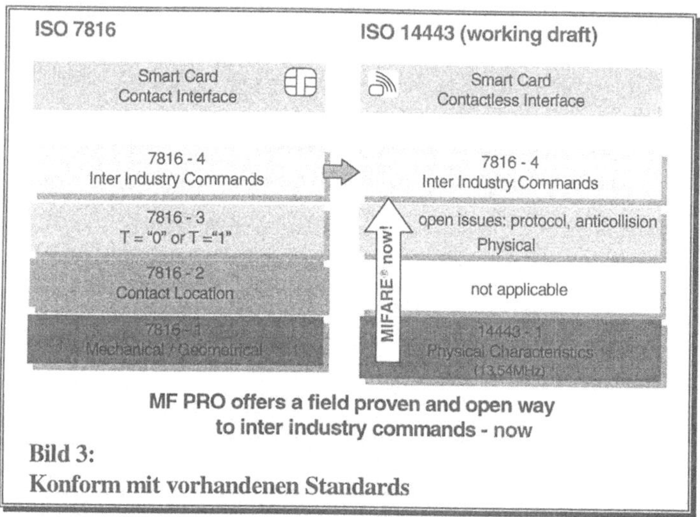

Bild 3:
Konform mit vorhandenen Standards

Der Realisierung eines elektronischen Fahrscheines, bei dem auch der komplette Zahlungsverkehr nicht nur elektronisch sondern auch mit der selben Karte abgewickelt werden kann, steht nichts mehr im Wege. Es kann so beispielsweise ein Tagesfahrschein mit der Elektronischen Geldbörse am Automaten gekauft werden, welcher automatisch als elektronisches Tikket auf der Karte gespeichert und später bei betreten des Bahnsteiges, des Busses oder auch einfach nur zeitgebunden elektronisch validiert wird.

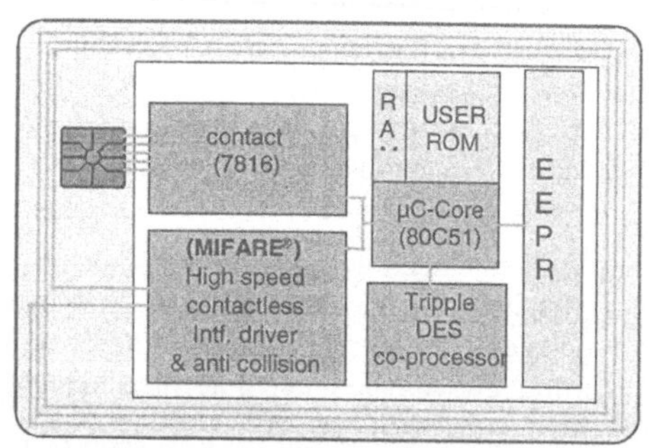

Bild 4:
Prinzip der Dual Interface Karte

Die gesamte Produktfamilie ist nach höchsten Sicherheitsmaßstäben designed und stellt alle in der Kontaktkartentechnik zur Selbstverständlichkeit gewordenen Sicherheitsmaßnahmen zur Verfügung. Darüber hinaus werden physikalische Attacken durch fortgeschrittene Designmethoden erheblich erschwert. Ein speziell auf die Anforderungen der kontaktlosen Kartentechnik optimierter DES-Co-Prozessor verkürzt die Rechenzeit für einen Tripple DES auf bis zu 130-Millionstel Sekunden. Das ist bis zu 400 mal schneller als mit vergleichbaren Lösungen in Software. Damit wird die für die Datensicherheit benötigte Rechenzeit innerhalb einer Transaktion vernachlässigbar klein und stellt keinen Engpaß für den Einsatz kontaktloser Karten im Zahlungsverkehr mehr dar.

In einer der möglichen Konfigurationen steht ein Emulationsmodus zur Verfügung, der Funktionskompatibilität zum MIFARE® S und MIFARE® PLUS Chip gewährleistet.

Das erste Mitglied der MIFARE® PRO Familie stellt dem Anwender 20KByte ROM und 8KByte EEPROM zur Verfügung.

Die MIFARE PRO Familie wird kurzfristig um eine kostenoptimierte Version mit einem auf 2K Byte EEPROM verkleinerten Speicher und einer "high-end" Variante welche den Einsatz von asymmetrischen Verschlüsselungsverfahren zuläßt, erweitert.

Die physikalischen Abmessungen der MIFARE® PRO Karte sind identisch mit kontaktbehafteten Karten. Zusätzlich zum Kontaktblock ist eine einfache Antenne in die Plastikkarte integriert, um das IC über die kontaktlose Schnittstelle sowohl mit Spannung zu versorgen als auch Schreib/Lese- Operationen bis zu einer Entfernung von bis zu 10 cm zu ermöglichen. Die Antenne besteht aus 4 Windungen einer gedruckten, geätzten oder gewickelten Spule, die mit dem MIFARE® PRO Chip verbunden ist.

Sichere Transaktionen mit kontaktlosen Chipkarten

Thilo Zieschang

SECUNET
Security Networks GmbH
Mergenthalerallee 77-81, D-65760 Eschborn
zieschang@acm.org

Zusammenfassung

Der Nutzung komfortabler, kontaktloser Chipkarten stehen in vielen Anwendungsgebieten nach wie vor ernsthafte Sicherheitsbedenken entgegen. Die Ausführung starker kryptographischer Mechanismen wie beispielsweise RSA im kontaktlosen Betrieb scheitert bei den meisten heutigen Kartentypen noch an technischen Hürden. Neuere kryptographische Protokolle erlauben jedoch die Verlagerung aufwendiger Teilberechnungen in eine Offline-Phase vor der eigentlichen Transaktion. Hier wird eine Variante vorgestellt, die während der Transaktionsphase keinerlei arithmetische Berechnungen mehr von der Chipkarte verlangt. Darüber hinaus werden einfache Möglichkeiten dargestellt, wie kontaktlose Chipkarten ohne große Rechenleistung eine sichere verschlüsselte Datenübertragung realisieren können. Beispiele für den gewinnbringenden Einsatz der beschriebenen kryptographischen Mechanismen in der Praxis erläutern die Vorzüge der dargestellten Verfahren.

1 Einführung

Intelligente Chipkarten erobern täglich weitere Anwendungsbereiche. Alle Experten stimmen darin überein, daß Chipkartenapplikationen während der kommenden Dekade einen gigantischen Wachstumsmarkt darstellen. Die hohen Sicherheitsanforderungen bei vielen Einsatzgebieten, beispielsweise bei Zutrittsschutz oder Digital Cash, implizieren die Verwendung intelligenter Prozessorchipkarten, gegebenenfalls sogar mit kryptographischem Koprozessor. Solche Karten sind zumeist erforderlich, um einerseits die benötigten kryptographischen Algorithmen implementieren zu können, und andererseits die in diesem Zusammenhang anfallenden sensiblen Schlüsselinformationen hinreichend gut vor unbefugtem Zugriff in der Chiphardware abzusichern.

In der Praxis sind der Verwendung dieser Kartentypen jedoch oftmals Grenzen gesetzt: kryptographische Koprozessoren und erhöhter Speicherbedarf verteuern die Hardware, Kompatibilitätsprobleme zu den bisherigen Infrastrukturen auf Basis symmetrischer Verfahren wie DES müssen überwunden werden, und vieles mehr.

Neben technischen und finanziellen Erwägungen findet sich freilich ein weiterer nicht zu vernachlässigender Aspekt beim Einsatz kontaktbehafteter Chipkarten, beispielsweise im Zutrittsschutz, in der Frage hinreichenden Komforts bzw. Praktikabilität im täglichen Betrieb. Die Vorteile kontaktloser Chipkarten bei vielen Einsatzzwecken liegen hier auf der

Hand. Bei Zutrittskontrollsystemen lassen sich mit kontaktloser Technik mehr Personen pro Zeiteinheit durchschleusen als bei Verwendung kontaktbehafteter Chipkarten. Ferner sind kontaktlose Karten, ebenso wie deren zugehörige Leseeinheiten, oftmals weniger anfällig gegen Störungen in rauhen Betriebsbedingungen. Überdies erschließen solche Karten (oder allgemeiner Tokens) mit kontaktloser Übertragung eine Fülle weiterer Anwendungsgebiete, beispielsweise in der Lagerhaltung, Viehzucht, Container- und Fahrzeugüberwachung, Personenlokalisierung in abgesicherten Räumen, etc. Ein zentrales Problem liegt bei Verwendung kontaktloser Chipkarten allerdings im reduzierten Sicherheitsniveau aufgrund einer Beschränkung auf zumeist schwächere kryptographische Mechanismen.

2 Kontaktlose Karten und Kombikarten

2.1 Verschiedene Risiken

Diverse Hersteller bieten kontaktlose Chipkarten unterschiedlichster Funktionalität an. Bei den einfacheren Versionen erfolgt die Kommunikation zwischen Karte und Lesegerät völlig unverschlüsselt, und bei den einfachsten wird hierbei lediglich eine individuelle Karten-Seriennummer übertragen. Modernere Varianten werben mit verschlüsselter Datenübertragung von und zur Chipkarte. Einige, teils gravierende Nachteile in Sachen Sicherheit sollten jedoch beachtet werden:

- Abhören und gegebenenfalls Analyse der Kommunikation zwischen Karte und Leser ist hier wesentlich einfacher (und unauffälliger) als bei leitungsgebundenen Verfahren.

- die verwendeten Verschlüsselungsverfahren sind im allgemeinen völlig unsicher und leicht zu knacken. Grundschema solcher Verfahren ist häufig ein leicht modifiziertes, blockweises XOR des Klartextes mit Datenblöcken, die in einfacher Weise vom geheimen Schlüssel abgeleitet wurden. Solche Verfahren kennen wir beispielsweise von bekannten Textverarbeitungsprogrammen, wo eine immergleiche Byte-Folge blockweise nacheinander mit dem zu „verschlüsselnden" Dokument per XOR verknüpft wurde. Begründbar ist diese Vorgehensweise jedoch zumindest bei den obengenannten, kontaktlosen Chipkarten: zur Ausführung aufwendigerer kryptographischer Algorithmen fehlt den meisten Karten dieses Typs nämlich schlicht die erforderliche Energie.

- ist die Karte erst geknackt, so gestaltet sich auch die weitere Vorgehensweise für Eindringlinge vergleichsweise einfach: zwar unterstützen einige Kartentypen die Verwendung differenzierter Schlüssel für unterschiedliche Applikationen bzw. Speicherbereiche. Doch diese Schlüssel sind häufig für alle beteiligten Karten dieselben.

- ein unauffälligeres Einbringen gefälschter oder simulierter Leser bzw. Karten ist im kontaktlosen Fall ein weiteres Problem. Ferner ist nicht immer gewährleistet, daß beim Versuch Unbefugter, die Karte heimlich auszulesen, keinerlei persönliche Daten preisgegeben werden. Überdies sind manche Karten zwar recht gut gegen unrechtmäßiges wiederaufbuchen von Geldbeträgen geschützt (s.u.), doch man stelle sich die Reaktion eines Anwenders vor, dessen Kartengeld sich, ohne sein Einverständnis, quasi im Vorbeigehen verflüchtigt hat...

2.2 Bisherige Lösungsansätze

Die momentan vorherrschende Philosophie bei der Implementierung von Sicherheitsmechanismen auf kontaktlosen oder Kombikarten besteht im wesentlichen darin, die konventionellen kryptographischen Mechanismen zur gegenseitigen Authentisierung von Chipkarte und Hostrechner nebst (optional) verschlüsselter Datenübertragung möglichst ähnlich wie bei der Verwendung kontaktbehafteter Mikrokontrollerchipkarten zu realisieren. Dies führt zu den oben bereits genannten Angriffspunkten durch teils unzureichende Kryptoalgorithmen infolge nur schwacher Rechenleistung.

2.3 Kontaktlose Karten

Im Falle kontaktloser Chipkarten ohne hinreichend leistungsfähigen Mikroprozessor ist die Anfälligkeit durch schwache Verfahren systembedingt und kaum zu beheben. Verbesserungspotential findet sich allerdings noch in der Weiterentwicklung solcher kryptographischer Verfahren, die mit geringem Rechenaufwand und stark eingeschränktem Umfang an Recheninstruktionen dennoch eine unter diesen Randbedingungen möglichst gute Resistenz gegenüber kryptoanalytischen Angriffen bieten. In der kryptographischen Forschung ist dies leider ein kaum beachtetes Thema.

2.4 Kombikarten

Unter sogenannten Kombikarten versteht man die Kombination einer kontaktbehafteten Mikrokontrollerchipkarte mit einem kontaktlosen Interface. Dies ist dann sinnvoll, wenn beide auf gemeinsame Ressourcen zugreifen können, beispielsweise ein gemeinsames EEPROM. Beispiel einer solchen Karte ist die Siemens Combi Card SLE 44R42, deren Blockdiagramm wie folgt darstellbar ist:

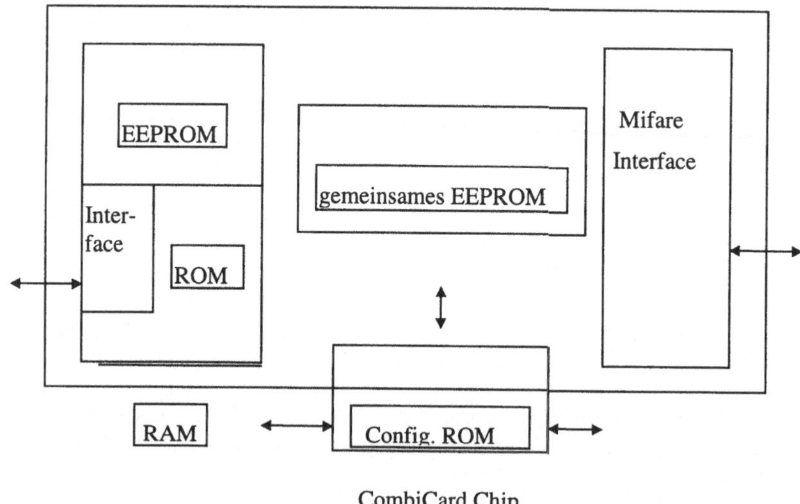

CombiCard Chip

Wie lassen sich hier nun sichere Anwendungen implementieren? Eine derzeit realisierte Lösung sieht vor, daß obiger Chip im Einsatz als Debitkarte ein Aufladen von Werteinheiten nur über das kontaktbehaftete Interface zuläßt, unter Verwendung starker Kryptofunktionen. Ein Abbuchen von Werteinheiten hingegen ist über das kontaktlose Interface möglich, bei Zugriff auf einen gemeinsamen EEPROM-Speicheranteil. Der EEPROM-Speicher des Kontrollers hingegen kann nicht über die kontaktlose Schnittstelle angesprochen werden. Diese Vorgehensweise stellt zwar einen gewissen Schutz des Kartenbetreibers dar, doch der Kartenbesitzer ist nicht im gleichen Maße vor Verlusten durch unrechtmäßiges (oder fehlerhaftes) Abbuchen geschützt. Wünschenswert wären hier zuverlässige Mechanismen, mit denen der Kartenbesitzer seine Transaktionen bzw. Abbuchungen sicher und nachweisbar autorisieren kann. Da solche Verfahren nicht ohne beträchtlichen Rechenaufwand auszuführen sind, besteht ein Ausweg darin, rechenintensive Bestandteile Offline bereits vor der jeweiligen, via kontaktlose Schnittstelle abzuwickelnden, Transaktion vorzubereiten.

Online/Offline Signaturen

Die in der Praxis oftmals gegebenen Einschränkungen hinsichtlich der Rechenkapazität von Chipkarten und anderen Hardware Tokens führten bereits zu diversen Überlegungen, einen Teil jener Berechnungen anderweitig auszuführen. Hierbei existieren unterschiedliche Möglichkeiten:

zum einen ist die Möglichkeit zu untersuchen, ob Teilberechnungen zum betreffenden Zeitpunkt (Online) von der ursprünglichen Hardware (Chipkarte, Hardware Token) ohne Sicherheitsrisiko ausgelagert werden können zwecks Ausführung durch andere beteiligte Hardwarekomponenten, beispielsweise im Kartenlesemodul oder auf dem zugehörigen Hostrechner.

Zum anderen lassen sich einige Methoden finden, wie bereits vor einer jeweiligen Transaktion, Offline - unter Ausnutzung größerer Ressourcen an Zeit und/oder Rechenkapazität, mittels Vorberechnungen und speziell abgestimmter Algorithmen ein Teil der Rechenarbeit erledigt werden kann.

Ein Beispiel für letztere Vorgehensweise ist ein digitales Signaturverfahren von Even, Goldreich und Micali namens On-Line/Off-Line Digital Signatures. Diese besteht aus zwei Phasen: die erste wird offline durchgeführt, bevor die später zu signierende Nachricht überhaupt bekannt ist. Die zweite Phase wird sodann online ausgeführt und ist mit im Vergleich zur Offline-Phase relativ wenig Rechenaufwand versehen. Benötigt werden hier trotz allem jedoch noch beispielsweise einige hundert DES-Verschlüsselungen, sowie eine modulare Exponentiation. Somit eignet sich ein solches Signaturverfahren unter anderem überall dort, wo zwar auf Prozessorchipkarten zurückgegriffen werden kann, diese Karten allerdings ohne Hilfe des genannten Verfahrens in der betreffenden Applikation überlastet wären.

Die Praxisanforderungen an gängige Kombikarten bei Verwendung der kontaktlosen Schnittstelle für Online-Transaktionen sind jedoch im allgemeinen noch deutlich restriktiver. Zwar werden in Zukunft auch Karten verfügbar sein, die starke Kryptoalgorithmen im kontaktlosen Betrieb abarbeiten können. Doch die heute weltweit im Einsatz befindlichen Kartentypen sind den Anforderungen beispielsweise des obigen Signaturverfahrens von Even, Goldreich und Micali nicht gewachsen.

One Time Digital Signatures

Bestandteil der obigen Online/Offline-Algorithmen von Even, Goldreich und Micali sind sogenannte One Time Digital Signatures, wie beispielsweise von Ralf Merkle vorgeschlagen. Diese One Time Signaturen benötigen allerdings noch immer die Ausführung umfangreicher Berechnungen (online), so daß solche One Time Signaturen nur für manche Anwendungen in Frage kommen. Zusätzlich fällt ein hoher Datentransfer an und für manche Protokolle ist zudem die Integration einer Trusted Third Party erforderlich.

Eine im vorliegenden Kontext hochinteressante Variante findet sich in einem Electronic Cash Protokoll von Chaum, Fiat und Naor, welches in modifizierter Form auch für One Time Signaturen zu gebrauchen ist (siehe Barry Hayes). Bevor wir die Vorteile dieses Verfahrens darstellen, soll zunächst das Protokoll selbst beschrieben werden.

Unsere beiden Teilnehmer im Protokoll seien Anna und Bert, wobei Anna eine One Time-Signatur beantragt bei Teilnehmer Bert. Die Bezeichnungen seien wie folgt: f und g bezeichnen zwei kollisionsfreie Einwegfunktionen mit jeweils zwei Argumenten. Das Symbol $\oplus$ bezeichne XOR, und $\|$ bezeichne Konkatenation. Der Sicherheitsparameter k werde gewählt in Abhängigkeit von der Anzahl der $k/2$ zu signierenden Bits (also beispielsweise $k = 200$). Bert besitzt einen Public Key n_{Bert}, dessen Faktorisierung nur Bert bekannt ist.

Anna wählt Zufallszahlen a_i, c_i, d_i, r_i, für $i = 1, ..., k$. Ferner konstruiert Anna Werte $m_1, ..., m_k$, von denen jeder unter anderem die Identität von Anna beschreibt und nicht leicht von Dritten zu erraten ist. Anna sendet nun an Bert die folgenden sog. Blinded Candidates $B_i = r_i^3\, f(g(a_i, c_i), g(a_i \oplus m_i, d_i))$.

Bert wählt aus jenen Werten $B_1, ..., B_k$ nun $k/2$ zufällig gewählte Kandidaten aus und sendet die betreffende Teilmenge R an Anna. (Zur einfacheren Notation nehmen wir im folgenden an, es sei $R = B_1, ..., B_{k/2}$.)

Anna offenbart nun ihre Parameter a_i, c_i, m_i, d_i, r_i für alle $i \notin R$.

Bert prüft, ob die von Anna offen gelegten Werte tatsächlich die korrekte Berechnung der zugehörigen B_i ermöglichen; ferner verifiziert Bert, daß die von Anna übermittelten Werte m_i, $i \notin R$, wirklich alle die Identität von Anna beinhalten.

Ist diese Überprüfung erfolgreich beendet, so übermittelt Bert an Anna den von ihm signierten Wert

$$(\Pi\, B_i)^{1/3} \bmod n_{Bert}.$$
$$\scriptstyle i \in R$$

Dies ist Bert's Signatur unter Verwendung des Signaturschlüssels n_{Bert}.

Anna entfernt aus dem obigen Produkt nunmehr die Blinding Factors und erhält den Wert

$$S_A = (\Pi\, f(g(a_i, c_i), g(a_i \oplus m_i, d_i)))^{1/3} \bmod n_{Bert}.$$
$$\scriptstyle i \in R$$

Mit diesem Wert S_A kann Anna nun folgendermaßen signieren:

Bert sendet die zu signierende Nachricht $\varepsilon_1, ..., \varepsilon_{k/2}$ an Anna.

Anna antwortet hierauf mit den folgenden Daten:

(i)   Falls $\varepsilon_i = 1$, so sendet sie an Bert die Werte $g(a_i \oplus m_i, d_i)$, a_i, und c_i.

(ii)   Falls $\varepsilon_i = 0$, so sendet sie an Bert die Werte $g(a_i, c_i)$, $a_i \oplus m_i$, und d_i.

Bert verfügt nun über alle erforderlichen Daten, um die korrekte Form von S_A zu verifizieren.

Verwendet Anna ihren von Bert ausgestellten Signaturstring S_A ein weiteres Mal, um eine andere Nachricht zu signieren, so läßt sich obiger Beschreibung entnehmen, daß sie in diesem Fall nicht mehr länger anonym bleibt und durch die Rekonstruierbarkeit (mindestens) eines Wertes m_i ihre Identität preisgibt. (Für weitere Funktionalitäten des obigen Protokolls verweisen wir auf Hayes.)

Wichtig zu vermerken ist beim beschriebenen Verfahren, daß weder die Erzeugung der Signaturen noch der Signaturvorgang selbst irgendeiner vertraulichen Datenübertragung bedürfen. Ferner sind zum Zeitpunkt der Signierung keine arithmetischen Berechnungen erforderlich, sondern lediglich die Übertragung der obigen Bits ε_i.

Somit empfiehlt sich dieses Verfahren möglicherweise beim Einsatz von Kombikarten, sofern im Vorfeld der jeweiligen Transaktion eine Vorberechnung gemäß obigem Protokoll zwischen Smart Card und beteiligtem Hostrechner durchgeführt werden kann. Der noch relativ hohe Speicherbedarf zur Speicherung der benötigten Parameter auf der Chipkarte läßt sich durch zusätzliche Maßnahmen noch reduzieren.

Ein simples Anwendungsszenario wäre beispielsweise der Einsatz im öffentlichen Nahverkehr, wo via kontaktbehaftete Karte vorausbezahlt wird, und sodann im Verkehrsmittel selbst via kontaktlose Karte gezahlt werden kann, wobei die Zahlung anonym erfolgt und nicht notwendig sofort eine Online-Verbindung zur Überprüfung hergestellt werden muß bzw. kann. Der obige Algorithmus erlaubt den Einsatz der One Time Signaturen analog zu beispielsweise Traveller Checks oder ähnlichem, wobei der Besitz einer One Time Signatur (in Abhängigkeit vom jeweiligen Modul n) vergleichbar wäre mit einem Traveller Check über einen bestimmten Geldbetrag. Das Einlösen (resp. Unterschreiben) dieses Traveller Checks entspricht dann dem digitalen Signieren unter Verwendung des Strings S_A. Eine Übertragung dieses Prinzips in den Bereich der Zutrittskontrolle bestünde beispielsweise in der Entsprechung unterschiedlicher Schecks mit unterschiedlichen Zutrittsrechten. Ebenso denkbar ist die Analogie zwischen einer bereitgestellten One Time Signatur und einem von der Bank zur Verfügung gestellten Euroscheckformular. So könnte man mit seiner One Time Signatur zum Beispiel einen Warenkauf signieren, wobei der Kaufwert bis zu einem bestimmten Höchstbetrag von der ausstellenden Bank gedeckt wird.

Weitere One Time-Signaturverfahren

Grundidee:

Preprocessing-Phase: wähle Zahlenfolge $X_1, ..., X_k$, und übermittle diese Zahlenfolge sowie ihren signierten Hashwert. (Hierbei genügt es, den ersten Wert zufällig zu wählen, und die weiteren Werte beispielsweise sukzessive in aufsteigender Reihenfolge festzulegen.) Der Host signiert letzteren Hashwert sodann ebenfalls und gewährleistet somit die Gültigkeit der später ausgeführten One-Time-Signatur. Der Teilnehmer berechnet nun für jedes X_i sowohl $X_i^{1/3}$ als auch $X_i^{1/5}$ und speichert diese Werte ab.

Online-Phase: die zu signierende Nachricht bzw. ihr Hashwert sei $m_1, ..., m_k$;

Falls $m_i = 0$ $\Rightarrow$ übermittle den Wert $X_i^{1/3}$

Falls $m_i = 1$ $\Rightarrow$ übermittle den Wert $X_i^{1/5}$

Ist der Public Key des (nicht anonymen) Teilnehmers bekannt, so läßt sich dessen Signatur durch Potenzbildung unmittelbar nachprüfen. Signiert der Teilnehmer zwei verschiedene Strings unter Verwendung derselben Zahlenfolge X_1, ..., X_k, so offenbart er zu mindestens einem X_i sowohl die dritte als auch die fünfte Wurzel und läßt sich daraufhin als Betrüger anklagen.

Eine anonyme Variante könnte folgendermaßen aufgebaut werden:

Wähle statt obiger Zahlenfolge X_1, ..., X_k im Protokoll die nachfolgenden Zahlenstrings Y_1^{15}, ..., Y_k^{15}, wobei jedes Y_i die folgende, spezielle Form hat: Y_i = „ID-Nummer $\parallel$ X_i", wobei X_i wie zuvor Pseudozufallszahl ist. In der Online-Phase übermittelt der Teilnehmer als Signatur:

Falls $m_i = 0$ $\Rightarrow$ sende den Wert Y_i^5

Falls $m_i = 1$ $\Rightarrow$ sende den Wert Y_i^3

Signiert der Teilnehmer nun zwei verschiedene Strings unter Verwendung derselben Zahlenfolge Y_1, ..., Y_k, so offenbart er zu mindestens einem Y_i sowohl die dritte als auch die fünfte Wurzel von Y_i^{15}, wodurch sich nunmehr Y_i selbst errechnen läßt. Dieses enthält die ID-Nummer des betreffenden Teilnehmers und erlaubt es, diesen daraufhin als Betrüger anzuklagen.

Verschlüsselung auf der Luftschnittstelle

Eine sichere Verschlüsselung der Kommunikation zwischen Karte und Kartenleseeinheit läßt sich sehr einfach realisieren. Hierzu muß die Karte lediglich über hinreichend großen Speicher verfügen zur Speicherung eines vor der Kommunikation offline berechneten Key Streams, d.h. einer Folge von (pseudo-) zufälligen Bits. Steckt die Karte im Kartenleser (kontaktbehaftetes Interface), so wird dieser Key Stream entweder extern erzeugt und auf die Karte geladen, oder (Kombikarte) durch den karteneigenen Prozessor beispielsweise auf Basis von DES auf der Karte berechnet und sodann abgespeichert. In der Online-Phase während der Datenübertragung ist dann nur noch eine bitweise XOR-Verknüpfung mit dem zu verschlüsselnden Text durchzuführen.

Wird der Key Stream auf der Karte erzeugt, so muß selbstverständlich der zugrunde liegende Initialisierungswert zur Berechnung dieses Key Streams auch auf Hostseite verfügbar sein. Dieser Initialisierungswert könnte beispielsweise zuvor (Offline) mit den Public Keys der fraglichen Hostrechner verschlüsselt und in dieser gesicherten Form später übertragen werden.

Anwendungen

Es gibt eine Fülle verschiedener Praxisanwendungen, in denen die in den vorherigen Kapiteln beschriebenen kryptographischen Methoden bei Verwendung kontaktloser Chipkarten gewinnbringend zum Einsatz kommen können. Neben der im vorherigen Abschnitt bereits erwähnten Übermittlung verschlüsselter Daten besteht insbesondere Bedarf an der Funktionalität digitaler Signaturen. Offensichtliche Beispiele finden sich bei Implementierungen elektronischer Zahlungsverfahren bzw. digitalen Geldes, bei der Signatur bzw. Autorisierung verschiedenster Transaktionen, etc.
Eine weitere Anwendung besteht in der Realisierung sicherer Zutrittskontrollsysteme unter Verwendung von Challenge-Response-Verfahren mit Hilfe der One-Time-Signatur einer übermittelten Zufallszahl. In diesem Kontext ließe sich durch passende Zusatzmechanismen

beispielsweise die Anzahl erlaubter Zutritte begrenzen, mittels Ausstellung einer begrenzter Anzahl von One-Time-Signaturen; für unterschiedliche, individuelle Zutrittsrechte ließen sich überdies verschiedene Public Key-Moduln verwenden. Jeder Modul korrespondiert hierbei mit einer bestimmten Sicherheitsstufe oder einer Klasse von Benutzer-Zugangsrechten.

3 Wünschenswerte Features für die Zukunft

Neben den vielfältigen Anforderungen an die Schutzmechanismen der Karte aus Sicht der IT-Sicherheit besteht zudem ein berechtigtes Interesse am Schutz personenbezogener Daten auf der Chipkarte vor unbefugtem Auslesen. Dies beinhaltet alle kartenindividuellen Informationen, insbesondere also beispielsweise auch die Seriennummer der Chipkarte. Viele kontaktlose Karten können sich einem lesenden Zugriff von außen bislang nur bedingt entziehen.

Einige technische Voraussetzungen für höhere Sicherheit beim Einsatz kontaktloser Karten werden momentan bereits in den verschiedenen Entwicklungslabors der Hersteller geschaffen. Viele der zuvor geschilderten Probleme wären als weitgehend gelöst zu betrachten, sobald im kontaktlosen Betrieb lauffähige Mikroprozessoren mit Fähigkeit zur Berechnung starker Kryptoalgorithmen für Massenanwendungen verfügbar sind. Ebenso erforderlich als Voraussetzung für den Einsatz komplexerer kryptographischer Protokolle ist die Realisierung schnellerer Datenübertragungsraten auf der Luftschnittstelle. Komplexe Algorithmen und hinreichende Schlüssellängen erfordern überdies mehr Speicherplatz, - ein Problem, das sich in Anbetracht der zu erwartenden weiteren Fortschritte in der Speichertechnologie vermutlich von alleine lösen wird.

Die Erfahrung hat gezeigt, daß sich in nahezu allen Anwendungsbereichen die komfortablere gegenüber der sichereren Lösung durchsetzt. Aus diesem einfachen Grund ist mit einem wachsenden Anteil kontaktloser Chipkarten bei Smartcard-basierten Systemen zu rechnen.

Im vorliegenden Beitrag wurde versucht, einige relevante Sicherheitsprobleme hierbei aufzuzeigen und mögliche neue Lösungsansätze zu skizzieren. Noch reichlich Forschungsarbeit ist zu leisten, bis kontaktlose Chipkarten eine vergleichbar hohe Sicherheit bieten werden wie herkömmliche Chipkarten mit kontaktbehaftetem Interface.

Bibliographie

Chaum, David, and Fiat, Amos, and Naor, Moni, *Untraceable Electronic Cash*, Advances in Cryptology - CRYPTO'88, Springer-Verlag, 19880, Lecture Notes in Computer Science

Even, Shimon, and Goldreich, Oded, and Micali, Silvio, *On-Line/Off-Line Digital Signatures*, Journal of Cryptology, 1996, Vol. 9, Nr. 1, pages 35 - 67

Hayes, Barry, *Anonymous One-Time Signatures and Flexible Untraceable Electronic Cash*, Advances in Cryptology - AUSCRYPT'90, Springer-Verlag, 1990, Lecture Notes in Computer Science 453, pages 294 - 305

Merkle, R. C., *A Certified Digital Signature*, Advances in Cryptology - CRYPTO'89, Springer-Verlag, 1990, Lecture Notes in Computer Science 435, pages 218 - 238

Rabin, M. O., *Digitalized Signatures*, in: R. DeMillo, D. Dobkin, A. Jones, R. Lipton (Editors), Foundations of Secure Computation, Academic Press, 1978, pages 155-168

Zieschang, Thilo, *Combinatorial Properties of Basic Encryption Operations*, Advances in Cryptology - EUROCRYPT'97, Springer-Verlag, 1997, Lecture Notes in Computer Science 1233, pages 14 – 26

Zieschang, Thilo, *Differentielle Fehleranalyse und Sicherheit von Chipkarten*, 5. Deutscher IT-Sicherheitskongreß des BSI, 1997, SecuMedia-Verlag, Ingelheim, Seiten 227 - 240

Erweitertes API zum TCOS[a]-Betriebssystem[b]

Stefan Pütz[1] · Torsten Henn[2]

[1]T-Mobil, Postfach 30 04 63, D-53184 Bonn
stefan.puetz@t-mobil.de

[2]secunet, Weidenauer Str. 217, D-57076 Siegen
T.Henn@secunet.de

Zusammenfassung

Dieser Beitrag zeigt die Anwendung von TCOS, Version 1.2, als Teilnehmerchipkarte bei der Realisierung eines neuen Sicherheitsprotokolls für UMTS (Universal Mobile Telecommunications System), der dritten Generation von Mobilfunksystemen. Dabei steht allerdings nicht das Protokoll im Vordergrund der Ausführungen, sondern die Schnittstelle, über die die User-Applikation auf die verfügbaren Chipkartenroutinen zugreift. Um diese Schnittstelle komfortabel zu gestalten, wurde ein erweitertes TCOS-API implementiert und darauf aufbauend eine Personalisierungsinstanz realisiert. Abschließend folgt eine Analyse der Ausführungszeit, die eine RSA *Secret-Key* Operation auf der verwendeten Chipkarte erfordert. Hier zeigt sich deutlich, daß nicht die RSA-Prozeßzeit der Karte, sondern die Bandbreite des Übertragungskanals die Ausführungszeit im wesentlichen begrenzt.

1 Einleitung

Aus dem GSM (Global System for Mobile Communication)-Standard, nach dem unsere heutigen digitalen D- und E-Netze der zweiten Mobilfunkgeneration arbeiten, ist die Trennung zwischen Endgerät und teilnehmerbezogener Chipkarte bekannt [Pütz_97]. Zur Nutzung der Mobilfunkdienste benötigt ein Teilnehmer neben einem geeigneten Endgerät eine Mobilfunkkarte. Diese Chipkarte hält sicherheitsrelevante Schlüssel und berechnet symmetrische kryptographische Operationen. Für UMTS (Universal Mobile Telecommunications System), dem Standard der nächsten Mobilfunkgeneration, wird derselbe Ansatz verfolgt. Allerdings basiert UMTS im Vergleich zu GSM auf einem veränderten Rollenmodell. Zudem wird erwartet, daß viele Netzbetreiber und Diensteanbieter dieses weltumspannenden UMTS-Netzes miteinander kooperieren und konkurrieren. Um die veränderten Sicherheitsanforderungen an UMTS zu erfüllen, werden vermehrt Protokolle diskutiert, die auf asymmetrischen Kryptoverfahren basieren.

---

[a] TCOS ist ein Produkt des Produktzentrums (PZ) Telesec der Deutsche Telekom GmbH, Netphen.

[b] Diesem Beitrag liegen Forschungsaktivitäten aus dem Projekt 'Mobilfunk-Authentikation' zugrunde, welches im Schwerpunktprogramm 'Mobilkommunikation' der Deutschen Forschungsgesellschaft angesiedelt ist und am Institut für Nachrichtenübermittlung der Universität Siegen bearbeitet wird.

Ausgangspunkt der hier dokumentierten Arbeit war es, anhand eines Demonstrators die Einsatzfähigkeit eines neuen Sicherheitsprotokolls zu zeigen. Dies erforderte die Auswahl und anschließende Einbindung einer entsprechenden Chipkarte.

2 Chipkarten-Betriebssysteme

Da die Mikrokontroller-Entwicklung für Chipkarten in den letzten Jahren große Fortschritte gemacht hat, sind die Anforderungen an leistungsfähige Betriebssysteme gestiegen. Die Vorzüge von Mikroprozessorkarten (ICC, Integrated Circuit Card) gegenüber Chipkarten mit festverdrahteter Logik sind sichere und flexible Datenspeicher unter der Verwendung von Kryptoalgorithmen.

Allerdings bedürfen Prozessorchipkarten zusätzlicher technischer Systeme, um Daten aus der Karte zu lesen, Daten auf die Karte zu schreiben oder um bestimmte Funktionalitäten des zugrunde liegenden Betriebssystems zu aktivieren. Als Systeme zur Erschließung der Daten und Funktionen einer Prozessorchipkarte werden Kartenterminals (CT, Chipcard Terminal) eingesetzt. Kartenterminal und Karte selbst sind wiederum in ein informationstechnisches Gesamtsystem eingebettet, welches die verschiedenen Anwendungsmöglichkeiten für Chipkartensysteme bereitstellt. Die Aufgabe des Kartenterminals besteht hier vor allem in der Annahme, Aufbereitung und Weiterleitung von Daten, die von einer Anwendung zur Chipkarte gesendet bzw. von der Chipkarte zur Anwendung übertragen werden. Insgesamt realisiert das Kartenterminal im Kontext mit einer Chipkartenanwendung die Mensch-Maschine-Schnittstelle einer Prozessorkarte. Vor dem Hintergrund, eine Prozessorchipkarte als UIM bei der Realisierung eines neuen Sicherheitsprotokolls für UMTS einzusetzen, ist bei der Wahl der Karte auch die Möglichkeit der Ansteuerung bzw. Kommunikation über das Kartenterminal zu berücksichtigen, d.h. es ist das gesamte Chipkartensystem zu betrachten.

Für die Auswahl einer geeigneten Prozessorchipkarte, welche einerseits als sicherer Datenträger fungiert und andererseits kryptographische Algorithmen zur Verfügung stellen muß, wurden die folgenden Kriterien aufgestellt. Sie definieren die Mindestanforderungen an das zugrundeliegende Betriebssystem:

- Steuerung der Zugriffs- und Nutzungsberechtigungen durch die Chipkarte selbst

- Benutzung allgemein anerkannter, veröffentlichter und verbreiteter Algorithmen für Verschlüsselungsfunktionen sowie zur Erzeugung von Zufallszahlen

- Unterstützung eines Dateisystems gemäß ISO 7816

- Public-Key Operationen

- Möglichkeit einer gegenseitigen Authentisierung von Chipkarte und Anwendung

- Realisierung aktiver Sicherheitsfunktionen zur Übertragungssicherung
  (Secure Messaging)

- Detaillierte Dokumentationen bzgl. Befehlsvorrat und Ansteuerung der Karte sowie eines geeigneten Kartenterminals

Um eine Auswahl vornehmen zu können, wurden Prozessorchipkarten mit folgenden Beriebssystemen in Erwägung gezogen:

- CardOS-Karte der Firma Siemens

- STARCOS-Karte der Firma Giesecke & Devrient

- TCOS-Karte der Deutsche Telekom AG

Der Bereich der Chipkartenbetriebssystem-Entwicklung ist noch relativ jung. Ein grundsätzliches Problem während der Suche nach einem geeigneten Betriebssystem bestand daher in der Verfügbarkeit entsprechend ausgestatteter Karten sowie notwendiger Detailinformationen zum Betriebssystem. CardOS stellt wie das Telesec Chipcard Operating System (TCOS) asymmetrische Kryptoverfahren zur Verfügung und kann für die Erschließung der Kartenfunktionalität und -daten über ein Kartenterminal nach B1-Spezifikation [CTB1_96] und die universelle Applikationsschnittstelle CT-API [CTAPI_96] angesprochen werden. STARCOS unterstützt nach derzeitigem Kenntnisstand ebenfalls Public Key Operationen. Für die Entwicklung von STARCOS-Applikationen steht ein eigenes Development Kit zur Verfügung.

Prinzipiell sind die meisten der oben genannten Chipkartenbetriebssysteme als Teilnehmerkarte im Kontext eines Sicherheitsprotokolls geeignet. Ausschlaggebend für den Einsatz des TCOS-Betriebssystems waren letztlich die folgenden Aspekte.

- Verfügbarkeit von TCOS-Karten und ausführlicher Dokumentation [TCOS_95a, TCOS_95b] (Bereitstellung durch das PZ Telesec der Deutsche Telekom AG)

- Evaluierung von TCOS nach den [ITSEC]-Sicherheitskriterien (Stufe E4)

- Flexibilität von TCOS

- Verfügbarkeit von B1-Leser, Programmierschnittstelle CT-API 1.1 und ausführlicher Dokumentation [CTB1_96, CTAPI_96, HTSI_96]

- Für Fragen wurde seitens des PZ Telesec ein Ansprechpartner bereitgestellt, so daß eine Implementierung performant durchgeführt werden konnte.

TCOS ist ein Betriebssystem für Prozessor-Chipkarten, das auf verschiedenen Hardwareplattformen einsetzbar ist [TCOS_95a]. Optional unterstützt TCOS Arithmetikprozessoren, die geeignet sind, modulare Rechenoperationen für Public-Key Anwendungen durchzuführen. Unter Ausnutzung der Hardwareschutzmöglichkeiten bietet TCOS noch zusätzliche Softwareschutzmöglichkeiten [KDSB_97, WoFo_97]. Die Daten werden von TCOS in einem Filesystem verwaltet, dessen Flexibilität TCOS multifunktionalen Einsatz in verschiedenen Anwendungen gestattet. Zum Schutz der Daten bietet TCOS für jedes File verschiedene Zugriffsrechte und kryptographische Sicherheitsmechanismen. Weiterhin schützt TCOS die Daten und Informationen durch Authentikation des Kommunikationspartners. Die Authentizität und Integrität der Befehle und Meldungen überprüft TCOS durch MAC-Berechnung.

TCOS erzeugt mit Hilfe kryptographischer Verfahren Zufallszahlen und symmetrische Schlüssel in verschiedenen Längen (z.B. für DES/DES3) und setzt diese z.B. als Session Keys ein. Mit diesen Schlüsseln unterstützt TCOS einerseits die gesicherte Datenübertragung von und zur Chipkarte (Encryption, MAC) und andererseits die Eigenschaft, Daten durch die Karte symmetrisch zu ver- bzw. entschlüsseln [NBS_77, NBS_80, ISO_9797]. Unter Anwendung asymmetrischer Kryptoalgorithmen, wie RSA [RSA_78], können mit TCOS digitale Signaturen generiert sowie Informationen asymmetrisch ver- und entschlüsselt werden.

Zur Kommunikation stellt TCOS verschiedene Chipkarten-Übertragungsprotokolle zur Verfügung [RaEf_95], die nach Bedarf eingesetzt werden können.

Ein Nachteil von TCOS besteht darin, daß auf die Kommandoschnittstelle zwecks Anlegen und Auslesen von Daten und Aktivierung bestimmter Operationen nicht komfortabel zugegriffen werden kann. Hierfür müssen die Codierungen der einzelnen Kommandos und deren Parameter herangezogen und in Form von Byte-Sequenzen über die CT-API an TCOS gesendet werden. Daher ergab sich die Notwendigkeit, ein erweitertes API zum TCOS-Betriebssystem zu implementieren.

3 Erweiterte Chipkarten-Schnittstelle

Der verwendete B1-Chipkartenleser stellt die Schnittstelle CT-API 1.1 zur Verfügung. Sie bietet neben der Initialisierung (CT_init) und dem Abbau (CT_close) einer HOST – CT Kommunikation eine weitgehend transparente Datenübertragung (CT_data) mit der Chipkarte selbst. Interface-Parameter dieser Schnittstelle werden durch den implementierten Befehl SetInterfParamCT beeinflußt.

Aufbauend darauf wird die Schnittstelle CT – ICC vorgestellt. Es wird dokumentiert, daß die elementaren TCOS-Befehle in komfortable C-Funktionen eingebettet wurden. Somit entstehen für den Anwendungsprogrammierer einheitliche und handliche Funktionsaufrufe mit einer übersichtlichen Parameterliste. TCOS Elementarfunktionen [ISO_7816-4, TCOS_95a] sind zuerst auf einer Zwischenstufe realisiert und anschließend, falls erforderlich, modular zu erweiterten Funktionen zusammengefaßt worden. Auf der untersten Ebene werden alle TCOS-Befehle mit dem Befehl DataCT (bzw. CT_data der CT-API 1.1) an die ICC abgesetzt.

Ist die Kommunikationsbeziehung einmal initialisiert, hat der Anwendungsprogrammierer keine Last mit der Adressierung seiner Befehle an die ICC, da die erweiterte API die Adreßanpassung und -aufbereitung übernimmt. Ebenso wird der Übertragungsmodus in den Befehlen an die ICC bzw. für die Meldungen von der ICC je nach den Erfordernissen des abgesetzten, speziellen Befehls angepaßt. Eine Vorwahl erfolgt durch den Befehl SetTransMode. Zur Adreßanpassung muß zwischen CT und ICC, beim Übertragungsmodus zwischen ungeschützt und geschützt (Encryption, MAC) unterschieden werden, wobei die jeweilige Übertragungsrichtung zu berücksichtigen ist [TCOS_95a].

```
CT-API 1.1 (Grundfunktionen)
    char __far __pascal CT_init        Kommunikationsbeziehung initialisieren (aktivieren)
    char __far __pascal CT_close       Kommunikationsbeziehung beenden (deaktivieren)
    char __far __pascal CT_data        Befehl/Daten an CT/ICC senden

Extended CT-API Libary
    char SetTransMode                  Übertragungsmodus wählen
    char InitCT                        Siehe CT_init (verwendet Default-Werte)
    char CloseCT                       Siehe CT_close
    char DataCT                        Siehe CT_data, Anpassung von Adresse und Übertra-
                                       gungsmodus (bzgl. Encryption, MAC), Fehlerbehandlung
    char ResetCT                       CT aktivieren, (Neu-) Start
    int  SetInterfParamCT              Interface-Parameter der Schnittstelle HOST – CT setzen
    int  ResetICC                      ICC aktivieren, (Neu-) Start
    int  DeactivateICC                 ICC deaktivieren
```

TCOS Version 1.2 (Grundfunktionen)

char Status	Kartenstatus der ICC abfragen
char VerifyPassword	PIN/Paßwort überprüfen
char ChangePassword	PIN/Paßwort ändern
char UnblockPIN	PIN mit PUK zurücksetzen
char SelectApp	Applikation auswählen
char CloseApp	Applikation schließen
char CreateFile	Datei erzeugen
char SelectFile	Datei auswählen
char DeleteFile	Datei löschen
char ReadBinary	Leseoperation, Lesebereich variabel
char ReadRecord	Leseoperation, Lesebereich blockorientiert
char UpdateBinary	Schreiboperation, Schreibbereich variabel
char UpdateRecord	Schreiboperation, Schreibbereich blockorientiert
char AskRandom	Zufallszahl ausgeben
char GetSessionKey	Sitzungsschlüssel ausgeben, geschützte Kommunikation
char GetResponse	Antwort ausgeben
static char InternalAuth	ICC authentisieren
static char ExternalAuth	Anwendung authentisieren
static char SetKey	Schlüssel für Verschlüsselung u. MAC-Berechnung setzen
static char Crypt	Daten ver-/entschlüsseln
static char Signature	Daten signieren, Einbinden von Signaturzähler u. ICC-ID

Extended TCOS-API Libary

Get_Session_Key

char GetSessionKeyDESEncrypted	Sitzungsschlüssel DES/DES3-verschlüsselt ausgeben, Kryptofunktionen HOST-seitig aufrufen
char GetSessionKeyRSAEncrypted	Sitzungsschlüssel RSA-verschlüsselt ausgeben, Kryptofunktionen HOST-seitig aufrufen

Authentication

int IntAuth	ICC authentisieren, Authentikations- u. Verschlüsselungsverfahren auswählen, Kryptofkt. HOST-seitig aufrufen
int ExtAuth	Anwendung authentisieren, Authentikations- u. Verschl.-verfahren auswählen, Kryptofkt. HOST-seitig aufrufen

Encryption/Decryption

int EncryptDES_ECB	Daten mit DES verschlüsseln, EBC-Mode
int EncryptDES3_ECB	Daten mit DES3 verschlüsseln, EBC-Mode
int EncryptDES_CBC	Daten mit DES verschlüsseln, CBC-Mode
int EncryptDES3_CBC	Daten mit DES3 verschlüsseln, CBC-Mode
int DecryptDES_ECB	Daten mit DES entschlüsseln, EBC-Mode
int DecryptDES3_ECB	Daten mit DES3 entschlüsseln, EBC-Mode
int DecryptDES_CBC	Daten mit DES entschlüsseln, CBC-Mode
int DecryptDES3_CBC	Daten mit DES3 entschlüsseln, CBC-Mode
char PublicKeyRSA	Daten mit RSA verschlüsseln
char SecretKeyRSA	Daten mit RSA entschlüsseln
char SignRSA	Daten signieren (interne RSA-Verschlüsselung)

Werden zur Ausführung einzelner Befehle (`GetSessionKeyDES/RSAEncrypted`, `In-tAuth/ExtAuth`) oder für eine geschützte Übertragung Kryptooperationen benötigt, rufen die erweiterten TCOS-Funktionen HOST-seitig entsprechende Routinen einer RSAREF-Implementation auf. So werden beispielsweise für die Authentikationen [ISO 9798-2] ebenso wie für eine verschlüsselte bzw. MAC-geschützte Übertragung von und zur ICC DES/DES3-Operationen benötigt. Wird ein Session-Key verschlüsselt ausgegeben, entschlüsseln ihn geeignete Algorithmen (RSA/DES), so daß er nach dem Funktionsaufruf im Klartext vorliegt.

4 Personalisierungsinstanz

Um TCOS-Chipkarten für ihren Einsatz als UIM vorzubereiten, ist eine Peronalisierungsinstanz entwickelt und realisiert worden, die einerseits die individuelle Personalisierung dieser Chipkarten erlaubt und andererseits deren Umkonfiguration oder Überpüfung unterstützt. Im Rahmen dieser Personalisierungsinstanz sind folgende Befehle implementiert.

<u>Befehle der Personalisierungsinstanz</u>

`Verify   PIN/Password`	PIN/Paßwort überprüfen
`Change   PIN/Password`	PIN/Paßwort ändern
`Unblock PIN`	PIN mit PUK zurücksetzen
`Select   App`	Applikation auswählen
`Create   App`	Applikation erstellen (Masterfile MF aktualisieren)
`Delete   App`	Applikation löschen (Masterfile MF aktualisieren)
`Close   App`	Applikation schließen
`Select   File`	Datei auswählen
`Create   File`	Datei erstellen, Dateiname, -größe und -rechte anlegen
`Delete   File`	Datei löschen
`Show     File`	Dateiinhalt ausgeben
`Check on all Files`	Liste aller existierenden Dateien ausgeben
`Write    Binary`	Dateiinhalt schreiben, Schreibbereich variabel
`Write    Record`	Dateiinhalt schreiben, Schreibbereich blockorientiert
`Copy     File`	Dateiinhalt kopieren (innerhalb TCOS)
`Modify   File-Header`	Dateikopf ändern, Dateiname, -größe und -nutzungsrechte aktualisieren
`Init     Password File`	Paßwort-Datei initialisieren, Fehlbedienungszähler setzen
`Create   RSA File-System`	Dateien erstellen, die ein RSA-Schlüsselsystem aufnehmen (Modulus, öffentl. und geheimer Exponent)
`Init     RSA File-System`	Schlüsselsystem in entsprechenden Dateien ablegen, Schlüsselsystem wird eingelesen oder aktuell generiert
`Show     RSA Public-Key/Modulus`	Modulus und öffentlichen Exponent ausgeben
`Check    RSA Key-System`	RSA-Operationen mit öffentlichem und geheimem Schlüssel nacheinander ausführen
`Authenticate Host`	HOST authentisieren
`Authenticate ICC`	ICC (TCOS-Chipkarte) authentisieren
`Init       CT`	Kommunikationsbeziehung zum CT initialisieren
`Reset      CT`	CT aktivieren
`Deaktivate CT`	CT deaktivieren
`Reset      ICC`	ICC (TCOS-Chipkarte) aktivieren
`Deaktivate ICC`	ICC (TCOS-Chipkarte) deaktivieren

Um ein RSA-Schlüsselsystem auf der ICC abzulegen, muß dieses entweder aus einer Datei eingelesen oder aber aktuell generiert werden. Auch hierzu ruft die Personalisierungsinstanz HOST-seitig entsprechende Funktionen einer RSAREF-Implementation auf.

5 Analyse der Ausführungszeit

Im Rahmen der Testaktivitäten des neuen Sicherheitsprotokolls ist eine ausführliche Betrachtung der Protokoll-Ausführungsgeschwindigkeit durchgeführt worden. Es hat sich herausgestellt, daß die Gesamtausführungszeit einer RSA *Secret-Key* Operation in hohem Maße durch die Datenübertragung von und zur ICC beeinflußt wird. Daher werden nachfolgend die Ausführungszeiten für RSA *Secret-Key* Operationen in Verbindung mit TCOS Version 1.2 und Version 2.0 analysiert. Laut Datenblatt [TCOS_96] wird mit dem Betriebssystem TCOS Version 1.2 auf einer geeigneten Smart Card, z.B. 44CR80S vom Hersteller Siemens, eine reine *Secret-Key* Operation bei einer Moduluslänge von 512 bit in nur 65 ms möglich, falls das Chinese/Remainder Verfahren angewendet wird. Eine *Public-Key* Operation erfolgt sogar in nur 10 ms, falls als öffentlicher Exponent F_4, die 4. Fermat-Zahl [Rula_93], verwendet wird. Allerdings unterstützt TCOS Version 1.2 an der Schnittstelle CT – ICC lediglich eine Bandbreite von 9.600 Baud, die CT-API Version 1.1 des B1-Chipkartenlesers an der HOST – CT Schnittstelle unter MS Windows 57.600 Baud (und unter DOS 115.200 Baud). Insgesamt ergibt sich rein rechnerisch eine Gesamtzeit (*Secret-Key* Operation inkl. Übertragungszeit) unter MS Windows von mindestens 221,5 ms (zzgl. der Prozeßzeit der *Set-Key* Operation), die der in Versuchsreihen gemessenen Zeit von 235 ms sehr gut entspricht. Die Versuchsreihen basieren auf dem Einsatz der Smart Card 40C200 von Siemens und dem Betriebssystem TCOS Version 1.2. Beide Zahlen veranschaulichen jedoch deutlich, daß die tatsächliche Gesamtzeit mit etwa 235 ms ungefähr der 3,5 fachen *Secret-Key* Prozeßzeit entspricht. Dieses Verhältnis zeigt, daß ein wichtiger Ansatzpunkt zur Verringerung der Gesamtzeit darin liegt, die Übertragungsrate speziell auf der CT – ICC Schnittstelle signifikant zu erhöhen, wie dies in [ISO_7816-3] vorgeschlagen und ab TCOS Version 2.0 mit max. 38.400 Baud vorgesehen ist. Theoretisch ergibt sich dann eine Gesamtzeit von 121 ms (zzgl. der *Set-Key* Prozeßzeit). Zudem verzichtet TCOS Version 2.0 auf die Übertragung des separaten *Set-Key* Befehls, wo-

Tabelle 1: Ausführungszeiten einer *Secret-Key* Operation inkl. Übertragungszeit

	TCOS Version 1.2		TCOS Version 2.0
Datenvolumen in [Byte]	161	161	144
Übertragungszeit in [ms] bei – 9.600 Baud	134,16		
– 38.400 Baud		33,54	30
– 57.600 Baud	22,36	22,36	20
Prozeßzeit in [ms] einer *Secret-Key* Operation	65	65	65
Gesamtzeit in [ms]	**221,52**	**120,90**	**115**

durch sich weitere Einsparungen ergeben. Bleibt der *Crypt* Befehl unverändert, sinkt die Gesamtzeit auf 115 ms.

Bei diesen Betrachtungen wurde der Aufwand für einen *Set-Key* und einen *Crypt* Befehl berücksichtigt sowie ebenfalls der Overhead durch die Verwendung des T = 1 Protokolls auf beiden Schnittstellen [RaEf_95]. Der Umfang der zu übertragenden Daten beläuft sich insgesamt auf 17 Byte für einen *Set-Key* Befehl sowie auf 144 Byte für einen *Crypt* Befehl (RSA-Algorithmus) je Schnittstelle.

6 Zusammenfassung

Der Beitrag zeigt die Funktionalität von TCOS auf. Die ursprüngliche Programmierschnittstelle gab Anlaß zur Implementation der erweiterten TCOS-API, die eine komfortable Einbindung der Kartenfunktionalitäten in neue Applikationen ermöglicht. Mit dieser Schnittstelle konnte die Karte sehr leicht als Teilnehmerchipkarte in den Demonstrator für ein neues UMTS-Sicherheitsprotokoll eingebettet werden. Zur Konfiguration der Chipkarten und zum anschließenden Funktionstest wurde eine Personalisierungsinstanz implementiert. Der auffallend hohe Einfluß der Übertragungszeit in die Gesamtausführungszeit einer RSA *Secret-Key* Operation und damit in die Protokollausführungsgeschwindigkeit bewirkte eine genaue Untersuchung dieses Punktes.

Literatur

ISO_7816-3   ISO/IEC DIS 7816: *Identification Cards – Integrated Circuit(s) Cards with Contacts – Part 3: Electronic Signals and Transmission Protocols. 1996.*

ISO_7816-4   ISO/IEC 7816: *Identification Cards – Integrated Circuit(s) Cards with Contacts – Part 4: Inter-industry Commands for Interchange. 1994.*

ISO_8372   ISO/IEC 8372: *Modes of Operation for a 64 Bit Block Cipher Algorithm. 1987.*

ISO_9797   ISO/IEC 7979: *Data Integrity Mechanism Using a Cryptographic Check Function Employing a Block Cipher Algorithm. 1994.*

ISO_9798-1   ISO/IEC 9798: *Information technology – Security techniques – Entity authentication – Part 1: General model. 1991.*

ISO_9798-2   ISO/IEC 9798: *Information technology – Security techniques – Entity authentication – Part 2: Mechanisms using symmetric encipherment algorithms. 1994.*

CTAPI_96   CT-API (CardTerminal Application Programming Interface) Version 1.1: *Schnittstellenspezifikation der anwendungsunabhängigen Kartenterminalfunktionen.* PZ Telesec der Deutsche Telekom AG, GMD Forschungszentrum Informationstechnik GmbH, RWTÜV Anlagentechnik GmbH, TeleTrust Deutschland e.V., Juli 1996.

CTB1_96   Intelligenter, multifunktionaler B1 Chipkartenleser (CardTerminal): *Betriebsanleitung Version 4.0*. Siemens Nixdorf AG, Mannheim, Mai 1996.

HTSI_96   B0/B1 HTSI (Host Transport Service Interface): *Programmierhandbuch Version 2.1*. PZ Telesec der Deutsche Telekom AG, Siegen, März 1996.

ITSEC   *Kriterien für die Bewertung der Sicherheit von Systemen in der Informationstechnik*. EGKS-EWG-EAG, Brüssel, Luxemburg 1991, ISBN 92-826-3003-X.

KDSB_97   Konferenz der Datenschutzbeauftragten des Bundes und der Länder: *Anforderungen zur informationstechnischen Sicherheit bei Chipkarten*. Datenschutz und Datensicherheit (DuD), 5/97, S. 254-259.

NBS_77   National Bureau of Standards (NBS): *Data Encryption Standard (DES)*. Federal Information Processing Standards Publication (FIPS-PUB) 46-1, US Department of Commerce, Jan. 1977.

NBS_80   National Bureau of Standards (NBS): *DES modes of Operations*. Federal Information Processing Standards Publication (FIPS-PUB) 81, US Department of Commerce, Dec. 1980.

Pütz_97   Pütz, Stefan: *Zur Sicherheit digitaler Mobilfunksysteme*. Datenschutz und Datensicherheit (DuD), 6/97, S. 321-327.

RaEf_95   Rankl, Wolfgang; Effing, Wolfgang: *Handbuch der Chipkarten: Aufbau – Funktionsweise – Einsatz*. Carl Hanser Verlag, München, Wien, 1995.

RSA_78   Rivest, Ronald L.; Shamir, Adi; Adleman, Leonard: *A Method for obtaining Digital Signatures and Public Key Cryptosystems*. Communications of the ACM, Bd. 21, Nr. 2, 1978, S. 120-126.

Rula_93   Ruland, Christoph: *Informationssicherheit in Datennetzen*. DataCom-Verlag, Bergheim, 1993.

TCOS_95a   TCOS (Telesec Chipcard Operating System) Betriebssystem für Chipkarten Version 1.2: *Spezifikation für Entwickler*. Produktzentrum (PZ) Telesec der Deutsche Telekom AG, Siegen, Januar 1995.

TCOS_95b   TCOS (Telesec Chipcard Operating System) Betriebssystem für Chipkarten Version 1.2: *Beispielapplikationen*. Produktzentrum (PZ) Telesec der Deutsche Telekom AG, Netphen, August 1995.

TCOS_96   TCOS (Telesec Chipcard Operating System) Betriebssystem für Smart Cards Version 1.2: *Datenblatt*. Produktzentrum (PZ) Telesec der Deutsche Telekom AG, Netphen, 1996.

WoFo_97   Wohlmacher, Petra; Fox, Dirk: *Hardwaresicherheit von Smartcards*. Datenschutz und Datensicherheit (DuD), 5/97, S. 260-265.

Wesentliche Kriterien beim Kryptochipkarten-Entwurf

Holger Bock · Wolfgang Mayerwieser
Karl C. Posch · Reinhard Posch

Institut für Angewandte Informationsverarbeitung
und Kommunikationstechnologie
Technische Universität Graz, Österreich
{hbock,wmayer,kposch,rposch}@iaik.tu-graz.ac.at

Zusammenfassung

Bei der Beschreibung von Chipkarten-Entwürfen mit kryptografischen Public-Key-Funktionen stand bisher meist deren Architektur im Vordergrund. Dies wohl deshalb, da Public-Key-Funktionen angesichts der jeweils vorhandenen technologischen Möglichkeiten sehr anspruchsvoll waren und nur durch starke Kompromisse beim Thema Sicherheit überhaupt realisiert werden konnten. Durch bessere Technologien und durch die starke Verbreitung von Chipkarten kommt jedoch der Sicherheitsaspekt immer mehr in den Vordergrund.

Diese Arbeit beschreibt anhand des Entwurfs eines Krypto-Co-Prozessors die beim Entwurfsprozess notwendige ganzheitliche Sichtweise, welche nicht nur traditionelle Parameter wie Chipflächenbedarf, Leistungsaufnahme und Rechengeschwindigkeit beachtet, sondern zusätzlich Begriffe wie Sicherheit gegenüber unerwünschter Manipulation, Skalierbarkeit, und Energiemanagement in den Entwurfsprozess miteinbezieht.

1 Einleitung

Chipkartensysteme mit kryptografischen Funktionen sind auf Grund der stark limitierten Siliziumfläche von maximal 25 mm^2 zu den eher kleinen elektronischen Systemen zu zählen. Betrachtet man jedoch die an sie gestellten Anforderungen bezüglich Verlässlichkeit, Informationssicherheit, Energieumsatz und Rechengeschwindigkeit, so zeigt sich klar, dass diese Rahmenbedingungen neue Denkansätze beim Systementwurf erfordern.

Diese Arbeit zeigt einige wesentliche Kriterien beim Entwurf von Kryptochipkarten auf, welche in zwei Projekten der Europäischen Union im Rahmen von OMI[1], SOSCARD[2] und CRISP[3], erarbeitet wurden. In diesen Projekten wurde ein RISC-Prozessor mit einer Hardware-Einheit für asymmetrische und symmetrische kryptografische Funktionen samt einem

---

[1] The Open Microprocessor Systems Initiative, 4th Framework Programme.
[2] Secure Operating System Smart Card SOSCARD, ESPRIT III Project 9259.
[3] CRISP Cryptographic Reduced Instruction Set Processor Smartcard, 4th Framework Programme Project 20847.

Betriebssystem für Multiapplikations-Chipkarten entwickelt. Technische Details zur kryptografischen Einheit dieser Chipkarte wurden in [Bock97] beschrieben. In [Bock98] findet man eine marktorientierte Auseinandersetzung zum gleichen Thema. Die vorliegende Arbeit konzentriert sich auf die Wichtigkeit der ganzheitlichen Sicht beim Entwurf des Chipkartensystems, welche über den mittlerweile weitverbreiteten Begriff *Hardware/Software Co-design* hinausreicht.

In [Naccache96] findet man eine Zusammenstellung bisheriger Ergebnisse im Bereich der Public-Key-Co-Prozessoren für Chipkarten. Man erkennt daraus den Trend, daß nach anfänglichen Versuchen, dieses Thema technologisch überhaupt zu lösen, es jetzt an der Zeit ist, eine gesamtheitliche Sichtweise in den Vordergrund zu rücken. In der jüngeren Vergangenheit stand hauptsächlich das Problem der zu knappen Rechenleistung bei eingeschränkter Maximalfläche und limitierter Leistungsaufnahme zur Debatte. Verschiedene Co-Prozessor-Architekturen für Public-Key-Verfahren werden etwa in [Ferreira96] verglichen. In dieser Arbeit versuchen wir, die eben zitierte ganzheitliche Betrachtung der Sicherheitsproblematik bei Chipkarten zu betonen.

Eine effiziente Realisierung eines Chipkartensystems ist nur möglich, wenn es einerseits in all seinen Betrachtungsebenen von der Applikation bis zur Hardwaretechnologie gemeinsam optimiert wird und andererseits dabei eine Vorgangsweise gewählt wird, welche gegenüber Technologieänderungen weitestgehend immun ist. Die klassischen Dimensionen der Optimierung sind durch Energieumsatz, Geschwindigkeit und Flächenbedarf gegeben. Als vierte Dimension lässt sich Informationssicherheit darstellen, wenngleich hier die Metrik nicht einfach festzulegen ist.

In Kapitel 2 dieser Arbeit greifen wir den Aspekt des Energieumsatzes heraus und behandeln diesen im Detail. Geschwindigkeit und Flächenbedarf sind stark von der jeweilig zur Verfügung stehenden Technologie abhängig. Diese einerseits auszureizen und andererseits gegenüber raschen Technologieänderungen immun zu sein ist aus Sicht der Optimierung wesentlich. Damit beschäftigen wir uns in Kapitel 3.

Die Auseinandersetzung mit allen Systemebenen beginnend von der Applikation bis zum Layout auf Silizium ist auch aus Gründen der Informationssicherheit notwendig. Geht man davon aus, dass Chipkarten vor allem als Träger sensitiver Information, wie sie etwa Schlüssel oder Geld darstellen, verwendet werden, so ist der Sicherheitsaspekt evident. Eine Reihe von sicherheitsbedingten Problemstellungen lässt sich selbst (und insbesonders) bei Verwendung von starken kryptografischen Methoden nur durch technische Verfahren lösen, welche auf mehreren Betrachtungsebenen ansetzen. Attacken auf Systeme, wie etwa die in [Kocher95] beschriebene „Timing Attack" oder die auf differentieller Fehleranalyse basierende Attacke [Biham96a, Biham96b] können bei einer ganzheitlichen Sicht leichter abgewehrt werden.

Aus Gründen der Informationssicherheit und der Effizienz sind Full-Custom-Methoden beim Layout-Entwurf angeraten. Um beim raschen Technologiefortschritt trotzdem weitestgehend technologieunabhängig zu bleiben, schlagen wir die Verwendung von skalierbaren Systemmodulen vor. Das Konzept der Skalierbarkeit führt zu einer Neudefinition des Begriffs „Zellbibliotheken", auf welchen wir in Kapitel 3 genauer eingehen.

In Kapitel 4 dieser Arbeit umreissen wir in groben Zügen die Funktionsweise und Leistungsmerkmale der kryptografischen Einheit von CRISP, bei welcher die hier vorgestellten Entwurfskriterien eingeflossen sind.

2 Energiemanagement

Energiemanagement für VLSI-Bausteine ist ein Thema, das insbesondere durch die Verfügbarkeit von portablen elektronischen Geräten wie Notebook-Computern, Mobiltelefonen etc. in den letzten Jahren stark an Bedeutung gewonnen hat. Viel Entwicklungsarbeit in Bezug auf sogenanntes Low-Power-Design wird nicht zuletzt auch deswegen geleistet, weil moderne Hochleistungsprozessoren, die bei immer höheren Taktfrequenzen betrieben werden, in ihrer Entwicklung durch die anfallende Verlustleistung begrenzt werden. Die größten Herausforderungen für die Prozessor-Entwickler stellen heutzutage einerseits das durch die schnellen Prozessoren entstandene Problem der mangelnden Memory-Bandbreite und andererseits das Verhindern der zu großen Erwärmung des Prozessors durch die große Verlustleistung bei hohen Taktraten dar.

Bei Chipkarten ist die Grundproblematik ähnlich, wenn auch von anderer Seite her motiviert: Nicht thermische Grenzen, sondern die restriktiven Anforderungen bezüglich des Strombedarfs und das limitierte Energiebudget müssen beim Entwurf leistungsfähiger Smartcard-Chips von Grund auf in die Überlegungen miteinbezogen werden. Dies gilt besonders für jene Chipkarten, die Public-Key-Funktionen unterstützen und daher im Vergleich zu Standard-Chipkarten wesentlich höhere Rechenleistung erreichen können müssen. Wenn man bedenkt, dass trotz stetig steigenden Bedarfs eine hohe Marktakzeptanz längerfristig nur durch den Einsatz von versorgungsunabhängigen, kontaktlosen Chipkarten sichergestellt werden wird können, kommt die Wichtigkeit der Thematik Energiemanagement/Energieeffizienz erst richtig zum Tragen.

Grundsätzlich kann man die Reduktion des Energiebedarfs auf verschiedenen Ebenen in Angriff nehmen:

- Auf der höchsten Ebene können Einsparungen durch geeignete Software erreicht werden. Dieser Ansatz ist konsistent zu der – auch aus anderen Motivationen heraus begründeten – Notwendigkeit, Hardware und die darauf auszuführende Software aufeinander abzustimmen. Aus den oben genannten Gründen wird der Faktor Energie ein immer wichtigerer integraler Bestandteil des Hardware/Software-Co-Designs werden müssen.

- Einen signifikanten Einfluss auf den Leistungsbedarf einer VLSI-Schaltung hat deren Architektur. Die Entscheidungen beim Entwurf geeigneter Architekturen von Prozessoren und Co-Prozessoren werden von vielen Parametern bestimmt. Dabei ist eine möglichst gute Ausgewogenheit der drei Kriterien Rechenleistung, Flächenbedarf und Leistungsaufnahme in Hinblick auf die speziellen Anwendungen anzustreben. Im konkreten Zusammenhang mit dem besonderen Erfordernis, das Design und damit auch die Architektur in Bezug auf Sicherheitsfragen zu optimieren, ergibt sich die Einführung einer vierten Dimension, die all diese Abwägungen noch weiter beeinflusst. Ein zentraler Punkt beim Entwurf von kryptografischen Smartcard-Chips ist daher die Frage, wie man all diese Aspekte kombiniert, und zugleich noch die Skalierbarkeit des Designs sicherstellt, um rasch auf neue Anforderungen reagieren zu können. So ergibt sich eine Fülle von Freiheitsgraden, deren Beherrschung geeignete Design-Methodologien erforderlich macht.

- Der im Zusammenhang mit Low-Power-Design meist diskutierte Punkt betrifft die Schaltungstechnik, in der der Chip letztlich realisiert wird. Auf dieser Ebene sollte man offensichtlich Logik-Arten, die inhärent mehr Energie umsetzen als andere, vermeiden.

Auf diese Punkte soll nun im Zusammenhang mit der krytpografischen Einheit von CRISP näher eingegangen werden. Diese besteht vereinfacht gesagt im wesentlichen aus einem Multiplizierer mit Pipeline-Struktur, der die eigentlichen Berechnungen durchführt, sowie aus Schieberegistern für Zwischenresultate. Die Wahl dieser Architektur scheint vor dem Hintergrund herkömmlicher Überlegungen erklärungsbedürftig, insbesondere da die erhöhte Aktivität, sowie die höhere Anzahl von Speicherelementen dieser Struktur im allgemeinen erhöhten Strom bedingen, und normalerweise der bloßen Durchsatzerhöhung dienen. Zwei Überlegungen mögen helfen, Motivationen hierfür zu beleuchten:

1. Die gewählte Architektur bedeutet, dass die Daten permanent zirkulieren. Die Verwendung von dynamischer Logik[4] erlaubt das Pipelining möglichst platz-, flächen- und geschwindigkeitseffizient zu realisieren, da quasi Logik- und Speicherelemente miteinander verschmelzen und so Transistoren eingespart werden. Beide Tatsachen – zirkulierende Struktur und die Verwendung dynamischer Logik, die bei Stillstand ihre logischen Pegel verliert – haben eine wesentliche Implikation für die Realisierung eines sicheren Designs: Durch den stetigen Datenfluss ist die Beobachtbarkeit der Daten extrem erschwert; bei Anhalten (d.h. Stoppen des Datenflusses) jedoch gehen wegen der Verwendung dynamischer Logik die Daten verloren. Beide Argumente zusammen ergeben also eine Erhöhung der Sicherheit gegenüber Angriffen auf das System. Dass dies unter Zuhilfenahme von Techniken erreicht wurde, die a priori einen höheren Strom bedingen, ist ein typisches Beispiel für obige Behauptung: Für kryptografische Anwendungen muss Sicherheit, neben Fläche, Geschwindigkeit und Stromaufnahme, als gleichberechtigter Parameter erachtet werden. Man muß Sicherheitsaspekte, die in das Design einfließen sollen, potentiell durch Abstriche bei einem oder mehreren der anderen Parameter erkaufen und die daraus resultierenden Effekte durch anderweitige Anstrengungen zu kompensieren versuchen. Dieses Beispiel zeigt auch, dass es Sicherheitsaspekte gibt, die auf höheren Ebenen des Entwurfszyklus einfließen müssen, da sie nachträglich nicht mehr effizient eingebracht werden können.

2. Ein weiteres Argument für Pipeline-Strukturen besteht – auf den ersten Blick vielleicht überraschenderweise – auch vor dem Hintergrund des Energiemanagement. Bekanntlich ist bei CMOS-Schaltungen das effizienteste Mittel zur Verringerung der Leistungsaufnahme die Reduzierung der Versorgungsspannung, da dies weitestgehend quadratischen Einfluss hat. Daraus ergibt sich jedoch eine Einbuße hinsichtlich der Geschwindigkeit, die zum Erreichen hinreichend großer Verschlüsselungsraten potentiell kompensiert werden muß. Eine Möglichkeit wäre die Verwendung von speziellen Prozessen mit Transistoren mit niedriger Threshold-Spannung, die jedoch höhere Leckströme aufweisen, was für Power-Down-Modi problematisch sein kann. Zudem ist es nicht besonders zielführend, ein Design, das über verschiedenste Prozesse hinweg portabel sein soll, auf spezielle Prozesse hin abzustimmen. Eine weitere Möglichkeit, den Geschwindigkeitsverlust durch verringerte Betriebsspannung auszugleichen wäre ein höheres Maß an Parallelität in der Architektur. Dies bedeutete jedoch ein deutliches Mehr an Fläche. Pipelining hin-

---

[4] Vielfach wird in der Literatur der Begriff „dynamische" Logik als Synonym für „Precharge-Logik" verwendet, obwohl letztere eher als Untermenge derselben betrachtet werden müsste. Dynamische Logik soll hier ganz allgemein bedeuten, dass Information in geladenen Knoten unter Zuhilfenahme der inhärent vorhandenen parasitären Kapazitäten erfolgt. Da Precharge-Logik aufgrund ihrer höheren Aktivitiät als relativ stark stromaufnehmend erachtet werden muss, wird dies durch die erwähnte begriffliche Ungenauigkeit oft fälschlicherweise dynamischer Logik allgemein zugeschrieben.

gegen bietet die Möglichkeit, unter eher geringem Flächenmehraufwand die Geschwindigkeit durch erhöhten Durchsatz auf die benötigte Verschlüsselungsrate anzupassen, auch wenn die Betriebsspannung aus Gründen der Leistungsaufnahme signifikant abgesenkt wurde. Richtig dimensioniert, kann durch Pipelining in Zusammenhang mit geringerer Betriebsspannung insgesamt eine Verringerung der Leistungsaufnahme erreicht werden.[Larsson96]

Bei der Auswahl der Schaltungstechnik wurden Ansätze gemieden, die den Energieumsatz inhärent steigern. Dazu zählen unter anderem Precharge-Gatter, die durch ihre hohe Aktivität für Low-Power-Anwendungen unattraktiv sind und ihre Berechtigung im Bereich hoher Schaltfrequenzen haben. Weiters wurde darauf geachtet, im gesamten Design keine Schaltungselemente einzuführen, die reduzierte Logik-Pegel zur Folge haben, wie etwa einzelne N- oder P-Pass-Transistoren. Reduzierte Logik-Pegel führen nämlich einerseits zu Kurzschlußströmen in nachfolgenden Gattern, andererseits verbietet der dadurch verringerte Störabstand den Betrieb der Schaltung bei geringen Versorgungsspannungen, was im Hinblick auf Leistungsminimierung äußerst unklug wäre. Nach eingehenden Untersuchungen unterschiedlichster Gatter wurden dynamische True-Single-Phase Doubled-C^2MOS-Gatter gewählt [Yi-Ren87, Yuan89]. Diese Wahl erwies sich nicht nur für die Schieberegister als geeignet, es gibt auch eine besonders energieeffiziente, auf dieser Logik basierende Implementierung von Volladdierern [Schindler96], die für den Multiplizierer Verwendung finden. Eigens für dieses Design wurde ein quasi-statisches Schieberegister[5] entwickelt, das ebenfalls auf den genannten dynamischen Gattern basiert. Dieses arbeitet im Normalbetrieb gleich wie die rein dynamische Version mit allen damit verbundenen Vorteilen, kann jedoch nach erfolgter Berechnung seine Daten über lange Zeit halten und vereint somit die Vorteile von dynamischen und statischen Techniken.

Grundsätzlich wurden alle Transistoren in der Multiplizierer/Schieberegister-Struktur mit minimaler Breite dimensioniert. Dies verringert Kapazitäten und damit die Stromaufnahme, verringert die Fläche, und stellt auch für die Geschwindigkeit kein Problem dar, da diese über das Pipelining variiert werden kann. Die Verwendung von Minimaltransistoren hat den weiteren Vorteil, dass die damit verbundene minimierte Ausgangsbelastbarkeit das unerwünschte Auslesen von Knotenspannungen tendenziell erschwert, also einen weiteren Teilaspekt in Richtung Sicherheit darstellt.

Nachdem nun die prinzipielle Architektur und Schaltungstechnik festgelegt ist, stellt sich die Frage nach dem Zusammenfügen zur Gesamtschaltung und deren Skalierbarkeit. Prinzipiell ist das Design in weiten Bereichen skalierbar; durch Variation einiger weniger Parameter (wie etwa der maximalen Schlüssellänge und der Größe des Multiplizierers) kann das Design auf unterschiedliche Flächen- und Durchsatzerfordernisse angepasst werden. Ausgehend von der Annahme einer vorgegebenen maximalen Fläche und einer gewünschten Verschlüsselungsrate für eine gegebene Schlüssellänge können die Parameter des Designs unter Minimierung der Energieaufnahme pro Verschlüsselungsoperation optimiert werden. Falls Randbedingungen bezüglich der Versorgungsspannung zu beachten sind (was im allgemeinen aus Kompatibilitätsgründen der Fall sein wird), gehen diese natürlich ebenso in die Betrachtungen ein. Wie in praktisch jedem Low-Power-Design sollte man darauf Wert legen, die Ge-

---

[5] Dabei handelt es sich um ein in dieser Architektur notwendiges Schieberegister, das Ergebnisse längerfristig halten können muss und daher als einziger Teil der Multiplizierer/Schieberegister-Struktur nicht rein dynamisch sein kann.

schwindigkeitsanforderungen möglichst genau einzuhalten und nicht einfach großzügig zu übertreffen, da im allgemeinen eine Schaltung, die mit Frequenz f_1 arbeitet, aber mit $f_2 > f_1$ arbeiten könnte, wesentlich weniger energieeffizient ist als eine, die von vornherein für eine tatsächliche Maximalfrequenz f_1 ausgelegt wurde.

Die Gesamtgröße der Schaltung wird im Normalfall hauptsächlich durch den Multiplizierer bestimmt. Hier empfiehlt es sich, die Kenngröße k des Multiplizierers[6] möglichst groß zu wählen, solange die Flächenvorgaben dies erlauben: Die Multipliziererfläche wächst quadratisch mit k, und damit auch die durchschnittliche Stromaufnahme. Da die Anzahl der benötigten Taktzyklen für eine Modulo-Multiplikation quadratisch mit k fällt, ergibt sich kein wesentlicher Unterschied bezüglich des Energieumsatzes. Um für unterschiedliche k dieselbe Verschlüsselungsrate zu erzielen, sinkt jedoch die benötigte Frequenz quadratisch mit k. Die Leistung für kleineres k bei höherer Frequenz, und für größeres k bei entsprechend geringerer Frequenz ist im wesentlichen gleich, und energetisch tritt a priori kein gravierender Unterschied ein. Der wesentliche Vorteil geringerer Frequenz liegt aber darin, dass die Schaltung nicht für hohe Geschwindigkeit dimensioniert und ausgelegt sein muß, was global einen sehr günstigen Einfluss auf den Energieumsatz hat. Ist k auf diese Weise festgelegt, hängt die Wahl der Betriebsfrequenz von der zu erreichenden Verschlüsselungsrate und der minimalen Versorgungsspannung ab. Damit zusammenhängend ergeben sich Entscheidungen über Designaspekte, wie etwa die Tiefe der kombinatorischen Logik, d.h. den Grad des Pipelining.

Wie man aus obigen grundlegenden Überlegungen sehen kann, stellt eine allgemeingültige Skalierbarkeit eines Designs unter Berücksichtigung möglichst effizienter Optimierung eine große Herausforderung dar, da die Menge der zu behandelnden Freiheitsgrade beachtlich ist, und die Zusammenhänge der einzelnen Faktoren kompliziert sind. Es ist eine der wesentlichsten weiterführenden Aufgaben auf Basis des CRISP-Coprozessor-Designs, zum bereits weit ausgereiften Skalierungskonzept noch weitere, insbesondere den Energieumsatz optimierende Parameter in konsistenter Weise hinzuzufügen. Davon, und von Modifikationen betreffend der für den Leistungsumsatz so wichtigen Taktverteilung, können noch enorme Steigerungen bezüglich Energieeffizienz erwartet werden.

3 Entwurfsmethoden für kryptografische Chipkarten

Um einerseits rasch auf die sich veränderten Anforderungen an kryptografische Prozessoren reagieren zu können, und um andererseits Sicherheitsaspekte in den Design-Prozess für Chipkarten auf integrierte Art und Weise einfließen zu lassen, ist es unumgänglich, geeignete Design-Strategien zu entwickeln. Einer dieser Aspekte betrifft die Notwendigkeit eines homogenen, sauberen und die Sicherheits- und Zuverlässigkeitsanforderungen erfüllenden Design-Flusses, und zwar von der Definition der grundlegenden Spezifikation bis hin zum realen Krypto-Chip. Ein weiterer Aspekt, der hier näher beleuchtet werden soll, betrifft den Nutzen von skalierbaren Custom-Zellbibliotheken, die unter besonderer Berücksichtigung von Sicherheitsaspekten entworfen wurden.

Zell-Bibliotheken sind ganz allgemein weitgehend fixer Bestandteil in einer Vielzahl von ASIC-Entwicklungsumgebungen. Um jedoch den speziellen Erfordernissen für kryptografi-

---

[6] k = Bitlänge der Operanden des Multiplizierers

sche Chipkarten genügen zu können, wurde eigens eine dynamische, skalierbare Custom-Zellbibliothek entworfen.

„Dynamisch" kann in diesem Zusammenhang auf zweierlei Art interpretiert werden: Einerseits handelt es sich bei diesen Zellen um solche, die in dynamischer Logik entworfen wurden. Dies bedeutet, dass die Information in jenen Kapazitäten gespeichert wird, die in jeder CMOS-Schaltung inhärent vorhanden sind. Mit dynamischer Logik ist es möglich, signifikant Transistoren einzusparen, was offensichtlich eine deutliche Reduktion an Flächen- und Strombedarf bei gleichzeitig höherer Verarbeitungsgeschwindigkeit bedeutet. Diese Ersparnisse sind in Anbetracht der sehr restriktiven Flächen- und Stromvorgaben auf Chipkarten von größter Bedeutung. Ein weiterer Vorteil der Verwendung dynamischer Logik kommt unter Berücksichtigung von Sicherheitsbetrachtungen hinzu: Die in geladenen Knoten gespeicherte Information ist „flüchtig", d.h. sie geht nach Anhalten des Systemtaktes verloren, beziehungsweise wird sie – etwa durch nachträglich aufgebrachte Messnadeln – durch geänderte Ladungsumverteilungsbedingungen leichter gestört. Diese Eigenschaft kann als eines der Mosaiksteinchen in den Bemühungen um die Erschwerung von Attacken betrachtet werden.

Die zweite Bedeutung des Wortes „dynamisch" deutet auf die Möglichkeit hin, die Zellbibliothek auf sehr einfache und effiziente Art und Weise auf neue Technologien anpassen zu können. Die Idee von technologieunabhängigen Zellbibliotheken als solche ist nicht neu, doch im Hinblick auf den Entwurf von kryptografischen Chipkarten wurde eine eigene Methode zur Erstellung und Handhabung von Zellbibliotheken entwickelt. Im folgenden werden die Grundzüge dieser Zellbibliothek skizziert.

Oft wird bei technologieunabhängigen Zellbibliotheken ein Kompromiss zwischen allgemeiner Formulierbarkeit der Zellbeschreibung auf höherer Ebene und dafür geringerer Flächeneffizienz eingegangen. Dies wurde hier ganz bewusst vermieden, um das Layout in Hinblick auf die restriktiven Flächenrandbedingungen so „dicht" wie möglich entwerfen zu können. Hier kommt der Begriff „Custom"-Zellbibliothek zum Tragen: Ziel war es, dem Designer eine möglichst einfache und komfortable, aber doch mächtige Schnittstelle zu bieten, um die Transistorschaltung und deren Geometrie in weiten Bereichen flexibel definieren zu können, sodass das Endergebnis jenem eines optimierten „Hand-Layouts" gleichkommt, das lästige und zeitaufwendige Plazieren der Elemente unter Einhaltung der jeweiligen Design-Rules jedoch vollkommen zu automatisieren. Realisiert wurde diese Idee durch einen Ansatz auf funktionaler Programmierung beruhend: Nachdem der Designer sich ein genaues Bild über den Aufbau der Zelle gemacht hat, erfolgt die Definition der Schaltungselemente und ihrer geometrischen Position und der Verbindungen über Listen. Diese Listen werden als Argumente an eine Generator-Funktion übergeben, die schließlich das Layout der Zelle generiert. Auch hier wurden wieder Detailaspekte in Bezug auf Sicherheitsfragen integriert. So etwa endet ganz bewusst die Freiheit für den Designer dort, wo es um den grundsätzlichen Stil des Aufbaus der Zelle geht. Zum Beispiel wurde aus Sicherheitsfragen heraus entschieden, die Zellen möglichst großflächig mit dem obersten Metall-Layer, über den keine relevante Information übertragen wird, zu bedecken, um den Zugang zu den darunterliegenden Layern möglichst zu erschweren. Dies ist ein weiterer Mosaikstein in Richtung sicheres Design. Wesentlich ist hier, dass die strikte Einhaltung dieses Design-Stils für die Zellen durch die Generatorfunktion selbsttätig bewerkstelligt wird, und er Designer keinen Einfluss darauf hat.

Wenn erst einmal eine spezielle Zelle über diese Listen definiert wurde, kann das Portieren auf neue Technologien quasi per Knopfdruck erfolgen: Die Generatorfunktion erzeugt unabhängig von der Technologie in Sekundenschnelle stets sehr kompaktes und den Design-Rules konformes Layout. Es sei hier angemerkt, dass diese Vorgangsweise nur unter Verwendung komplexer Design-Werkzeuge implementierbar ist, die sowohl derartige funktionale Formulierungsmöglichkeiten zur Verfügung stellen, einen objektorientierten Ansatz für die Zusammenfassung von Layoutelementen zu Gruppen bieten, als auch über mächtige Kompaktierungsoptionen verfügen.

Zusätzlich zu den hier beschriebenen Werkzeugen zur Implementierung der physikalischen Domäne deckt die Zellbibliotheksumgebung auch die strukturelle und funktionale Domäne ab, inklusive automatisierter Charakterisierungs-Werkzeuge für dynamische Logik auf SPICE-Ebene.

Bei den auf diese Art erzeugten Basis-Zellen wurde darauf Wert gelegt, dass deren Ein- und Ausgänge so plaziert sind, dass es möglich ist, sie per „Abutment" im Stile von Kacheln anzuordnen, sodass keine expliziten Verbindungen mehr zwischen den Zellen gezogen werden müssen. Dies ist einerseits deshalb von Vorteil, weil die Schaltungen, die aus den Basiszellen aufgebaut werden, auf einfache Weise skalierbar sind, indem z.B. beim Aufruf des Generators für einen Multiplizierer bloß als Parameter dessen Bit-Breite übergeben wird, und durch simple Aneinanderreihung der entsprechenden Anzahl von Basiszellen der Multiplizierer aufgebaut werden kann. Zudem unterstützt die Möglichkeit dieses bloßen Aneinanderfügens der Basiszellen die Fortführung des in diesen Basiszellen eingeführten Konzepts des Erschweren der Zugreifbarkeit auf Verbindungsleitungen auch auf höherer Strukturebene.

Zusammenfassend kann gesagt werden, dass die integrierte Betrachtungsweise der Erfordernisse an sich schnell ändernde Anforderungen an Chipkarten, der einschränkenden Flächen- und Stromvorgaben, sowie der speziellen Sicherheitsanforderungen an Chipkarten es notwendig macht, traditionelle VLSI-Designstrategien zu überdenken und in etlichen Punkten alternative Ansätze zu finden.

4 Die kryptografische Einheit von CRISP

Die Krypto-Einheit der CRISP-Smartcard mit dem Akronym CLU (Cryptographic Logic Unit) basiert auf einem neuen Makrozellen-Konzept, welches symmetrische und asymmetrische kryptografische Funktionen unterstützt. Die CLU ist gleichzeitig eine mit Full-Custom-Methoden entworfene Zelle, die auf Schaltkreisen in True-Single-Phase-Technik aufgebaut ist. Damit sind minimale Fläche, niedrige Leistungsaufnahme und bestmögliche Hardware-Security zu erreichen. Die wesentlichen Eigenschaften der CLU sind:

- Flexibilität bei der Ausführung all jener kryptografischer Algorithmen, die auf Moduloarithmetik oder DES basieren.
- Skalierbarkeit bezüglich Energieumsatz, Geschwindigkeit und Schlüssellänge.
- Minimaler Energieumsatz auf Grund der verwendeten Schaltkreistechniken.
- Hohe Effizienz durch den Einsatz spezieller Hardware-Algorithmen.

Die CLU ist als Makrozelle erhältlich. Das Layout steht in GDSII-Format zur Verfügung und funktionale Beschreibungen liegen in VHDL und Verilog® vor. Mehrere Prototypen wurden auf Silizium implementiert und erfolgreich getestet.

4.1 Funktionsweise der CLU

Die CLU ist ein mikroprogrammierbarer Krypto-Coprozessor. Zu ihren typischen kryptografischen Funktionen gehören

- RSA-Schlüsselgenerierung
- RSA-Exponentiation
- Modulo-Multiplikation und Inverse
- Symmetrische Kryptofunktionen die auf DES basieren

Mit Hilfe einer zweistufigen Kompression der Mikrobefehle für die Langzahlenarithmetik bietet die CLU ein flexibles Interface nach aussen. Da für die Ausführung dieser Befehle gleichartige Basisoperationen mehrmals hintereinander ausgeführt werden müssen, bietet sich eine Code-Struktur an, die aus Basiskontrolldaten und Wiederholungszählerständen besteht. Die Auswertung dieser Zählerstände und das wiederholte Anlegen der Kontrolldaten an die CLU übernimmt im Regelfall die Interface-Logik. Dadurch können die Kontroll-Skripts selbst sehr verkürzt werden, was die CPU, welche die CLU ansteuert, je nach Interface-Variante mehr oder weniger stark entlastet.

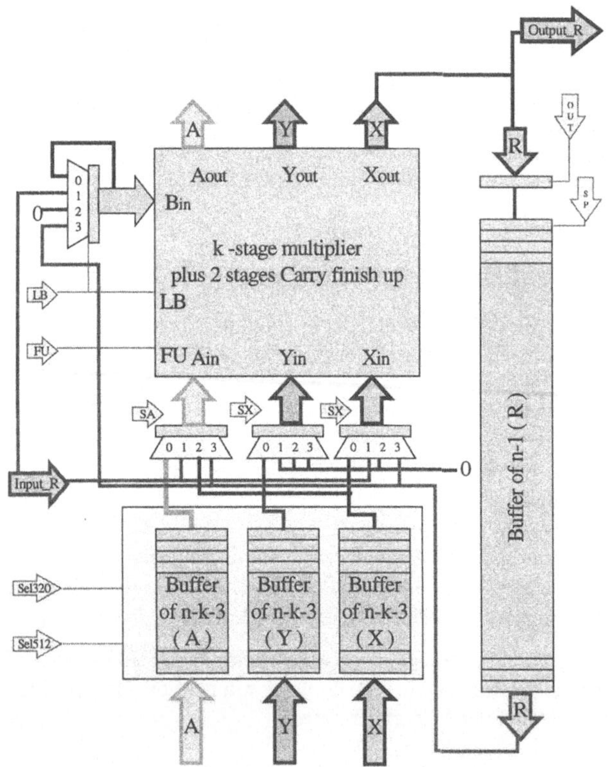

Abb. 1: Blockschaltbild der CLU (ohne DES)

4.2 Skalierbarkeit der CLU

Das Makrozellen-Konzept der CLU erlaubt eine Reihe von Implementierungsoptionen:

- Verschiedene Größen des Kernmultiplizierers, z.B. 8x8, 16x16, 24x24, 32x32.
- Einen großen Variationsbereich für die Schlüssellänge von 320 bis 2048 Bit.
- Mehrere Optionen hinsichtlich der Taktfrequenz für die Makrozelle.

Um die Schlüssellänge für die asymmetrischen Kryptofunktionen bei verschiedenen Implementierungen der CLU variabel zu gestalten, sind die Registerlängen konfigurierbar. Nach der Konfiguration sind durch die Verwendung von Layoutgeneratoren Full-Custom-Layouts weitestgehend automationsunterstützt herstellbar. Damit wird die Zeit zwischen Spezifikation der Zelle und dem fertigen Silizium auf ein Minimum reduziert. Zum Zweck der Einbettung in Systemmodelle und zur Simulation im Gesamtkontext gibt es funktionale Modelle mit exaktem Eingangs-/Ausgangszeitverhalten in VHDL und Verilog®. Zusätzlich sind Architekturmodelle in C++ vorhanden.

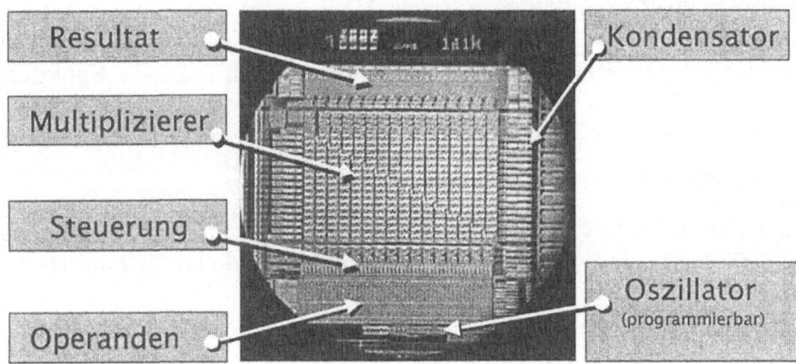

Abb. 2: Mikroskopische Aufnahme eines CLU-Prototyps

4.3 Leistungsmerkmale

Die CLU besteht aus folgenden Blöcken:

- Multiplizierer
- Register für Zwischenresultate und Ergebnis
- Programmierbarer Taktgenerator von 3 bis 140 MHz
- DES-Modul

Durch die Kombination von programmierbarer Taktrate, einer Versorgungsspannung zwischen 2.2 und 5.6 Volt, der Vielfalt an Algorithmen und der variablen Schlüssellänge entsteht ein breites Spektrum an Anwendungsmöglichkeiten. Die zu erreichende Leistung der asymmetrischen Option der CLU hängt vom verwendeten Prozess und von den ausgewählten Parametern ab. Ein Beispiel ist in der folgenden Tabelle dargestellt:

Die Leistungsaufnahme hängt stark von dem tatsächlich verwendeten Prozess ab. Als ein Beispiel sei die Implementierung einer 640-Bit-CLU mit einem 16x16-Bit-Multiplizierer auf einem 0.8µ-CMOS-Prozess der Firma AMS erwähnt, bei dem die Stromaufnahme im Betrieb von 60 MHz um 60 mA liegt.

Prozess	0.6μ CMOS
Größe des Multiplizierers	16x16
Schlüssellänge	1024
Taktrate	100 MHz
RSA unter Einsatz des Chinesischen Restesatzes (CRT)	ca. 75 ms
RSA ohne Verwendung von CRT	ca. 215 ms
Größe der CLU	2.61 mm^2

4.4 Die CLU in Chipkarten

Die CLU wurde unter besonderer Berücksichtigung ihres Einsatzes in Chipkarten entworfen. Dabei wurde speziell auf Aspekte der Informationssicherheit Wert gelegt. Einige wesentliche Merkmale sind die folgenden:

- Multiplikation und Quadrieren verhalten sich identisch.
- Kontroll-Skripts gewährleisten ein schlüsselunabhängiges Profil bei der Exekution
- Daten werden meist nicht statisch gespeichert
- Der Einsatz von Transistoren minimaler Größe erschwert das Auslesen von elektrischen Werten durch Messspitzen, da Schaltungsknoten dadurch minimale Kapazitäten besitzen.

4.5 Die DES-CLU

Die CLU unsterstützt bei Bedarf auch eine symmetrische Option. Auf einer Fläche von etwa 0.5 mm^2 (bei 0.6μ-CMOS-Technologie) lässt sich ein DES-Modul für folgende DES-Modi unterbringen:

- DES
- Triple-DES
- Two-key DES

Im Triple-DES-Modus ist eine Verschlüsselungsrate von über 150 Mbit/s möglich. Pro 1 mA Stromaufnahme sind 30 000 Triple-DES-Operationen pro Sekunde erhältlich.

5 Zusammenfassung

Auf Grund der skalierbaren Eigenschaften der CLU ist es möglich, jeweils vorhandene Silizium-Technologien optimal auszunutzen. Damit kann ohne Änderung der Architektur auf den Bedarf nach längeren Schlüsseln eingegangen werden. Kleinere Siliziumstrukturen erlauben die Verwendung von größeren Kernmultiplizierern, womit der durch die längeren Schlüssel bedingte erhöhte Rechenbedarf egalisiert werden kann.

Es gibt derzeit im wesentlichen drei verschiedene Interface-Konzepte für die CLU. Einerseits läßt sich die CLU eng an den Host-Prozessor binden, sodass die CLU wie eine spezielle arithmetische Einheit der Host-CPU betrachtet werden kann.

Darüber hinaus gibt es eine Interface-Logik zur Anbindung der CLU an den von der Open Microprocessor Systems Initiative vorgeschlagenen und standardisierten PI-bus (*OMI Standard 324 PI-BUS Peripheral Interconnect Bus*). Dieser Bus unterstützt ein Makrozellenkonzept, bei dem verschiedene Hersteller von Modulen in einfacher Weise ein „System auf Silizium" assemblieren können. In dieser Variante ist die CLU über die ebenfalls von OMI initiierte *Eurocell Library and Interfaces* (ELI) erhältlich.

Schließlich ist es möglich, die CLU in einfacher Weise als Co-Prozessor zusammen mit einer CPU auf einer Platine zu einem System zu verbinden.

Literatur

[Biham96a] Biham E., A. Shamir: A new cryptoanalytical attack on DES. Draft paper, 1996.

[Biham96b] Biham E., A. Shamir: The Next Stage of Differential Fault Analysis: How to Break Completely Unknown Cryptosystems. Draft paper, 1996.

[Bock97] Bock H., W. Mayerwieser, K.C. Posch, R. Posch, V. Schindler: An Integrated Co-processor Architecture for a Smartcard. Erscheint in: Journal of Network and Computer Applications, Academic Press Inc., San Diego.

[Bock98] Bock H., W. Mayerwieser, K.C. Posch, R. Posch, V. Schindler: Scalability Concepts in Cryptographic Co-processing. In: J. Y. Roger et al.: Advances in Information Technologies: The Business Challenge - Proceedings of the EMMSEC '97 Conference, 1998, S. 922-929.

[Ferreira96] Ferreira R., R. Malzahn, P. Marissen, J.-J. Quisquater, T. Wille: FAME: A 3rd Generation Coprocessor for Optimising Public Key Cryptosystems in Smart Card Applications. In: Hartel P.H., P. Paradinas, J.-J Quisquater: Smartcard Research and Advanced Applications; Proceedings of CARDIS 96, Stichding Mathematisch Centrum, Amsterdam, 1996, S. 59-72.

[Ji-Ren87] Ji-Ren Y., I. Karlsson, and C. Svensson: A True Single-Phase-Clock Dynamic CMOS Circuit Technique. IEEE Journal of Solid-State Circuits, vol.sc-22, no.5, 1987, S. 899-901.

[Kocher95] Kocher P.C.: Cryptanalysis of Diffie-Hellman, RSA, DSS, and Other Systems Using Timing Attacks. Zu finden im World Wide Web unter URL <http://www.cryptography.com/timingattack/paper.html>, 1995.

[Larsson96] P. Larsson-Edefors P.: Reducing Power Consumption by using Optimally Pipelined Circuits. Proceedings of Sixth International Workshop on Power and Timing Modelling, Optimization and Simulation, 1996, S. 103-112.

[Naccache96] Naccache D., D. M'Raihi: Arithmetic co-processors for public-key cryptography: the State of the Art. In: Hartel P.H., P. Paradinas, J.-J Quisquater: Smartcard Research and Advanced Applications; Proceedings of CARDIS 96, Stichding Mathematisch Centrum, Amsterdam, 1996, S. 39-58.

[Schindler96] Schindler V.: A Low-Power True Single Phase Clocked (TSPC) Full-Adder. Proc. of the 22nd European Solid-State Circuits Conference, Neuchâtel, 1996.

[Yuan89] Yuan J. and C. Svensson: High-Speed CMOS Circuit Technique. IEEE Journal of Solid-State Circuits, Vol.24, No.1, 1989, S. 62-70.

Dublettenfreie Schlüsselgenerierung durch isolierte Instanzen

Patrick Horster

Universität Klagenfurt
Institute für Informatik - Systemsicherheit
pho@ifi.uni-klu.ac.at

Zusammenfassung

In zahlreichen Anwendungsbereichen ist es wünschenswert oder gar zwingend erforderlich, daß bei der Verwendung kryptographischer Verfahren die zum Einsatz kommenden Schlüssel beziehungsweise Schlüsselparameter einzigartig und zudem individuell (frisch) generiert werden. Ergänzend wird zudem gefordert, daß die Schlüssel zufällig oder zumindest pseudozufällig sind. Betrachtet man geschlossene Systeme, so können die aufgeführten Anforderungen durch geeignete Sicherheitsmaßnahmen relativ leicht erfüllt werden. In verteilten offenen Systemen ergeben sich allerdings Probleme, deren Lösung nicht so einfach realisiert werden kann. Wird darüber hinaus auch noch gefordert, daß alle betroffenen Instanzen zur Schlüsselgenerierung beitragen sollen, was insbesondere für den Endbenutzer von Bedeutung sein kann, so sind spezielle Sicherheitsüberlegungen anzustellen. Infolge verschiedener Lösungsansätze in unterschiedlichen Applikationen kann der vorliegende Beitrag nicht als Lösung oder gar als Universallösung angesehen werden, vielmehr soll er dazu dienen, zugrundeliegende Probleme zu beschreiben und anregen, neue und vielleicht akzeptable Lösungen im Umfeld chipkartenbasierter Sicherheitslösungen aufzuzeigen. Der Schlüsselgenerierung im Kontext digitaler Signaturen kommt dabei eine besondere Bedeutung zu.

1 Einleitung

Schlüsseldubletten kennen wir nicht nur in Form autorisiert oder unautorisiert angefertigter mechanischer Nachschlüssel, auch vorhandene Originalschlüssel können durchaus identisch sein. In der Vergangenheit konnte es beispielsweise vorkommen, daß ein fremdes Auto mit dem eigenen Autoschlüssel geöffnet werden konnte. Entsprechend konnte der Besitzer des anderen Autos ebenfalls beide Autos öffnen.

Die Wahrscheinlichkeit, eine solche Situation zu erleben, ist zwar in den letzten Jahrzehnten deutlich geringer geworden, jedoch nicht völlig auszuschließen. Dies gilt auch für mangelhaft gestaltete elektronische Schließsysteme und Wegfahrsperren. Da ein Kraftfahrzeug einen erheblichen Wert darstellt, ist es leicht einzusehen, daß jedes Auto mit einem individuellen elektronischen Schloß und zugehörigen Schlüsseln ausgestattet sein sollte. Dasselbe gilt in analoger Weise für elektronische Wegfahrsperren. Wird eine Autotüre mit einem Funk- oder Infrarot-Schlüssel gesichert, so sind zusätzliche Aspekte zu betrachten, insbesondere sollten sich die gesendeten Daten bei jeder Transaktion von bisher gesendeten Daten unterscheiden.

Was für den Diebstahlschutz von Autos gilt, daß muß erst recht für Daten der Informations- und Kommunikationstechnik gelten. Werden Daten über einen öffentlichen Kommunikationskanal übertragen und enthalten die Daten sensible Informationen, so ist es zwingend erforderlich, die Daten geeignet zu verschlüsseln. Hierzu wird zumeist ein symmetrisches (klassisches) Verschlüsselungssystem eingesetzt, das durch einen Sitzungsschlüssel k parametrisiert werden muß. Wird jeweils derselbe Sitzungsschlüssel verwendet, so könnte ein Angreifer unter gewissen Umständen bereits gesendete (verschlüsselte) Daten wiederholt senden, ohne daß der Empfänger einen Verdacht schöpft. Auch hier sollten sich die verwendeten Sitzungsschlüssel möglichst nicht innerhalb der Lebensdauer der verwendeten Komponenten wiederholen. Zudem ist es wünschenswert, daß unterschiedliche Benutzer auch unterschiedliche Schlüsselfolgen generieren und benutzen.

In zahlreichen Anwendungen ist es somit erstrebenswert, daß die zum Einsatz kommenden kryptographischen Schlüssel bzw. Schlüsselkomponenten genau einmal existieren oder genau einmal eingesetzt werden. Im Kontext der Wegfahrsperren sollte es dabei selbstverständlich sein, daß jedem Automobil zumindest ein individueller Schlüssel zugewiesen werden kann. Bei der Zuweisung sind in der Regel verschiedene Instanzen betroffen, angefangen vom Chiphersteller über den Automobilhersteller bis zum Kunden. Würde jede dieser Instanzen einen Teilschlüssel generieren, so könnte dem resultierenden Gesamtsystem ein größeres Vertrauen entgegengebracht werden, als wenn nur eine Instanz, etwa der jeweilige Chiphersteller, die Schlüssel aller Automobile erzeugen würde.

Die zugrundeliegenden Probleme und deren Lösung können als Teil eines, wie auch immer gearteten Schlüsselmanagement aufgefaßt werden. Dabei ist es – was die Schlüsselerzeugung betrifft – unerheblich, ob die Schlüssel in einer Wegfahrsperre, einem funkgesteuerten Sicherheitsmechanismen oder einem System der modernen Kommunikationstechnik eingesetzt werden. Wir beschränken uns im folgenden daher auf den allgemeinen Fall der Kommunikationssysteme, wobei nochmals darauf hingewiesen sei, daß die vorgestellten Aspekte Beispielcharakter haben und keinesfalls als generelle Lösungen angesehen werden können.

2 Individualisierung von Schlüsselparametern

Werden in einem Kommunikationssystem zur Realisierung von Vertraulichkeit und Integrität Sitzungsschlüssel verwendet, so ist es in zahlreichen Anwendungen erforderlich, daß die verwendeten Schlüssel einmalig und zufällig gewählt sind. Ist der verwendete Schlüsselraum groß genug, so können (echte) Zufallsschlüssel dazu beitragen, das gewünschte Resultat zu erreichen. Die jeweils zur Anwendung kommenden Sitzungsschlüssel werden dann frisch generiert und entsprechend verwendet, etwa in einem Envelop System, in dem der frische Schlüssel k mit dem öffentlichen Schlüssel OSA eines Empfängers A verschlüsselt und gemeinsam mit einem (unter Verwendung des Schlüssels k) verschlüsselten Klartext gesendet wird. Der autorisierte Empfänger A ist dann in der Lage, den Schlüssel k mittels des nur ihm bekannten geheimen Schlüssels GSA zu rekonstruieren und somit den Klartext zu ermitteln. Würde dieses System ohne weitere Maßnahmen zur Unterstützung der Integritätssicherung (etwa geeignete Timestamps) eingesetzt werden, so kann durch wiedereinspielen alter, bereits zu einem früheren Zeitpunkt gesendeter Daten eine einfache Replay Attacke durchgeführt werden.

Werden die verwendeten Sitzungsschlüssel im vorgestellten Beispiel zufällig aus einer endlichen Menge S von möglichen Schlüsseln generiert, so zeigt die Wahrscheinlichkeitsrechnung, daß wir nach etwa $\sqrt{|S|}$ Schlüsselgenerierungen mit 50% Wahrscheinlichkeit zumindest zwei gleiche Schlüssel generiert haben. Dieses als Geburtstagsparadoxon bekannte Phänomen wird im vierten Kapitel noch näher betrachtet.

Werden $|S|$+1 Schlüssel generiert, so ist es unvermeidbar, daß zwei gleiche Schlüssel erzeugt werden. Zufallsgeneratoren können aber – im Gegensatz zu Pseudozufallsgeneratoren – nicht garantieren, daß $|S|$ generierte Schlüssel unterschiedlich sind, obwohl das theoretisch möglich wäre. Als einfaches Beispiel wäre eine kanonische Aufzählung der Schlüssel anzusehen, wobei der Zufallscharakter allerdings verloren geht.

Werden geeignete Pseudozufallsgeneratoren verwendet, so kann garantiert werden, daß eine erste Schlüsselwiederholung erst nach der Periodenlänge des Generators auftritt. Im günstigsten Fall tritt dies erst nach $|S|$ Schlüsselgenerierungen auf. Bezeichnet R einen Pseudozufallsgenerator, und bezeichnet r_0 einen Startwert, so können die generierten Folgewerte durch $r_{i+1} = R(r_i)$ beschrieben werden. Hierbei ist insbesondere darauf zu achten, daß beim Neustart des Generators der aktuelle Wert r_{i+1} und nicht etwa r_0 verwendet wird. Denken wir dabei an kontaktlose Chipkarten oder elektronische Autoschlüsseln, in denen Pseudozufallsgeneratoren integrierte sind, so sind die damit verbunden Probleme leicht zu lokalisieren.

Bezeichnet $E: M \times K \to M$ ein geeignetes Verschlüsselungssystem, das durch einen Schlüssel $k \in K$ parametrisiert werden kann, so kann sichergestellt werden, daß die erzeugten Schlüssel $s \in S$ einmalig und quasi zufällig sind. Wir setzen zunächst voraus, daß $M = S$ ist. Als Startwert wählen wir einen Wert $s_0 \in S$ und berechnen den ersten Zufallswert $s_1 = E(s_0 + 1, k)$, der nächste Zufallswert wird durch $s_2 = E(s_0 + 2, k)$ bestimmt. Der Wert s_i, $i > 0$, ergibt sich somit als $s_i = E(s_0 + i, k)$, wobei die Addition modulo der Anzahl der möglichen Argumente $|M|=|S|$ durchzuführen ist: $s_i = E((s_0 + i) \text{ MOD } |M|, k)$. Die ersten $|S|$ generierten Schlüssel s_0, s_1, ..., $s_{|S|}$ sind somit alle unterschiedlich, da die Verschlüsselungsfunktion E für alle festen k bijektiv ist. Ist die Kardinalität der Menge M größer als die von S, $|S| < |M|$, und unterscheiden sich die Kardinalitäten nicht wesentlich, so ist eine Modifikation des Verfahrens erforderlich, da es sonst vorkommen kann, daß $E((s_0 + i) \text{ MOD } |M|, k) \notin S$ ist. Tritt ein solcher Fall auf, so wird die Schlüsselgenerierung solange fortgesetzt, bis ein Wert $E((s_0 + j) \text{ MOD } |M|, k) \in S$, $j > i$, gefunden wird. Auf diese Art werden wiederum zunächst alle Schlüssel $s \in S$ erzeugt, erst danach wird die erste Schlüsseldublette generiert.

Bei der Schlüsselgenerierung ist dabei darauf zu achten, daß vor der Ausgabe des aktuell generierten Zufallswertes s_i der neue Zählerstand $(s_0 + i + 1) \text{ MOD } |M|$ im Zufallsgenerator gespeichert und beim nächsten Aufruf verwendet wird. Dies ist erforderlich, damit bei einem ungewollten Abbruch garantiert werden kann, daß als nächste Pseudozufallszahl keine Schlüsseldublette erzeugt wird. Im Falle eines ungewollten Abbruchs kann es dann vorkommen, daß ein generierter Schlüssel nicht zur Anwendung kommt. Hierdurch könnte sich ein Angriff ergeben, der darin besteht, daß solange neue Schlüssel vom Zufallsgenerator angefordert werden, bis ein „genehmer" Schlüssel generiert wurde. Um diese Attacke zu verhindern beziehungsweise zu erschweren, kann im verwendeten Übertragungsprotokoll eine Quittierung eingeführt werden. Dies sollte so geschehen, daß eine angeforderte oder empfangene Zufallszahl durch eine entsprechende Response beantwortet werden muß. Geschieht dies nicht, so wird keine neue Zufallszahl generiert, die Verfügbarkeit muß dann durch geeignete organisatorische Maßnahmen gesichert werden.

Hierdurch kann der angedeutete Angriff allerdings noch nicht verhindert werden, lediglich die Anzahl der Schlüssel pro Zeiteinheit kann, bedingt durch die erforderlichen Interaktionen verringert werden. Erst die Beschränkung der Schlüsselanzahl in einem zu bestimmenden Zeitintervall könnte eine geeignete Maßnahme zur Risikominderung sein, dazu ist es allerdings erforderlich, daß ein geeigneter Zeitgeber zur Verfügung steht. Im Umfeld von Chipkarten könnte dies durch entsprechend ausgestattete Chipkartenterminals unterstützt werden. Vor einer Kommunikation zwischen Chipkarte und Terminal ist dann eine gegenseitige Authentifikation und eine darauf aufbauende authentische Kommunikation erforderlich. Durch geeignete Systemintegration kann die angedeutete Attacke vermieden werden.

Das beschriebene Verfahren zur Generierung von Sitzungsschlüssel besitzt zwei freie Parameter, den Generatorschlüssel k und den Startwert s_0. Durch geeignete organisatorische Maßnahmen könnten diese Parameter so eingestellt werden, daß sowohl Chiphersteller, Systemhersteller und Benutzer Einfluß auf diese Parameter haben. Der Chiphersteller (bzw. Personalisierer) könnte die Initialwerte $(k'', s'') = (a, b)$ in die entsprechenden Speicherplätze so laden, daß sie nicht ausgelesen werden können. Zusätzlich werden ein Transport- und ein Initial-PIN festgelegt. Der Systemhersteller könnte nun die ihm unbekannten Initialwerte durch Eingabe von Werten (c, d) zu $(k', s') = (a, b) \oplus (c, d)$ modifizieren. Dies kann durch Eingabe des Transport-PIN ermöglicht und erzwungen werden. Bei der Erstbenutzung wird der Besitzer (Benutzer) nach Eingabe seiner Initial-PIN aufgefordert, eine (neue) Benutzer-PIN und zwei weitere Parameter (e, f) zu wählen, aus denen dann die zum Einsatz kommenden Werte $(k, s_0) = (a, b) \oplus (c, d) \oplus (e, f)$ ermittelt werden. Weder Chiphersteller, Systemhersteller noch Besitzer kennen die verwendeten Initialwerte, wodurch maximale Sicherheit und insbesondere Vertrauen gewährleistet werden kann. Voraussetzung ist hier natürlich, daß die beteiligten Instanzen und die eingesetzten Komponenten genau das leisten was sie leisten sollen – nicht mehr und nicht weniger.

Sind in den eingesetzten Chipkarten geeignete Algorithmen implementiert, so lassen die unterschiedlichen Eingabeparameter zahlreiche Alternativen zu. Werden hierzu Verschlüsselungsverfahren eingesetzt, so kann sichergestellt werden, daß die generierten Werte (k, s_0) für alle Beteiligten Instanzen eine zufriedenstellende Zufälligkeit aufweisen. Solche grundlegenden Verfahren könnten dann sogar veröffentlicht werden, was in vielen Fällen zur breiten Akzeptanz beitragen könnte.

Werden nun Chipkarten an unterschiedliche Benutzer ausgeliefert und auf die beschriebene (oder angedeutete) Art initialisiert, so kann es durchaus vorkommen, daß verschiedene Benutzer identische Schlüssel erhalten oder identische Schlüsselfolgen generieren. Bei der dublettenfreien Schlüsselgenerierung gilt es aber gerade das zu vermeiden. Zudem kann gefordert werden, daß zumindest der Benutzer nachvollziehen kann, daß seine Eingaben zur Initialisierung entscheidend beigetragen haben. Ist der Schlüsselraum $|S|$ groß genug, so können die Mengen M und K dazu genutzt werden, daß eine geeignete dublettenfreie Individualisierung der zu generierenden Schlüssel stattfinden kann.

Wir betrachten ein Beispiel mit den Instanzen Chiphersteller, Systemhersteller und Benutzer. Dabei muß nun unterschieden werden, ob die Benutzer nach erfolgreicher Initialisierung eine dublettenfreie Schlüsselfolge oder dublettenfreie Schlüssel erzeugen sollen. Unter dublettenfreien Schlüsselfolgen sind dabei Folgen zu verstehen, die nicht unmittelbar miteinander korrelieren. Insbesondere soll dabei vermieden werden, daß aus der Kenntnis einer gegebenen Teilfolge auf Teilfolgen derselben oder einer anderen Schlüsselfolge geschlossen werden kann. Dies wird im wesentlichen dadurch realisiert, daß zur Generierung der Schlüsselfolgen unterschiedliche Schlüssel und geeignete Verschlüsselungsverfahren verwendet werden.

Im beschriebenen Fall muß also sichergestellt werden, daß alle zur Anwendung kommenden Benutzerschlüssel unterschiedlich sind. Setzen wir voraus, daß alle Benutzer dasselbe Verschlüsselungsverfahren einsetzen und nehmen wir an, daß ein Schlüssel K eine Länge von 128 bit besitzt, so können theoretisch 2^{128} unterschiedliche Schlüssel zum Einsatz kommen.

Betrachten wir einen Schlüssel K als Konkatenation (Hintereinanderschaltung) von Teilschlüsseln K1, K2 und K3 mit K=K1∥K2∥K3, so kann bei geeignete Wahl der jeweiligen Längen der Teilschlüssel eine Individualisierung leicht erreicht werden. Der Teilschlüssel K1 könnte dem Chiphersteller etwa dazu dienen, den Systemhersteller eindeutig festzulegen. Der Teilschlüssel K2 könnte durch den Systemhersteller den Benutzer eindeutig festlegen, während der Teilschlüssel K3 vom Benutzer gewählt werden könnte. Ein eventuell erforderlicher Startwert könnte wiederum durch alle Beteiligten festgelegt werden. Tatsächliche Realisierungen könnten hiervon etwa dadurch voneinander abweichen, daß die einzelnen Bits der Teilschlüssel auf den Gesamtschlüssel verteilt und nicht hintereinander abgelegt sind. Zudem können einem Systemhersteller auch mehrere Schlüssel zugewiesen werden, etwa in Form von Chargenschlüssel, hierbei ist dann allerdings sicherzustellen, daß niemals unterschiedliche Systemhersteller Lieferungen aus einer Charge mit demselben Chargenschlüssel erhalten.

Um weitgehend Mißtrauen auszuschließen, könnte zudem der erzeugte Schlüssel K selbst verschlüsselt werden, wodurch sichergestellt werden kann, daß selbst der Chiphersteller keinen Informationsvorsprung bei der Rekonstruktion einzelner Teilschlüssel nutzen kann. Hierzu ist allerdings ein (in einem Security Module untergebrachter) Master Key erforderlich, mit dessen Hilfe der Schlüssel K verschlüsselt wird. Aus Sicherheitsgründen sollte die Verschlüsselung dabei in der Chipkarte ablaufen. Das Verschlüsselungsergebnis kann dann als individueller Schlüssel in der Chipkarte genutzt werden.

Bisher sind wir im wesentlichen davon ausgegangen, daß die erzeugten Schlüssel zur gültigen Schlüsselmenge gehören und direkt zur Anwendung kommen können, das ist jedoch nicht generell der Fall. Denken wir an spezielle (schwache oder halbschwache) DES-Schlüssel, so kann das Problem durch elementare Tests behoben werden. Im Kontext digitaler Signaturen ist es jedoch erforderlich, daß aus (pseudo-) zufälligen 0-1-Folgen Zahlen generiert werden müssen, die spezielle zahlentheoretische Eigenschaften besitzen. In der Regel handelt es sich dabei um Primzahlen oder Primitivwurzeln. Hierbei kann der bisher betrachtete Ansatz vom Prinzip her weiterverfolgt werden. Allerdings ist dann darauf zu achten, daß bei den folgenden Testverfahren sichergestellt wird, daß innerhalb vorgegebener Intervallgrößen auch tatsächlich ein Exemplar des gesuchten Zahlentyps ermittelt werden kann.

3 Individualschlüssel im Kontext digitaler Signaturen

Sollen Digitale Signaturen flächendeckend in offenen Systemen eingesetzt werden, so verlangt das Signaturgesetz (SigG) [IuKD97] bzw. die zugehörige Signaturverordnung (SigV) [SigV97] eine Schlüsselgenerierung, die es praktisch ausschließt, daß zwei Teilnehmer identische Schlüssel besitzen. Darüber hinaus sollte gefordert sein, daß die Schlüsselparameter so gewählt werden, daß sichergestellt ist, daß unterschiedliche Signaturschlüssel nicht dazu führen, daß sie isoliert betrachtet als sicher angenommen, gemeinsam jedoch zu beliebig unsicheren Systemen führen können.

Betrachten wir exemplarisch das RSA-Signaturverfahren [RiSA78], bei dem jeder Teilnehmer zwei Primzahlen, p und q, wählt, deren Produkt $n = p \cdot q$ bildet und zwei Schlüsselparameter e und d mit $1 = e \cdot d$ MOD $\varphi(n)$ wählen muß. Neben weiteren Anforderungen ist insbesondere sicherzustellen, daß alle gewählten Primzahlen unterschiedlich sind, da ansonsten die gewählten Moduln einen gemeinsamen Teiler besitzen könnten und daher von zumindest zwei Teilnehmern faktorisierbar wären. Betrachtet man die Größenordnung der verwendeten Primzahlen, die sich nach dem Ergebnis der bisherigen Diskussionen [Mkat97] zunächst in einer Größenordnung von 512 bit (pro Primzahl) bewegen, so ist bei einer zufälligen Wahl der Primzahlen bzw. der Startwerte für einen primzahlenerzeugenden Prozeß die Wahrscheinlichkeit extrem gering (aber nicht auszuschliessen), daß zwei Teilnehmer zumindest eine identische Primzahl wählen. Im Umfeld digitaler Signaturen ist es aber wünschenswert, daß Primzahldubletten generell verhindert werden [SigV97].

Es ist aus Sicherheitsgründen für den Besitzer einer SmartCard erstrebenswert, daß geheime und sicherheitsrelevante Schlüssel in der Karte selbst generiert werden und diese nie verlassen. Allerdings stellt sich dann das Problem, wie diese Schlüssel rekonstruiert werden können, wenn die SmartCard verloren geht oder funktionsunfähig wird. Dieses Problem muß im Einzelfall diskutiert und gegebenenfalls durch organisatorische Maßnahmen gelöst werden. Eine Möglichkeit dazu wäre, den Initialisierungswert des verwendeten Pseudozufallsgenerators von außen in die Chipkarte einzugeben und in ein spezielles Verzeichnis zu speichern. Dieser Initialisierungswert ist dann aber genauso sicher aufzubewahren wie die damit erzeugten geheimen Schlüssel selbst. Alternativ können die zur Generierung verwendeten Teilschlüssel an dafür vorgesehene Stellen gespeichert werden. Handelt es sich dabei um unterschiedliche Stellen, so muß ein eindeutiger Bezug zur entsprechenden Karte herstellbar sein, außerdem müssen alle betroffenen Algorithmen deterministisch ablaufen.

Initialisierung von Startwerten:

Am Beispiel der Schlüsselgenerierung beim RSA-Verfahren wird dargestellt, wie die prinzipielle Vorgehensweise sein kann, damit jeder Benutzer (pseudo-) zufällige Schlüsselkomponenten erhält. Hierbei muß besonderes Augenmerk darauf gelegt werden, daß jeder Benutzer zwei (wesentlich) verschiedene Primzahlen p und q zugewiesen bekommt. Insbesondere muß dabei garantiert werden, daß das Produkt $n = p \cdot q$ nicht durch einfache Verfahren (etwa durch die Methode der Differenz der Quadrate) faktorisierbar ist. Außerdem wird gefordert, daß alle generierten Primzahlen verschieden sind. Dies muß auch dann gelten, wenn die Primzahlen von unabhängigen Instanzen, insbesondere in Chipkarten, generiert werden.

Ziel der Initialisierung ist es, zwei Initialisierungswerte P und Q derart festzulegen, daß sich Primzahlen p und q der Form P$\|X$ und Q$\|Y$ finden lassen, wobei die Längen von p und q vorgegeben sind. Unsere Vorgehensweise wird dabei so sein, daß P und Q feste Längen besitzen und die Werte X und Y in Abhängigkeit zufälliger (ungerader) Werte $X0$ und $Y0$ bestimmt werden. Ein probabilistischer Primzahltest ermittelt dazu die jeweils „nächsten" Primzahlen. Sind $X0$ und $Y0$ ungerade Binärzahlen der Länge l, so werden zunächst Werte der Form $X = X0 + 2 \cdot k \leq 2^l - 1$ betrachtet, um zu entscheiden, ob P$\|X$ eine Primzahl ist. Wurde keine Primzahl gefunden, so werden Werte der Form $X = 1 + 2 \cdot k \leq X0$ betrachtet.

Wir wollen zwei unterschiedliche Primzahlen p und q, jeweils von der Größenordnung 512 bit, in einer Chipkarte generieren, wobei wir nachweisen, daß bei Einhaltung der beschriebenen Vorgehensweise keine Primzahldubletten generiert werden können. Hierbei gehen wir natürlich davon aus, daß die Anzahl der benötigten Primzahlen (Anzahl der Chipkarten) wesentlich kleiner als die Anzahl der generierbaren Primzahlen ist.

Der im weiteren vorgestellte Lösungsansatz erhebt weder den Anspruch auf Universalität noch auf Praktikabilität. Eine praxisrelevante Lösung kann letztlich nur im Einklang mit allen Beteiligten erreicht werden. Hierzu ist gegebenenfalls ein standardisiertes bzw. harmonisiertes Verfahren erforderlich.

Die Zuweisung der erforderlichen einzelnen Parameter (für eine Chipkarte) kann durch unterschiedliche Instanzen geschehen. Im beschriebenen Beispiel gehen wir davon aus, daß es sich dabei um einen Chiphersteller, einen Systemhersteller und um einen Benutzer handelt. In einer konkreten Anwendung (etwa in einem geschlossenen System) könnte dies auch völlig anders aussehen. Als weitere Instanz ist ein Trust Center oder eine Trusted Third Party zu berücksichtigen. Das Trust Center soll einerseits dazu beiträgt, daß die Schlüsselgenerierung in eine dafür vorgesehene Umgebung durchgeführt wird. Andererseits ist es erforderlich, daß die Echtheit der öffentlichen Schlüssel durch ein Zertifikat bestätigt werden muß.

Um die unterschiedlichen Chip- und Systemhersteller zu unterscheiden ist jeweils eine Herstellerkennung erforderlich. Diese Herstellerkennungen könnten jeweils durch eine 20-stellige Binärzahl eindeutig festgelegt sein. Die Kennung des Chipherstellers wollen wir mit CH0, die des Systemherstellers mit SH0 bezeichnen. Die Festlegung sollte durch übergeordnete Instanzen (Registrierungsstellen) erfolgen, um zu garantieren, daß die Herstellerkennungen (innerhalb der Chip- und Systemhersteller) unterschiedlich sind. Daneben würde im vorliegenden Fall eine Kennzeichnung CH1 (1 bit) benötigt, um die Primzahlen p und q zu unterscheiden. Stehen dem Chiphersteller insgesamt 64 bit zur Verfügung, so könnte er die restlichen Stellen durch eine 43 bit Zufallszahl CHR ergänzen und zur Diversifikation der zu generierenden Schlüsselkomponente nutzen.

Erhält ein Chiphersteller einen Auftrag eines Systemherstellers mit der Herstellerkennung SH0, so werden in jeder Chipkarte zwei Parameter PCH (ein Wert für p) und QCH (ein Wert für q) geladen, die weder manipulierbar noch auslesbar sein dürfen. Der Parameter PCH ist dabei abhängig von den Werten CH0, CH1=0 und einer 43 bit Zufallszahl CHR1, QCH ist abhängig von den Werten CH0, CH1=1 und einer (weiteren) Zufallszahl CHR2. Wählen wir PCH=CH0||0||CHR1 und QCH=CH0||1||CHR2, so ist sichergestellt, daß sich PCH und QCH in wenigstens einer Stelle unterscheiden.

Die identischen Anfangsstücke könnten sich bei genauerer Betrachtung im später folgenden Prozeß der Primzahlgenerierung als nachteilig herausstellen. Daher betrachten wir folgende Modifikation, bei der die Parameter PCH und QCH als Resultat einer geeigneten Verschlüsselung E festgelegt werden: PCH=E(CH0||0||CHR1,K1) und QCH=E(CH0||1||CHR2,K1). Werden bei der Verschlüsselung identische Schlüssel K1 verwendet, so kann wiederum sichergestellt werden, daß sich PCH und QCH in wenigstens einer Stelle unterscheiden. An dieser Stelle können die Zufallszahlen CHR1 und CHR2 zudem durch fortlaufende Nummern oder Pseudozufallszahlen ersetzt werden. Um zu garantieren, daß unterschiedliche Chiphersteller keine Werte PCH bzw. QCH erzeugen, die bereits ein anderer Chiphersteller erzeugt hat, müssen im vorliegenden Beispiel alle Chiphersteller den gemeinsamen Schlüssel K1 verwenden. Dieser Schlüssel sollte in einem Security Module aufbewahrt werden. Wird dagegen auf eine Verschlüsselung verzichtet, so muß sichergestellt werden, daß alle Chiphersteller eine einheitliche Reihenfolge der Parameter einhalten. Dabei kommt der gewählten Reihenfolge der Parameter eine besondere (zahlentheoretische) Bedeutung zu, auf die hier nicht weiter eingegangen werden soll.

Damit sichergestellt wird, daß unterschiedliche Systemhersteller niemals Parameter bereitstellen können, aus denen dann möglicherweise identische Primzahlen generiert werden, ist eine zusätzliche Parametrisierung bezüglich der Systemhersteller erforderlich. Hierzu könnten zwei weiterer 64 bit Blöcke dienen (einer für p und einer für q), die insbesondere von der Herstellerkennung SH0 der Systemhersteller abhängen. Zusätzlich könnten sich Chip- und Systemhersteller optional auf einen weiteren Parameter innerhalb diese Blocks verständigen. So könnte an dieser Stelle beispielsweise eine Kennzeichnung SH1 der zum Einsatz kommenden Anwendung einfließen. Die verbleibenden Stellen sollten wieder mit einer Zufallszahl SHR aufgefüllt werden. Die 64 bit Parameter PSH=SH0||SH1||SHR1 und QSH=SH0||SH1||SHR2 können direkt oder durch PSH=E(SH0||SH1||SHR1,K2) und QSH=E(SH0||SH1||SHR2,K2) bestimmt werden. Auch hier ist wieder dafür Sorge zu tragen, daß alle Chiphersteller denselben Verschlüsselungsprozeß und gegebenenfalls denselben Schlüssel K2 verwenden, da nur dann sichergestellt werden kann, daß verschiedene Systemhersteller keine identischen Parameter erhalten. Die Werte PSH und QSH werden wiederum manipulationssicher und unauslesbar in die dafür vorgesehenen Speicherplätze einer Chipkarte geladen. Hierbei ist es zunächst nicht notwendig, daß sich die erzeugten Werte PSH und QSH unterscheiden. Die Möglichkeit der Unterscheidung sollte aber auch hier genutzt werden. Im Verschlüsselungsfall kann dazu anstelle eines Zufallsgenerators wieder ein geeigneter Zähler (etwa ein Pseudozufallsgenerator) verwendet werden.

Chiphersteller	Systemhersteller	Chipkarte	Benutzer
CH0||0||CHR1	SH0||SH1||SHR1	CC0||CCR1	PB0
CH0||1||CHR2	SH0||SH1||SHR2	CC0||CCR2	PB0

Abb.1: Beispiel von Schlüsselparametern und Zugehörigkeiten

Betrachten wir die bisherigen Zwischenergebnis A=PCH||PSH und B=QCH||QSH, so können wir einerseits garantieren, daß in jeder einzelnen Chipkarte die Werte A und B unterschiedlich sind. Andererseits können unterschiedliche Systemhersteller weder identische Werte A noch identische Werte B erhalten. Dennoch ist es nicht ausgeschlossen, daß ein Systemhersteller identische Werte A oder identische Werte B erhält. Bei der weiteren Schlüsselgenerierung ist dies besonders zu beachten, da die zu erzeugenden Primzahlen alle verschieden sein müssen.

Die weitere Vorgehensweise kann mit der bisherigen verglichen werden: Zunächst werden Parameter individualisiert, dann durch Zufallszahlen verlängert und eventuell verschlüsselt. Auch an dieser Stelle ist eine Verschlüsselung angebracht. Im Gegensatz zur bisherigen Betrachtung können Systembetreiber aber frei gewählte (System-) Schlüssel KS1 und KS2 einsetzen. Einmal verwendete Schlüssel dürfen allerdings nicht mehr verändert werden. Sind die Schlüssel KS1 und KS2 verschieden, so muß sichergestellt sein, daß sie nicht vertauscht werden können. An dieser Stelle würde es ausreichen, wenn jeder Chipkarte eine 32 bit Folgenummer CC0 und eine 32 bit Zufallszahlen (CCR1 und CCR2) zugewiesen würden. Die so erzeugten Parameter PCC=E(CC0||CCR1,KS1) und QCC=E(CC0||CCR2,KS2) würden dann dazu beitragen, daß sich die Gesamtheit aller erzeugter Werte PCH‖PSH‖PCC und QCH‖QSH‖QCC in wenigstens einer Stelle unterscheiden.

Bei den folgenden Betrachtungen gehen wir davon aus, daß sich die jeweils erzeugten Werte in einer Chipkarte befinden und weder verändert von ausgelesen werden können. Mit den bisher durchgeführten Parameterwahlen können 2^{20}=1.048.576 verschiedene Chiphersteller und 2^{20} verschiedene Systemhersteller integriert werden. Zusätzlich kann jeder Systemhersteller mindestens 2^{32}=4.294.967.296 Chipkarten eines Chipkartenherstellers ausgeben, bevor die Gefahr einer Schlüsseldublette droht. Jeder der 192 bit langen Schlüsselparameter ist zudem von einer (192–(20+20+32+1))=119 bit langen (Pseudo-) Zufallszahl abhängig. Durch eine geeignete Anpassung der Parameter lassen sich gewünschte Systemeigenschaften leicht und transparent realisieren. Einmal getroffene Entscheidungen lassen sich allerdings nicht ohne weitere Vorkehrungen zurücknehmen.

Aber selbst eine Modifikation der Parametergrößen zu einem späteren Zeitpunkt kann auf Kosten einiger Binärstellen realisiert werden. Gehen wir davon aus, daß nach einer ersten Verständigung zwischen allen beteiligen Instanzen eine Revision erforderlich werden könnte, so kann beispielsweise ein geeignet langer Versionsparameter dazu beitragen, daß mögliche Revisionen Berücksichtigung finden können. Hierbei ist es durchaus empfehlenswert, daß solche Versionsparameter Bestandteil der Parameter der einzelnen Instanzen sind.

Verfolgen wir unser Beispiel weiter, so stehen für jede Primzahl noch 512–192=320 bit (40 Byte) zur freien Verfügung, um Startwerte für einen probabilistischen Primzahltest festzulegen. Diese Werte könnten unter Einbeziehung des Benutzers festgelegt werden. Hierzu könnte ein Benutzer etwa aufgefordert werden, jeweils 8-stellige alphanumerische (64 bit) Zeichenketten PB0 und QB0 in ein geeignetes Chipkartenterminal einzugeben. Dadurch kann sichergestellt werden, daß die erzeugten Startwerte von allen Beteiligten (im Beispiel sind dies bisher die Chiphersteller, die Systemhersteller und die Benutzer) abhängig sind. Als wesentlich ist zudem anzusehen, daß die einzelnen Schlüsselparameter an unterschiedlichen Stellen unabhängig voneinander generiert werden können.

Um den Benutzer noch stärker in die Schlüsselgenerierung einzubinden, könnten durch den Benutzer weitere 64 bit Parameter PB1 und QB1 eingegeben werden oder durch eine geeignete (chipkarteninterne) Verkettung der Werte PB0, PCH, PSH und PCC ermittelt werden; entsprechendes gilt für die zu Q gehörenden Werte. Die verbleibenden 24 Byte (192 bit) könnten dazu dienen, die niederwertigsten Stellen der Startwerte P0 und Q0 festzulegen. Dies könnte durch ein Trust Center, eine Trusted Third Party oder geeignetes Sicherheitsequipment geschehen.

Berücksichtigt man die Tatsache, daß vor einer Schlüsselgenerierung alle Teilparameter der Initialisierung vorliegen müssen, so können die Teilparameter in beliebiger Reihenfolge festgelegt werden, davon könnte gegebenenfalls die Generierung der Teilparameter PB1 und QB1 ausgenommen sein, da sich hierzu die Teilparameter des Chip- und des Systemherstellers bereits in der Karte befinden müßten.

Bei den Betrachtungen gehen wir davon aus, daß sich für eine beliebig vorgegebene 320 bit Binärzahl x zwischen $x \cdot 2^{192}$ und $x \cdot 2^{192}+2^{192}-1=(x+1) \cdot 2^{192}-1$ wenigstens eine Primzahl befindet. Die Größe des Suchraums ist hier allerdings nur von untergeordneter Bedeutung.

Die naive Vorstellung, daß jede ungerade Zahl eine potentielle Primzahl ist, verschwendet bei dem folgenden Primzahltest in der Regel unnötige Ressourcen. Der intelligentere Ansatz, daß zunächst jeder Kandidat daraufhin getestet wird, ob er einen kleinen Primfaktor besitzt, ist ebenfalls aus Gründen der Performance nicht praktikabel. Berücksichtigt man jedoch den Umstand, daß gewisse zusammengesetzte Zahlen erst gar nicht auf Primalität getestet werden müssen, so kann die im Mittel erforderliche Testzeit auf einen Bruchteil der naiven Vorgehensweise reduziert werden. Hierzu bedarf es allerdings spezieller Chipkarten oder spezieller Interaktion zwischen Chipkarte und Chipkartenterminal.

4 Theoretische und algorithmische Aspekte

In den folgenden Ausführungen werden Verfahren und Algorithmen aufgeführt, die zur Realisierung der aufgezeigten Konzepte dienlich sein können. Außerdem werden Ergebnisse präsentiert, die im Umfeld unserer Betrachtungen in Zukunft noch genauer zu untersuchen sind.

4.1 Pseudozufallsgeneratoren und Zufallsgeneratoren

Pseudozufallszahlen können durch eine Funktion $f : X \to X$ beschrieben werden, wobei X eine endliche Menge darstellt. Ist ein Initialwert $x \in X$ vorgegeben, so stellt die Folge $x, f(x), f^2(x), f^3(x), \ldots$ die generierte Pseudozufallsfolge dar, wobei $f^{i+1}(x) = f(f^i(x))$ und somit etwa $f^3(x) = f(f^2(x)) = f(f(f(x)))$ ist. Da deterministisch arbeitende Pseudozufallsgeneratoren nur eine endliche Anzahl unterschiedlicher Werte generieren können, müssen zwangsläufig zwei Werte generiert werden, die identisch sind. Bei bestimmten Pseudozufallsgeneratoren können wir garantieren, daß diese Situation erst dann eintritt, wenn alle möglichen Werte bereits einmal generiert worden sind. Die kleinste positive ganze Zahl d für die die Bedingung $f^{d+i}(x) = f^i(x)$ für alle $i \geq l$ gilt, heißt Periodenlänge des Pseudozufallsgenerators, die ersten l Werte bezeichnen wir als Vorperiode.

Es existiert eine Vielzahl unterschiedlicher Klassen von Pseudozufallsgeneratoren [Knut97]. Im folgenden werden die drei am häufigsten eingesetzten Verfahren kurz charakterisiert. Zusätzlich wird darauf eingegangen, wie Verschlüsselungsverfahren zur Erzeugung von Pseudozufallszahlen genutzt werden können.

Lineare Kongruenzen: Bei der Methode der linearen Kongruenzen wird die Menge X durch den Restklassenring $Z_m = \{0, 1, \ldots, m-1\}$, und f durch $f(x) = (a \cdot x + b) \text{ MOD } m$ repräsentiert. Ist ein Initialisierungswert $x_0 \in Z_m$ vorgegeben, so wird die Pseudozufallsfolge $x_0, x_1, x_2, x_3, \ldots$ durch $x_{n+1} = (a \cdot x_n + b) \text{ MOD } m$ ermittelt. Die maximal mögliche Periodenlänge $d = m$ kann bei geeigneter Wahl der Werte a, b und m erreicht werden.

Schieberegister: Betrachten wir linear rückgekoppelte Schieberegister der Länge n, so wird die Menge X durch 0-1-Vektoren $x = (b_1, b_2, \ldots, b_n)$ dargestellt, und die Funktion f kann als lineare Transformation aufgefaßt werden: $f(x) = x \cdot T$, wobei T eine binäre $k{\times}k$-Matrix darstellt. Die anfallenden Operationen werden modulo 2 ausgeführt. Bezeichnet x den Initialvektor, so wird die Folge $x, x \cdot T, x \cdot T^2, x \cdot T^3, \ldots$ generiert. Wird die Binärmatrix T geeignet gewählt, so kann die maximale Periodenlänge $d = 2^n - 1$ erreicht werden.

Fibonacci-Generatoren: Fibonacci-Generatoren können als Erweiterung der Methode der linearen Kongruenzen betrachtet werden. Die Menge X wird durch Vektoren $x = (x_1, \ldots, x_r)$ der Länge r über einer endlichen Menge M, in der eine binäre Operation $\otimes$ definiert ist, festgelegt. Die Funktion f ist dabei durch $f(x_1,...,x_r) = (x_2,...,x_r,x_1 \otimes x_{r+1-s})$ festgelegt, wobei $1 \leq s < r$ ist. Liegen der Initialisierungswert $x = (x_1,...,x_r)$ vor, so können die Elemente der Pseudozufallsfolge $x_1, x_2,...,x_r,x_{r+1},...$ für alle $n > r$ durch $x_n = x_{n-r} \otimes x_{n-s}$ generiert werden. Solche Generatoren werden häufig durch die Bezeichnung $F(r,s,\otimes)$ gekennzeichnet. Weitere Einzelheiten und Analysen können beispielsweise [Knut97] entnommen werden.

Verschlüsselungsfunktionen: Verschlüsselungsfunktionen $E: X \times K \to X$ realisieren für alle festen Schlüssel $k \in K$ bijektive (Parameter-) Funktionen auf einer endlichen Menge X. Betrachten wir $f = E_k$, so kann diese Funktion auf unterschiedliche Weise als Pseudozufallsgenerator eingesetzt werden. Wird zur Initialisierung der Wert $x \in X$ gewählt, so kann durch $x, f(x), f^2(x), ...$ eine Pseudozufallsfolge bestimmt werden. Die Periodenlänge d ist dabei sowohl vom Initialwert x als auch vom Schlüssel k abhängig. Verbindliche Aussagen über $d = d(x,k)$ können hier in der Regel nicht gemacht werden. Betrachten wir dagegen die folgende Modifikation, so kann sichergestellt werden, daß die Periodenlänge der Kardinalität der Menge X entspricht und somit maximal ist. Hierzu nehmen wir an, daß X die Menge aller Binärzahlen der Länge n enthält, womit X auch durch die Menge $M = \{0,1,...,2^n - 1\}$ repräsentiert werden kann. Wählen wir den Startwert $x \in M$, und betrachten wir die Pseudozufallsfolge $x_0 = f(x)$, $x_1 = f(x \circ 1)$, $x_2 = f(x \circ 2)$, ... mit $x_i = f(x \circ i) = f((x + i) \,\mathrm{MOD}\, 2^n)$, so ist sichergestellt, daß alle Werte aus M durchlaufen werden, bevor der Wert x_0 wiederholt auftritt. Dies ist darin begründet, daß E_k und somit f bijektive Funktionen darstellen.

Besitzen Pseudozufallsgeneratoren eine bekannte Periodendauer, so können sie insbesondere zur „zufälligen Numerierung" eingesetzt werden. Der Startwert kann dann als Offset dienen, und die Elemente der generierten Folge zur Numerierung. Wird in einer Anwendung ein solches Numerierungssystem anstelle der natürlichen Numerierung 1, 2, 3, usw. verwendet, so kann insbesondere vermieden werden, daß auf die Anzahl der bisher vergebenen Nummern (oder die Anzahl der so gekennzeichneten Chipkarten) geschlossen werden kann. Diese Gefahr besteht bei der natürlichen Numerierung auch dann, wenn ein Beobachter nur von einigen (wenigen) der bereits vergebenen Nummern Kenntnis erhält. Das damit verbundene Taxiproblem behandelt die folgende Fragestellung: Welche Aussage kann man über die Anzahl der Taxis eines Unternehmens machen, wenn es seine Taxis durchnumeriert und jemand die Nummern zufällig beobachteter Taxis notiert? Werden die Nummern n_1, n_2, ..., n_k notiert und nehmen wir zusätzlich an, daß $n_i < n_{i+1}$ ist, so kann die Anzahl T der vorhandenen Taxis durch die Heuristik $T = n_k + \max_{1 \leq i < k} (n_{i+1} - n_i)$ relativ gut abgeschätzt werden.

Bei echten Zufallsgeneratoren ergibt sich eine völlig andere Situation. Zunächst können wir keine Voraussagen darüber machen, wie sich die generierten Werte verhalten. Ist die Anzahl der möglichen Werte groß, so können wir lediglich statistische Analysen und Betrachtungen durchführen. Im Kontext kryptographischer Anwendungen ist dabei das sogenannte Geburtstagsparadoxon von Bedeutung: Wählt man aus einer Menge von N unterschiedlichen Objekten m Objekte zufällig aus, wobei Wiederholungen möglich sind, so läßt sich die Wahrscheinlichkeit p, daß sich unter den ausgewählten Objekten mindestens zwei gleiche befinden leicht abschätzen. Sind die Werte p und N vorgegeben, so resultiert folgender Zusammenhang: $m = \sqrt{2 \cdot N \cdot ln(1-p)}$, womit sich für $p = 0,5$ die Abschätzung $m = 1,18 \cdot \sqrt{N}$ ergibt. Befinden sich beispielsweise in einer Klasse 23 Schüler, so tritt dieser Fall ($p = 0,5$)

bei der Fragestellung auf, wie groß die Wahrscheinlichkeit ist, daß sich in der Klasse mindestens zwei Schüler befinden die am selben Tag (an einem von 365 möglichen Tagen) Geburtstag haben.

Wählen wir von den 2^{56} möglichen Schlüsseln beim Data Encryption Standard (DES) 2^{28} zufällig aus und verwenden wir die aufgeführte Formel, so befinden sich darunter mit etwa 40% Wahrscheinlichkeit zwei identische Schlüssel. Beim Einsatz von Zufallsgeneratoren kann es dabei durchaus vorkommen, daß bereits die beiden ersten Schlüssel identisch sind, die Wahrscheinlichkeit dafür ist mit 2^{-56} allerdings verschwindend gering. Ebenso ist es möglich, daß die ersten 2^{56} Schlüssel alle verschieden sind, der nächste generierte Schlüssel ist dann natürlich eine Schlüsseldublette; die Wahrscheinlichkeit hierfür ist ebenfalls zu vernachlässigen.

Zufallsgeneratoren werden als integrierte Schaltungen von verschiedenen Herstellern angeboten. So bietet beispielsweise Newbridge Microsystems einen Random Bit Generator RBG 1220 an, der auf der Basis des sogenannten Johnson Phänomens eine Rauschquelle mit weißem Rauschen erzeugt, das anschließend durch einen speziellen Analog-Digital-Wandler so in zufällige 0-1-Folgen umgesetzt wird, daß die Wahrscheinlichkeit des Auftretens einer 0 oder einer 1 etwa gleichwahrscheinlich sind [Newb90]. Die Firmenunterlagen geben keine weitere Auskunft über die Qualität des Zufallsgenerators, bei Bedarf können jedoch detaillierte Testresultate angefordert werden. Mit dem Random Number Generator T7001 wird von der Firma AT&T Microelectronics ein wesentlich komplexerer Baustein angeboten [AT&T86]. Die verwendete Rauschquelle basiert auf einem frei laufenden Jitter-Oszillator. In einem Durchlauf wird dabei eine 536 bit (67 Byte) Zufallszahl erzeugt. Durch bestimmte Angaben, etwa minimales und maximales Gewicht (Anzahl der Einsen), kann dabei der akzeptierte Zufallswert eingeschränkt werden. Hierdurch wird die Zufälligkeit zwar in gewissen Grenzen beeinflußt, den Bedürfnissen der praktischen Anwendung wird dadurch aber in vielen kryptographisch relevanten Anwendungsfällen durchaus Rechnung getragen.

Die Überprüfbarkeit der korrekten Implementierung von Zufalls- und Pseudozufallsgeneratoren und deren korrekte Arbeitsweise ist bei Massenanwendungen de facto unmöglich. Hinzu kommt, daß im Fall der Generierung von Signaturschlüsseln der Zufallsprozeß in der Regel nur einmal einsetzt wird, wodurch sich die Frage der Verhältnismäßigkeit stellt. Von besonderer Bedeutung ist aber die nicht vorhandene Transparenz: Wie kann sichergestellt werden, daß die Kritiker nicht gerade im Zufallsprozeß das wesentliche Problem festmachen? Wünschenswert ist eine Lösung, bei der Manipulationsmöglichkeiten weitgehend ausgeschlossen, die individuelle Schlüsselgenerierung garantiert, eine gewissen Zufälligkeit gegeben und die jeweils Beteiligten einen unabhängigen Beitrag zur Schlüsselgenerierung leisten können.

4.2 Algorithmen der Zahlentheorie

Der Miller-Rabin-Primalitätstest: Beim Miller-Rabin-Test [Knut97] handelt es sich um einen probabilistischen Primzahltest. Der zugehörige Algorithmus überprüft, ob eine vorgegebene ungerade Zahl n eine Primzahl ist oder nicht. Hierzu werden Zufallszahlen b daraufhin überprüft, ob sie „Zeugen" für die Primalität von n sind. Entscheidet der Algorithmus „n ist zusammengesetzt", so ist dies immer richtig, lautet die Entscheidung jedoch „n ist eine Primzahl", so kann dies mit einer gewissen Wahrscheinlichkeit falsch sein. Diese Wahrscheinlichkeit ist beim Miller-Rabin-Test kleiner als 0,25. Da die so ermittelten Entscheidungen für verschiedene Zufallszahlen unabhängig voneinander sind, kann diese „Fehlerwahrscheinlichkeit" durch wiederholte Anwendung des Algorithmus beliebig klein gemacht werden. Lautet die Entscheidung bei k unabhängigen Versuchen immer „n ist eine Primzahl", so ist die Fehlerwahrscheinlichkeit kleiner als 4^{-k}. Ist beispielsweise $k = 10$, so ist die Fehlerwahrscheinlichkeit kleiner als 10^{-6}. Der Algorithmus läßt sich folgendermaßen darstellen: Ist n eine ungerade ganze Zahl, so läßt sie sich eindeutig in der Form $n = 2^k \cdot m + 1$ darstellen, wobei m eine ungerade ganze Zahl ist. Der Algorithmus berechnet die Werte $x_0 := b^m \bmod n$ und $x_j := x_{j-1}^2 \bmod n$ für $1 \leq j \leq k$, wodurch sich $x_k = b^{n-1} \bmod n$ ergibt. Hieraus kann nun gefolgert werden, daß n zusammengesetzt, also keine Primzahl ist, wenn $x_0 \neq 1$ und $x_j \neq n - 1$ für $1 \leq j < k$ ist. Anderenfalls ist n vermutlich eine Primzahl.

Primitivwurzeln modulo einer Primzahl: Für Verfahren auf Grundlage des diskreten Logarithmusproblems [KnH192, KnH292], wie etwa die Verfahren von Diffie-Hellman oder El-Gamal [ElGa85], ist eine Primitivwurzel α modulo einer Primzahl p erforderlich. Es ist zwar sichergestellt, daß für jede Primzahl p genau $\varphi(\varphi(p)) = \varphi(p-1)$ viele unterschiedliche Primitivwurzel modulo p existieren. Ist aber lediglich die Primzahl p gegeben, so ist kein Algorithmus bekannt, der auch nur eine dieser Primitivwurzeln effizient ermitteln kann. Ist p gegeben und die Faktorisierung von $p - 1$ unbekannt, so kann man zwar eine beliebige Zufallszahl α wählen und testen, ob $\alpha^{p-1} = 1 \bmod p$ und $\alpha^i \neq 1 \bmod p$ für $2 \leq i \leq p - 2$ ist; womit α eine Primitivwurzel modulo p wäre. Für kryptographisch relevante Primzahlen ist dies wegen der Größe von p jedoch unmöglich. Ist dagegen die Primfaktorzerlegung von $p - 1$ bekannt, $p - 1 = \prod_{i=1}^{k} p_i^{a_i}$, so kann gezeigt werden, daß α genau dann eine Primitivwurzel modulo p ist, wenn für alle $i = 1, 2, \ldots, k$ die Beziehung $\alpha^{(p-1)/p_i} \neq 1 \bmod p$ gilt.

Die beschriebenen Verfahren verlangen wegen des zugrundeliegenden diskreten Logarithmusproblems, daß $p - 1$ nachweisbar einen großen Primfaktor enthält. Ist beispielsweise q eine Primzahl und $p - 1 = 2 \cdot q$, so reduziert sich der Test darauf, ob

$$\alpha^2 \neq 1 \bmod p \quad \text{und} \quad \alpha^q \neq 1 \bmod p \text{ ist.}$$

Ist dies der Fall, so ist das betrachtete p eine Primzahl und außerdem ist α dann eine Primitivwurzel modulo p. Ist eine Primitivwurzel α modulo p bekannt, so können alle weiteren leicht berechnet werden. Dies ist beispielsweise dann von Interesse, wenn die ermittelte Primitivwurzel α aus irgendwelchen Gründen durch eine andere Primitivwurzel β ersetzt werden muß. Ist k eine zu $p - 1$ teilerfremde ganze Zahl, so ist $\beta = \alpha^k \bmod p \neq \alpha$ ebenfalls eine Primitivwurzel modulo p.

Simultanes Erzeugen von Primzahlen und Primitivwurzeln: Ist q eine Primzahl, $p = 2 \cdot q + 1$, $\alpha^2 \neq 1 \bmod p$ und $\alpha^q = -1 \bmod p$, so haben wir aufgeführt, daß dann p ebenfalls eine Primzahl und außerdem α eine Primitivwurzel modulo p ist. Diese Eigenschaft kann dazu verwendet werden, um aus einer vorgegebenen Primzahl q eine größere Primzahl p zu konstruieren. Anstelle von $p = 2 \cdot q + 1$ kann es dann aber auch sinnvoll sein, Werte der Form $p = k \cdot q + 1$ zu betrachten, wobei k als Produkt kleiner (möglicherweise gleicher) Primzahlen vorgeben ist.

Parameter für das RSA-Verfahren: Die Parameter eines RSA-Verfahrens sind der öffentliche Modul n, der öffentliche Schlüssel e sowie der geheime Schlüssel d und die geheimen Primzahlen p und q mit $n = p \cdot q$. Diese Parameter können für den jeweiligen Benutzer folgendermaßen ermittelt werden. Werden geeignete Startwerte vorgegeben, so können Primzahlen p und q mit dem Miller-Rabin-Test effizient bestimmt werden.

Zur Berechnung des RSA-Schlüsselpaares (e, d) zum Modul n ist zunächst die Ermittlung von $\varphi(n) = (p - 1) \cdot (q - 1)$ und die Erzeugung einer Zufallszahl t erforderlich. Teilt man t solange durch den größten gemeinsamen Teiler von t und $\varphi(n)$, bis das Ergebnis und $\varphi(n)$ den größten gemeinsamen Teiler 1 besitzen, so erhält man eine zu $\varphi(n)$ teilerfremde Zahl d, deren multiplikativ Inverses e eindeutig mittels des erweiterten Euklidschen Algorithmus ermittelt werden kann. Hierbei sind eventuell vorgegebene Zusatzbedingungen zu erfüllen, so muß etwa gewährleistet werden, daß $d \neq 1$ ist. Als Schlüsselparameter d kann aber auch eine geeignet Primzahl gewählt werden, um die aufgeführten Schritte zu sparen. Selbstverständlich kann d auch abhängig von e berechnet werden. Im Anhang von [CCIT89] wird für diese Zwecke die (Fermat-) Primzahl $65537 = (10.00000.00000.00001)_2$ vorgeschlagen.

Aus Sicherheitsgründen sollte man für den Einsatz in Verschlüsselungsverfahren allerdings kleine e (etwa $e = 3$) vermeiden. Sollte sich ein solches e ergeben, so ist der beschriebene Vorgang solange zu wiederholen, bis ein geeignetes e gefunden wurde.

4.3 Primzahlen und ihre Verteilung

Aus Gründen der Komplexität (insbesondere Rechenzeit in der Chipkarte) ist es erforderlich, Betrachtungen darüber durchzuführen, wie groß der Suchraum (das zu durchsuchende Intervall) sein muß, damit mit einer gewissen Wahrscheinlichkeit eine Primzahl schnell gefunden werden kann. Zur Motivation seien hier nur einige der bekannten Ergebnisse zusammengestellt, detaillierte Betrachtungen können der Literatur [Ries85, Shan85, Bres89, Ribe91, Knut97] entnommen werden. Kann das Suchintervall klein gestaltet werden, so stehen mehr Stellen für die Qualität (Zufälligkeit) der generierten Werte zur Verfügung.

Bezeichnet $\pi(n)$ die Anzahl der Primzahlen die kleiner oder gleich einer ganzen Zahl n sind, so kann gezeigt werden, daß $\pi(n)$ für große n gegen $n / \log(n)$ strebt. Bertrand äußerte 1845 die Vermutung, daß zwischen einer positiven ganzen Zahl n und der Zahl $2 \cdot n$ wenigstens eine Primzahl liegt, womit $\pi(2 \cdot n) - \pi(n) \geq 1$ wäre. Diese Vermutung wurde erstmals 1852 durch Tschebyschew bewiesen, er wies für $n \geq 5$ folgende Beziehung nach:

$$\frac{n}{3 \log n} < \pi(2n) - \pi(n) < \frac{7n}{5 \log n}.$$

Erdös generalisiert 1934 das Resultat, indem er zeigte, daß für jedes $\lambda > 1$ ein $C = C(\lambda) > 0$ und ein $x_0 = x_0(\lambda) > 1$ existiert, so daß für alle $x > x_0$ die folgende Aussage gilt:

$$\pi(\lambda x) - \pi(x) > C \frac{x}{\log x}.$$

Das Ergebnis von Ishikawa (1934) ist ebenfalls eine Konsequenz des Satzes von Tschebyschew: Sind $x \geq y \geq 2$ und $x \geq 6$, dann gilt $\pi(x \cdot y) > \pi(x) + \pi(y)$. Angeregt durch Ergebnisse von Rosser und Schoenfeld (1975) zeigte Vaughan, daß $\pi(x+y) \leq \pi(x) + \dfrac{2y}{\log y}$ ist, womit insbesondere $\pi(x+y) \leq \pi(x) + 2\pi(y)$ folgt. Rosser und Schoenfeld haben zudem gezeigt, daß für alle $x \geq 11$ die Relation $\pi(2x) < 2\pi(x)$ gilt. Eine weitere Beziehung, die 1882 von Opperman angegeben wurde und in unserem Kontext durchaus von Interesse sein kann, konnte bis heute weder bewiesen noch widerlegt werden: $\pi(n^2+n) > \pi(n^2) > \pi(n^2)$.

5 Resümee

Es wurde gezeigt, daß in vielen Anwendungsfällen eine (pseudo-) zufällige Generierung von Schlüsselparametern ermöglicht werden kann. Geschieht dies mit der erforderlichen Sorgfalt, so können die abgeleiteten Schlüssel als Sitzungsschlüssel verwendet werden. Außerdem gewinnen individuelle Schlüsselparameter zunehmend an Bedeutung. Als Paradebeispiel können private Signaturschlüssel angesehen werden. In diesem Zusammenhang ist zudem in der Regel zu garantieren, daß alle sicherheitsrelevanten Schlüsselparameter dublettenfrei generiert werden. Die Länge der verwendeten Parameter lassen ergänzend eine weitgehend zufällige Initialisierung der Parameter zu. Da mehrere unabhängige Instanzen bei der Schlüsselgenerierung mitwirken können, kann den resultierenden Schlüsselparametern ein großes Vertrauen entgegengebracht werden. Dies gilt um so mehr, wenn sogar der Benutzer einen angemessenen Anteil zur Schlüsselgenerierung beiträgt.

Im Kontext digitaler Signaturen wurden am Beispiel der Primzahlgenerierung konkrete Ideen vorgestellt, die es grundsätzlich ermöglichen, die benötigten Schlüsselparameter quasi zufällig zu generieren. Allen beteiligten Instanzen kann dabei eine „Schlüsselrolle" eingeräumt werden, indem sie zur Schlüsselgenerierung aktiv beitragen. Das vorgestellte Konzept verfolgt dabei aber das vorrangige Ziel der individuellen dublettenfreien Schlüsselgenerierung, wodurch die diesbezüglichen Forderungen des Signaturgesetzes erfüllt werden könnten. Da die Schlüsselkomponenten in einer Chipkarte generiert werden sollen, erfordert das vorgestellte Konzept eine geeignete Sicherheitsinfrastruktur und ein in weiten Bereichen standardisiertes Verfahren. Im Umfeld der digitalen Signatur könnte dies mit geringem Mehraufwand leicht realisiert werden.

Die vorgestellte Konzeption ist (insbesondere im Kontext digitaler Signaturen [Hors96, HoKr96]) zu diskutieren und zu präzisieren, auch ist zu überprüfen, ob sich hierdurch nicht die Möglichkeit der sicheren Rekonstruktion von Schlüsselkomponenten ergeben kann. Hierzu wäre es erforderlich, daß alle beteiligten Instanzen ihre Parameter in Verbindung mit einem individuellen Merkmal der jeweiligen Chipkarte speichern, um bei berechtigtem Anlaß eine Wiederherstellung der Schlüsselkomponenten zu ermöglichen.

Vor einem praktischen Einsatz sind allerdings noch zahlreiche Detailfragen zu klären, wobei Aspekte der Zahlentheorie und der Umsetzung in chipkartenrelevante effiziente Algorithmen von besonderer Tragweite sind. Um der Bedeutung sicherheitsrelevanter Anwendung und daraus abgeleiteten Sicherheitsanforderungen entsprechend zu begegnen, sind insbesondere Forschungsaktivitäten in den aufgeführten Bereichen erforderlich.

Literatur

[AT&T86] AT&T: Communication Device Data Book, T7010 Random Number Generator (1986) 3.29-3.42.

[Bres89] Bressoud,D.M.: Factorization and Primality Testing, Springer (1989).

[CCIT89] CCITT, Recommendation X.509. The Directory - Authentication Framework, Blue Book - Melbourne 1988, Fasscicle VIII, International Telecommunication Union, Geneva (1989) 48-81.

[ElGa85] ElGamal,T.: A Public Key Cryptosystem and a Signature Scheme Based on Discrete Logarithms, IEEE Transactions on Information Theory 31 (1985) 469-472.

[Hors96] Horster,P. (Hrsg.): Digitale Signaturen, DuD-Fachbeiträge, Vieweg (1996).

[HoKr96] Horster,P., Kraaibeek,P.: Grundüberlegungen zu digitalen Signaturen, Tagungsband Digitale Signaturen, Vieweg (1996) 1-14.

[IuKD97] Gesetz zur Regelung der Rahmenbedingungen für Informations- und Kommunikationsdienste (Informations- und Kommunikationsdienste-Gesetz - IuKDG), Bundesgesetzblatt 1869, Teil I G5702 (1997) 1869-1880.

[KnH192] Knobloch,H.-J., Horster,P.: Eine Krypto-Toolbox für Smartcard-Chips mit speziellen Calculation Units, Tagungsband 2. GMD-SmartCard Workshop (1992).

[KnH292] Knobloch,H.-J., Horster,P.: Eine Krypto-Toolbox für Smartcards, DuD - Datenschutz und Datensicherheit (1992) 353-361.

[Knut97] Knuth,D.: The Art of Computer Programming, Vol. 2: Seminumerical Algorithms, 3rd ed., Addison Wesley (1997).

[MKat97] Angaben des BSI zum Maßnahmenkatalog gemäß SigV §§ 12(2) und 16(6), Ausgabe Version 1.0 vom 18.11.1997, www.bsi.bund.de

[Newb90] Newbridge Microsystems: CMOS and Bipolar Products, RGB 1210 Random Bit Generator (1990) 3.28-3.31.

[Ribe91] Ribenboim,P.: The Little Book of Big Primes, Springer (1991).

[Ries85] Riesel,H.: Prime Numbers and Computer Methods for Factorization, Progress in Mathematics Vol. 57, Birkhäuser (1985).

[RiSA78] Rivest,R.L., Shamir,A., Adleman,L.: A Method for Obtaining Digital Signatures and Public Key Cryptosystems, Communications of the ACM (1978) 120-126.

[Shan85] Shanks,D.: Solved and Unsolved Problems in Number Theory, 3rd ed., Chelsea Publishing (1985).

[SigV97] Verordnung zur digitalen Signatur (Signaturverordnung - SigV), Stand: 8. Oktober 1997, www.iid.de/aktuelles/index.html/#iukdg

Implementierung Elliptischer Kurven auf Chipkarten

Detlef Hühnlein

Secunet Security Networks GmbH
Mergenthalerallee 79-81
D-65760 Eschborn
huehnlein@secunet.de

Zusammenfassung

Kryptosysteme auf Basis Elliptische Kurven über endlichen Körpern erfreuen sich wachsender Beliebtheit. So sieht beispielsweise das BSI im Anforderungskatalog für das Signaturgesetz u.a. auch die Verwendung Elliptischer Kurven vor. In dieser Arbeit soll die Implementierung Elliptischer Kurven über GF(p) auf Chipkarten diskutiert werden. Neben der offensichtlich möglichen Implementierung auf Karten mit kryptographischem Coprozessor, werden auch andere Ansätze betrachtet. So wird auch die Möglichkeit der Implementierung auf Karten ohne kryptographischem Coprozessor und Java-Karten beleuchtet.

1 Einleitung

Kryptosysteme auf Basis Elliptischer Kurven über endlichen Körpern, wie in [26] und [16] vorgeschlagen und momentan in [14] standardisiert, bieten gegenüber den heute gebräuchlicheren Verschlüsselungs- und Signatursystemen, wie RSA [34] , ElGamal [6] und DSA [30], den Vorteil, daß die zu verwendenden Parameter bei vergleichbarer Sicherheit deutlich kleiner gewählt werden dürfen und deshalb eine effizientere Implementierung ermöglichen.

Typischerweise hat heute z.B. ein RSA-Modulus eine Länge von 768 oder gar 1024 Bit. Man erwartet, daß die Faktorisierung (siehe z.B. [2] und [5]) von 512 Bit Moduli in wenigen Jahren mehr oder minder 'problemlos' möglich sein wird. Auch das Berechnen diskreter Logarithmen in $GF(p)^*$ (siehe z.B. [10], [36]) wird für 512 Bit Primzahlen p in wenigen Jahren leicht möglich sein. In beiden Fällen arbeitet man mit dem sog. Zahlkörpersieb [20]. Der Algorithmus besitzt eine subexponentielle Laufzeit

$$O(\exp(1.9223(\log m)^{1/3}(\log \log m)^{2/3})$$

wobei m der Modulus n bzw. die Primzahl p ist. Deshalb schlägt das BSI in [3] für Signaturen mit einer Lebensdauer von sechs Jahren die Verwendung von mindestens 1024 Bit Moduli bzw. Primzahlen vor.

Dagegen ist heute *kein Algorithmus subexponentieller Laufzeit* bekannt, der diskrete Logarithmen in der Punktegruppe Elliptischer Kurven berechnet. Die besten Algorithmen hierfür (Shank's Baby-Step-Giant-Step, Pollard-Rho und Pohlig-Hellman, [25], S. 103 ff.) besitzen exponentielle Laufzeit. Deshalb schlägt das BSI für Signaturen die Verwendung Elliptischer Kurven über *GF(q)* vor, wobei *q* entweder eine Primzahl mit $\log_2 q \geq 160$ oder $q=2^m$ mit $m \geq$ 160 ist. Das bedeutet, daß mit einer 160-Bit Körperarithmetik gearbeitet werden kann, was eine vergleichsweise effiziente Implementierung ermöglicht. Die Arithmetik für Körper der Charakteristik 2 kann in Hardware sehr effizient implementiert werden. Eine reine Software-Implementierung dieser Arithmetik (siehe [21]) erreicht eine ähnliche Effizienz, wie sie von gewöhnlichen 'Langzahlarithmetiken' her bekannt ist durch den speicherplatzintensiven Einsatz von vorausberechneten Tabellen bestimmter Elemente, z.B. eines Teilkörpers, wie in [44] und [11]. Ob bei einer Software-Implementierung auf Chipkarten auf diese speicherplatzintensiven Tabellen verzichtet werden kann muß näher untersucht werden.

In dieser Arbeit wollen wir in Kapitel 2 die nötigen *Grundlagen Elliptischer Kurven* wiederholen. In Kapitel 3 werden wir grob auf verschiedene *Karten-Arten* eingehen, die möglicherweise für die Implementierung von Kryptosystemen auf Basis Elliptischer Kurven in Frage kommen. Schließlich wollen wir in Kapitel 4 *implementierungstechnische Aspekte* beleuchten und die Implementierung mit den heute weiter verbreiteten Verfahren RSA und DSA vergleichen.

2 Elliptische Kurven

In dieser Arbeit wollen wir uns auf Kurven über *GF(p)* beschränken. Für die Diskussion von Kurven über *GF(2ⁿ)* sei auf [23] verwiesen.

Eine *Elliptische Kurve E(K)* über einem Körper *K* (*char(K)* $\neq$ 2,3) ist definiert als die Menge aller Punkte *P=(x,y)* der Ebene *K* $\times$ *K*, die die Gleichung

$$F(x,y): y^2 = x^3 + ax + b \tag{1}$$

erfüllen und kein Punkt ein singulärer Punkt ist, d.h.

$$4a^3 + 27b^2 \neq 0 \tag{2}$$

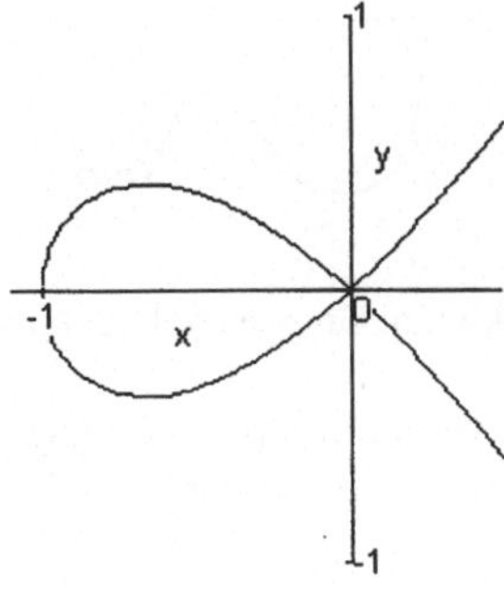
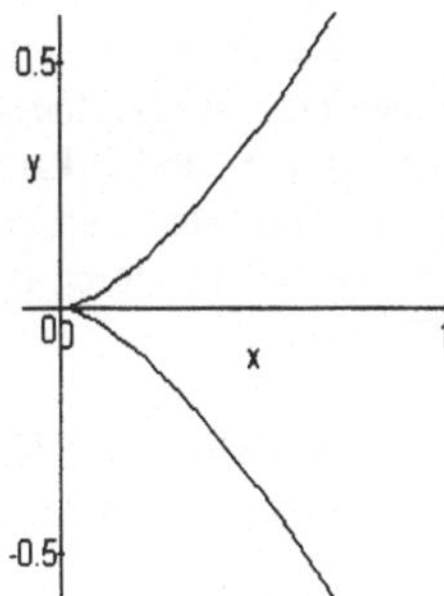

Abb. 1: $E_1{:}y^2{=}x^3{+}x^2$ **Abb. 2**: $E_2{:}y^2{=}x^3$

Die Forderung (2) stellt übrigens sicher, daß keine Knoten (Abb. 1[1]) oder Spitzen (Abb. 2) auftreten. Es läßt sich nämlich zeigen (z.B. [13], S. 77ff), daß die Punktegruppe einer singulären Kurve über einem endlichen Körper *GF(p)* bei einem Knoten isomorph zur multiplikativen Gruppe *GF(p)**, bzw. bei einer Spitze sogar isomorph zur additiven (!) Gruppe *GF(p)*[+] des zugrundeliegenden Körpers ist. Zusätzlich zu den so definierten Punkten nimmt man den Punkt $O = (\infty,\infty)$ im Unendlichen zur Elliptischen Kurve hinzu.

Beispiele für Elliptische Kurven über den reellen Zahlen sehen wir im folgenden (Abb. 3 und Abb. 4). Der Punkt $O = (\infty,\infty)$ im Unendlichen wäre hier anschaulich die unendliche Verlängerung der rechten Kurvenäste.

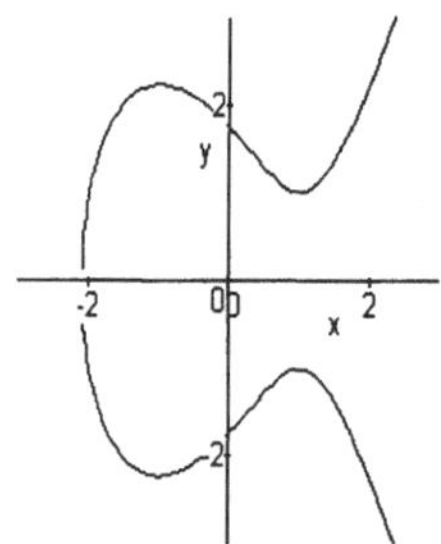

Abb. 3: $E_3: y^2 = x^3 - 3x + 3$

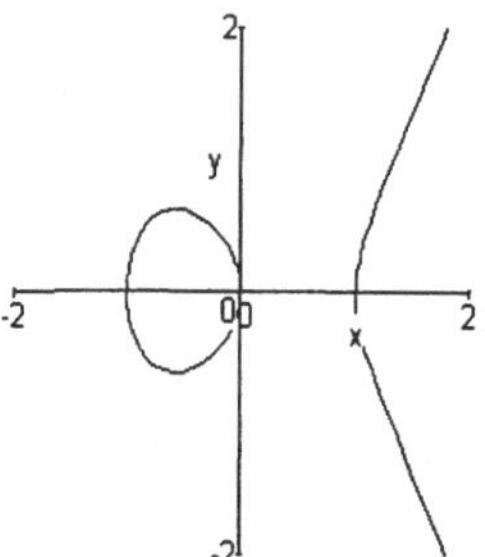

Abb. 4: $E_4: y^2 = x^3 - x$

Da sich auf diesen Punkten eine Gruppenverknüpfung definieren läßt, sind Elliptischen Kurven für die praktische Kryptographie interessant. Die als 'Sehnen-Tangenten-Regel' bekannte Gruppenverknüpfung sei an Abb. 5 veranschaulicht.

$P+Q$ berechnet sich als der an der *x*-Achse gespiegelte weitere Schnittpunkt *PQ* der Gerade *g* (Sehne) durch *P* und *Q*. Eine Punktverdopplung *2P* wird durch anlegen der Tangente an *P* erreicht. Das neutrale Element der Gruppe ist der Punkt $O=(\infty,\infty)$ im Unendlichen. Daraus ist klar, daß Invertieren einer Spiegelung an der *x*-Achse entspricht. Der Nachweis, daß es sich in der Tat um eine Gruppe handelt kann rein geometrisch geführt werden. Die folgenden Formeln für die Gruppenverknüpfung entstehen durch Berechnung der Steigung λ der Geraden *g*.

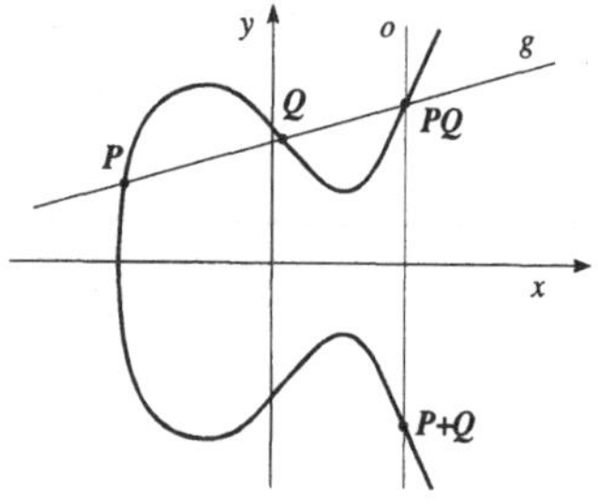

Abb. 5: Gruppenverknüpfung +

Es sei $E: y^2 = x^3 + ax + b$ eine Elliptische Kurve über einem Körper mit *char(K)* $\neq$ *2,3* und $P,Q,R \in E$, $R(x_3,y_3) := P(x_1,y_1) + Q(x_2,y_2)$, $P,Q \neq O$ und $P \neq -Q$. Dann erhalten wir für

$$x_3 = \lambda^2 - x_1 - x_2$$

---

[1] Aus Gründen der Anschaulichkeit sind alle abgebildeten Kurven über den reellen Zahlen definiert.

$$y_3 = -y_1 + \lambda(x_1 - x_3)$$

$$\text{wobei} \quad \lambda = \begin{cases} \dfrac{y_2 - y_1}{x_2 - x_1} \ \textit{für} P \neq Q \\[2mm] \dfrac{3x_1^{\,2} + a}{2y_1} \ \textit{für} P = Q \end{cases} \qquad (3)$$

Daraus sehen wir, daß die Addition zweier Punkte mit *drei Multiplikationen* und *einer Inversion (mod p)* möglich ist. Das Verdoppeln eines Punktes benötigt *vier Multiplikationen* und wiederum *eine Inversion (mod p)*. Hier sei angemerkt, daß die Addition *(mod p)* und die Multiplikation mit kleinen Konstanten vernachlässigt wurde.

Für das Inverse $-P$ des Punktes $P(x_1, y_1)$ erhält man einfach durch Negieren der y-Koordinate; d.h. -P(x_1,-y_1). Addition des neutralen Elementes O ändert nichts; d.h. $P+O = P$.

Diese Formeln sind sowohl für Kurven über den reellen Zahlen, als auch für Kurven über großen Primkörpern gültig. Der 'Übergang' von der reellen Zahlen zu *GF(p)* und die Rolle des Punktes im Unendlichen sei an folgender Abb. 6 veranschaulicht.

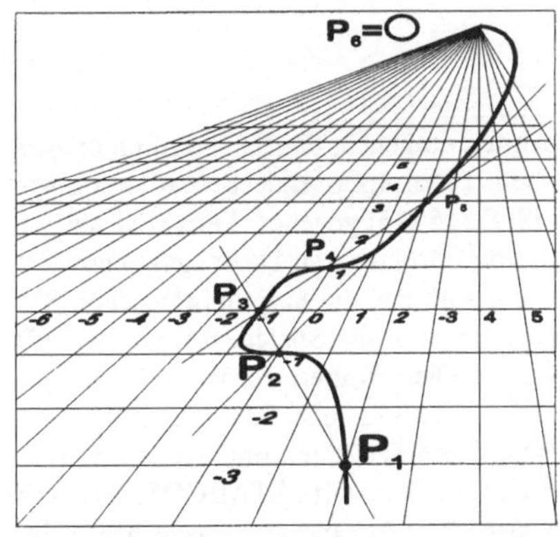

Abb. 6: *E5: y2=x3+1* in der Ebene

In dieser Abbildung sehen wir eine Elliptische Kurve $E_5\!:y^2=x^3+1$ über den reellen Zahlen (durchgezogene Linie). Das 'nette' an dieser Kurve ist nun, daß die ganzzahligen Kurvenpunkte genau die Punkte der Kurve über dem, für kryptographische Anwendungen zugegebenermaßen etwas kleingeratenen, endlichen Körper *GF(5)* sind. Die Kurve über *GF(5)* hat die sechs Punkte $P_1(2,-3)$, $P_2(0,-1)$, $P_3(-1,0)$, $P_4(0,1)$, $P_5(2,3)$ und $P_6(\infty,\infty)=O$. Es ist sofort zu sehen, daß $P_1=-P_5$, $P_2=-P_4$ und $P_3=-P_3$. Außerdem ist die Rolle des neutralen Elementes O gut ersichtlich. Schließlich sieht man, daß $2P_1= P_2$, $P_2+P_1=P_3$, $P_3+P_1=P_4$, $P_4+P_1=P_5$ und $P_5+P_1=P_6=O$. P_1 hat also Ordnung 5 und ist ein primitives Element der Punktegruppe.

Für den allgemeinen Fall einer Elliptischen Kurve über *GF(p)* ist mit dem *Satz von Hasse* (siehe z.B. [13], S. 243) bekannt, daß für die Anzahl der Punkte *N* gilt

$$N = p + 1 - t, \quad \textit{mit} \quad |t| \leq 2\sqrt{p}. \qquad (4)$$

In der Arbeit von Waterhouse [43] wurde Hasses Resultat etwas präzisiert. Er zeigt u.a., daß alle (natürlich ganzzahligen) Werte t, $ggT(t,p)=1$ in diesem Intervall auch auftreten. Außerdem gibt es sog. supersingulären Kurven mit Spur $t=0$. Eine Elliptische Kurve ist nach [40], Theorem 4.1, Seite 140 genau dann supersingulär, wenn ihre definierende Gleichung die Form

$$y^2 = x^3 + ax, \quad \textit{mit} \quad p \equiv 3 \bmod 4 \qquad (5)$$

oder

$$y^2 = x^3 + b, \quad mit \quad p \equiv 2 \bmod 4 \tag{6}$$

hat. Die Tatsache, daß die Anzahl der Punkte ($N=p+1$) bekannt ist, verschaffte den supersingulären Elliptischen Kurven zumindest in den frühen Jahren ihre Beliebtheit. Allerdings wurde in [24] gezeigt, daß man Elliptische Kurven über $GF(p)$ mit Hilfe der sog. Weil-Paarung in einen zugehörigen Erweiterungskörper $GF(p^k)$ einbetten kann. Ausgenutzt wird bei dieser Einbettung nur, daß die Punkteanzahl der Gruppe N die Gruppenordnung von $GF(p^k)$ teilt, also $N | (p^k -1)$. Die Berechnung diskreter Logarithmen erfolgt nach dieser Transformation nicht in der Punktegruppe sondern in der multiplikativen Gruppe $GF(p^k)^*$ mit subexponentiellen Algorithmen, wie dem Zahlkörpersieb [36]. Nun ist dieser Angriff nicht das Ende aller Kryptosysteme auf Basis Elliptischer Kurven, da der Grad der Körpererweiterung k im allgemeinen sehr groß ist. Allerdings ist klar, daß supersinguläre Kurven über $GF(p)$ nicht mehr verwendet werden dürfen, da diese Kurven bereits in den Erweiterungskörper $GF(p^2)$ eingebettet werden können, da $p^2-1=(p+1)(p-1)$.

Kürzlich wurde von N. Smart gezeigt, daß auch Kurven mit Spur $t=1$ nicht mehr verwendet werden dürfen, da bei diesen Kurven die Gruppenordnung gleich p ist und diskrete Logarithmen mittels p-adischer Logarithmen in polynomieller Zeit berechnet werden können. Welche Kurven für kryptographische Zwecke geeignet sind und wie man solche erhält wurde in [28] ausführlich diskutiert.

3 Chipkarten

In diesem Kapitel wollen wir etwas näher auf Chipkarten eingehen, die für die Implementierung Elliptischer Kurven potentiell in Frage kommen. Eine umfassende Spezifikation von Chipkarten mit Kontakten ist mit der Normenreihe ISO7816 [15] gegeben. Dort sind z.B. die elektrischen und mechanischen Eigenschaften, sowie die Struktur von Datenelementen festgelegt. Da Ziel der Implementierung in jedem Fall sein soll, daß die kryptographischen Protokolle selbst auf der Karte ablaufen, kommen natürlich keine sog. Speicherkarten, wie z.B. Karten mit Chips der Siemens SLE44XY-Reihe in Frage. Diese Karten werden z.B. für Telefon- und Krankenversicherungskarten eingesetzt. Die grundsätzliche Spezifizierung eines Chipkartenbetriebsystems ist in ISO7816-4 geschehen. Eines der ersten praktischen Betriebsysteme war das von Giesecke & Devrient und der GMD entwickelte STARCOS, das recht ausführlich z.B. in [33] beschrieben ist. Allerdings sieht STARCOS aus historischen Gründen (noch) nicht die Verwendung von Public-Key-Algorithmen vor. Ein anderes Chipkarten-Betriebssystem, das auch die Verwendung von Public-Key-Algorithmen unterstützt ist z.B. das von der Telesec entwickelte TCOS Betriebssystem [42]. Als Industriestandard für Chipkartenleser darf das B1-Konzept [41] angesehen werden. In dieser Arbeit wollen wir uns nun auf die verschieden Prozessortypen für eine mögliche Implementierung Elliptischer Kurven konzentrieren. Die meisten verfügbaren Prozessoren für Chipkarten (siehe [29]) basieren im Kern entweder auf den CISC-Intel-Chip 8051 oder den RISC-Chip 68HCO5 der Firma Motorola. Wir teilen die verfügbaren Prozessor-Karten in drei Kategorien ein: *Chipkarten mit kryptographischem Coprozessor, Chipkarten ohne* einen solchen *Coprozessor* und *Java-Interpreter-Karten*.

3.1 Chipkarten mit Coprozessor

Bei diesen Karten steht zusätzlich zur normalen CPU ein arithmetischer Coprozessor zu Verfügung, der die Berechnung der modularen Multiplikation bzw. Exponentiation übernimmt.

Da selbst die Implementierung von 512-Bit RSA nicht mit akzeptabler Performance ohne diese Coprozessoren möglich ist, sind alle heute verfügbare RSA-Karten mit einem solchen Coprozessor ausgestattet. In der folgenden Tabelle wollen wir die wichtigsten Daten der gängigsten Prozessoren auflisten. Für eine umfassendere Darstellung sei an [29] und Herstellerdatenblätter, wie [39], [27] oder [31] verwiesen.

Hersteller	Kern	Name	ROM in kByte	RAM in Byte	EEPROM in kByte	Max. Takt in MHz
Motorola	68HC05	SC29	13	512	4	5
Motorola	68HC05	SC49	20	896	4	5
Motorola	68HC05	SC50	20	896	8	5
Philips	8051	P83C852	6	256	2	6
Philips	8051	P83C855	20	512	2	8
Philips	8051	P83C858	20	640	8	8
Philips	8051	P83W858	20	672	8	8
Siemens	8051	SLE44C200	9	256	9	5
Siemens	8051	SL44CR80S	14	706	8	5
SGS	68HC05	ST16CF54	16	352	16	5
SGS	68HC05	ST16KL74	20	608	20	5

Tabelle 1:Karten mit Coprozessor

3.2 Chipkarten ohne Coprozessor

Da die Karten mit ohne Coprozessoren im allgemeinen etwa für den halben Preis zu haben sind, wie Karten mit Coprozessoren, wäre eine mögliche Implementierung von Elliptischen Kurven auf gewöhnlichen Karten natürlich besonders interessant. Auch hier sei für eine umfassendere Darstellung auf [29], [27], [31] und [39] verwiesen.

Hersteller	Kern	Name	ROM in kByte	RAM in Byte	EEPROM in kByte	Max. Takt in MHz
Motorola	68HC05	SC01	1,6	39	1	4
Motorola	68HC05	SC03	2	52	2	4
Motorola	68HC05	SC11	6	128	8	4
Motorola	68HC05	SC21	6	128	3	5
Motorola	68HC05	SC24	3	128	1	5
Motorola	68HC05	SC26	6	224	1	5
Motorola	68HC05	SC27	16	240	3	5
Motorola	68HC05	SC28	12,8	240	8	5
SGS	68HC05	ST1821	2	44	1	5
SGS	8048	ST1834	4	76	3	5

SGS	8048	ST16612	6	224	2	5
SGS	68HC05	ST16F48	6	224	3	5
SGS	68HC05	ST16CF54	16	512	8	5
SGS	68HC05	ST16301	3	160	1	5
Philips	8051	P83C864	16	256	4	8
Philips	8051	P83W864	20	256	4	8
Philips	8051	P83C868	20	384	8	8
Philips	8051	P83W868	20	384	8	8
Philips	8051	P83W8616	20	384	16	8
Siemens	8051	SLE44C10S	7	256	1	5
Siemens	8051	SL44C20S	7	256	2	5
Siemens	8051	SLE44C40S	7	256	4	5
Siemens	8051	SL44C42S	15	256	4	5
Siemens	8051	SLE44C80S	15	256	8	5
Siemens	8051	SLE44C160S	15	256	16	5‘

Tabelle 2: Karten ohne Coprozessor

3.3 Java™-Karten

Im Gegensatz zu den obengenannten Karten, die mit dem jeweiligen 8051- bzw. 68HC05 Assemblercode programmiert werden müssen, sind seit April 1997 erste Java-Interpreter-Karten verfügbar [38]. Das Konzept der Java-Karte ist in Abb. 7 ersichtlich. Der Unterschied zu konventionellen Karten ist die Interpreter-Schicht. Java ist bekanntlich eine von Sun Inc. entwickelte objektorientierte Programmiersprache, die einen großen Teil ihrer Popularität dem Internet verdankt. Dort können sog. Java-Applets durch den Web-Browser heruntergeladen und auf dem lokalen Rechner ausgeführt werden. Den dadurch entstehenden Sicherheits-Risiken wurden durch das Design der Java-API, d.h. dem stark eingeschränkten ‚Java-Sandkasten‘, bereits Rechnung getragen. Im Fall der Java-Karte können dann diese Applets mit entsprechenden Sicherheitsvorkehrungen auch auf die Karte geladen und dort ausgeführt werden. Für Entwickler ist Java auch besonders interessant, da sich durch den objektorientierten Ansatz der Quellcode gut wiederverwenden

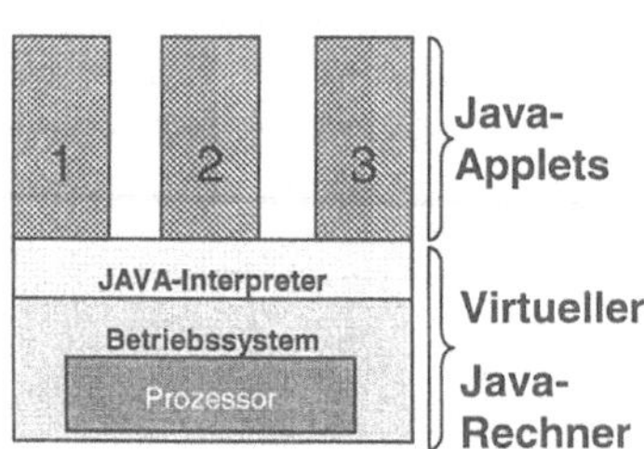

Abb. 7: Konzept der Java-Karte

läßt. Für unsere Zwecke würde das bedeuten, daß eine Java-Implementierung Elliptischer Kurven auf der Karte, sowie z.B. auf einem PC ablaufen kann. Die oben gemachte Unterscheidung zwischen den Prozessor-Typen wäre nicht nötig, da die Java-Implementierung prozessorunabhängig ist. Hier sind allerdings ein paar Wehrmutstropfen angebracht:

- Es ist bisher nur eine Java-Karte verfügbar - die *'Cyberflex'*-Karte der Firma Schlumberger [38]. Diese Karte besitzt einen 8-Bit-Motorola 68HC05-SC49 Prozessor mit Coprozessor. Allerdings wird dieser Coprozessor *nicht* genutzt. In der aktuellen Version der Java-Card-API wird bisher nur 8-Bit Arithmetik unterstützt.

- Es gibt (noch) keine Langzahl-Arithmetik, die als Applet genutzt werden kann

- Durch das Interpreterkonzept sind arithmetische Operationen etwa *100 mal langsamer*, als die gleichen Operationen auf dem gleichen Chip in Assembler.

Für die Zukunft, d.h. nach Erweiterung der Java-Card-API und bei Verfügbarkeit entsprechend leistungsstarker (16 oder gar 32-Bit) Chips, jedoch dürfte das Konzept der Java-Karte auch für die Implementierung von rechenintensiven kryptographischen Algorithmen sehr interessant werden.

4 Implementierung Elliptischer Kurven

In diesem Abschnitt wollen wir etwas näher auf die implementierungstechnischen Aspekte Elliptischer Kurven auf Chipkarten eingehen. Grundsätzlich existieren drei Möglichkeiten die Implementierung Elliptischer Kurven zu optimieren:

1. **Optimierung der Körperarithmetik**

2. **Varianten für Punktmultiplikation**

Die zeitaufwendigste Operation in einem kryptographischem DL-Protokoll ist die Berechnung eines Punktvielfachen nP. Ein Überblick über die verschieden Verfahren findet man z.B. in [12]. Die naive Methode diese Punktmultiplikation durchzuführen ist mit dem sog. Square&Multiply-Verfahren (z.B. [17]) gegeben. Der Aufwand für die Berechnung von nP, also die Länge der Additionskette beträgt

$$l(n) = \lambda(n) + \gamma(n) - 1 \tag{7}$$

Gruppenoperationen, wobei $\lambda(n) = \lfloor \log_2 n \rfloor$ die Länge und $\gamma(n)$ das Gewicht des Faktors. Unterstellt man Gleichverteilung für die Faktoren, so benötigen wir für 160 Bit Faktoren durchschnittlich etwa *239 Gruppenoperationen*. Allerdings kann dieser Aufwand durch eine geschicktere Auswertung des Faktors reduziert werden. Während der Aufwand für eine *m*-Bit-Fenstermethode (im Mittel, bei Gleichverteilung der Faktoren) nur etwas geringer[2] ist

$$l_m(n) = 2^{m-1} + \frac{2^m - 1}{2^m} \left\lceil \frac{\lfloor \log_2 n \rfloor}{m} \right\rceil + \lambda(n), \tag{8}$$

läßt sich der Aufwand durch Precomputation Varianten deutlich reduzieren. Bei der für 160 Bit Faktoren optimalen Fenstergröße *m*=4 entspricht das durchschnittlich etwa *206 Gruppenoperationen*. Da der verwendete permanente Speicherplatz im EEPROM der Karte u.U. stark eingeschränkt ist, kommen keine 'Intensiv'-Precomputation-Varianten für die 'Punktmultiplikation' wie z.B. [1] zum Einsatz. Allerdings stellen Precomputation, bei denen lediglich ein weiterer Kurvenpunkt in der Karte gespeichert werden muß einen guten Kompromiß zwischen Laufzeit und Speicherplatzbedarf dar. Aus [12] wissen wir, daß mit der *'(2 × 1) Additionskette mit Online Precomputation'* eine sehr effiziente Variante gegeben ist, bei der lediglich ein weiterer Kurvenpunkt im EEPROM der Karte gespeichert werden muß. Diese Methode verbindet die permanente Precomputation aus [1] und [22] mit der Online-

---

[2] Der Aufwand (8) ist für die statische Auswertung eines *m*-Bit-Fensters und läßt sich durch eine dynamische Auswertung weiter geringfügig verbessern.

Precomputation der Festervariante aus [17]. Der Aufwand für dieses Verfahren beträgt (wieder bei Gleichverteilung der möglichen Faktoren) durchschnittlich

$$l_{(2\times1)}(x) = \left\lceil \frac{\lambda(x)}{2} \right\rceil + \frac{2^4 - 1}{2^4} \left\lceil \frac{\lambda(x)}{4} \right\rceil + 10 \tag{9}$$

Gruppenoperationen. In unserem Fall sind das durchschnittlich *127.5 Gruppenoperationen.*

3. Punktdarstellung (affin - projektiv)

Die Frage, welche Punktdarstellung für die jeweilige Punktmultiplikationsvariante zu bevorzugen ist, wurde in [12] ausgiebig erörtert. Da die affine Punktdarstellung in jedem Fall die Implementierung des Erweiterten Euklidischen Algorithmus auf der Karte voraussetzt und selbst bei relativ effizienter Implementierung des EEA der projektiven Punktdarstellung unterlegen ist, wird die projektive Punktdarstellung verwendet.

Die *projektive Ebene* $\mathbf{P}^2(K)$ über einem Körper K ist die Menge aller Äquivalenzklassen der Tripel (A_1, A_2, A_3), $A_i \in K$ bezüglich der Äquivalenzrelation $(X_1, X_2, X_3) \sim (Y_1, Y_2, Y_3) :\Leftrightarrow \exists \lambda \in K \setminus \{0\}: X_i = \lambda Y_i$, ohne der Äquivalenzklasse des Tripels $(0,0,0)$. Der Wechsel von affiner zu projektiver Darstellung geschieht durch Anhängen der dritten Koordinate $A_3 = 1$. Der umgekehrte Weg kann durch eine Inversion und zwei Multiplikationen erreicht werden. Der Punkt im Unendlichen, was in der projektiven Ebene dem Horizont entspricht hat übrigens die Koordinaten $O(0,1,0)$.

Das projektive Pendant zu (1), d.h. die definierende Gleichung der elliptischen Kurve in der projektiven Ebene ist nun

$$E(X,Y,Z): Y^2 Z = X^3 + aXZ^2 + bZ^3 \tag{10}$$

Es sei wieder $P(X_1, Y_1, Z)$, $Q(X_2, Y_2, Z)$ und $R(X_3, Y_3, Z_3)$, $P,Q,R \in E$, $P,Q \neq O(0,1,0)$ und $P \neq -Q$. Dann ist $R := P+Q$

für $P \neq Q$

$$\begin{aligned}
X_3 &= -su \\
Y_3 &= t(u + s^2 X_1) - s^3 Y_1 \\
Z_3 &= s^3 Z \\
&\textit{wobei} \\
u &= s^2(X_1 + X_2) - t^2 Z \\
&\textit{und} \\
s &= X_2 - X_1, \quad t = Y_2 - Y_1
\end{aligned} \tag{11}$$

und für $P = Q$

$$X_3 = 2sh$$
$$Y_3 = w(4f - h) - 8e^2$$
$$Z_3 = 8s^3$$
$$mit$$
$$s = Y_1 Z$$
$$w = 3X_1^2 + aZ^2$$
$$e = Y_1 s$$
$$f = X_1 e$$
$$h = w^2 - 8f$$

$$(12)$$

Mit diesen beiden Formeln aus [18] benötigen wir für die Punktaddition mit (11) zehn Körpermultiplikationen, falls die beiden Punkte bereits die gleichen Z-Koordinaten besitzen. Im allgemeinen muß jedoch vor Anwendung von (11) kreuzweise multipliziert werden (fünf Multiplikationen), womit wir insgesamt bei *15 Körpermultiplikationen* angelangt wären. Für das Verdoppeln eines Punktes benötigen wir mit (12) *zehn Körpermultiplikationen.*[3]

4.1 Implementierung auf Karten mit Coprozessor

Die naheliegendste Plattform für die Implementierung Elliptischer Kurven auf Smartcards sind Chips mit integrierten kryptographischen Coprozessoren. Diese Coprozessoren sind je nach konkreter Ausprägung in der Lage Modulomultiplikationen und -exponentiationen mit einer Modul-Länge bis zu 1024 Bit auszuführen. Das bedeutet, daß die Implementierung Elliptischer Kurven über *GF(p)* mit 160 Bit p problemlos möglich ist, da die Modulo-Arithmetik des Coprozessors verwendet werden kann. Für die Kurvenarithmetik müssten lediglich die folgenden Routinen (allerdings im jeweiligen Assembler) implementiert werden:

- PADD(P_1,P_2) Punktaddition mit (11)

- PDBL(P) Punktverdopplung mit (12)

- PSUBNORM(P_1,P_2) Punktsubnormalisierung, d.h. kreuzweise Multiplikation
  $P_1(x_1,y_1,z_1) \rightarrow P_1(x_1 x_2, y_1 y_2, z_1 z_2)$
  $P_2(x_2,y_2,z_2) \rightarrow P_2(x_2 x_1, y_2 y_1, z_2 z_1)$

- PNORM(P_1) Punktnormalisierung, d.h. Transformation in affine Koordinaten, durch eine Inversion und zwei Multiplikationen

- PEQUAL(P_1,P_2) Gleichheitstest

- PMUL(P_1,P_2) Punktmultiplikation mit einer der oben geschilderten Varianten

Zur Performanceabschätzung und dem Vergleich der EC-DSA-Signatur mit DSA- und RSA-Signaturen wollen wir davon ausgehen, daß die RSA- und DSA Exponentiation mit einer 5-Bit-Fenstermethode durchgeführt wird, das Gewicht des Exponenten die Hälfte der Bitlänge beträgt und daß die Modulo-Multiplikation quadratische Komplexität besitzt. Das bedeutet, daß (nach (8)) bei der gewöhnlichen Signatur (ohne CRT) durchschnittlich 1239 modulare

---

[3] Es sei noch einmal angemerkt, daß wir Additionen, Subtraktionen sowie die Multiplikation mit kleinen Konstanten (z.B. a in den Kurvengleichungen (1) und (10)) bei dieser Betrachtung vernachlässigen können.

1024-Bit Multiplikationen nötig sind. Bei Anwendung des chinesischen Restsatzes (CRT) [32] benötigen wir rund 2·628+2=1258 modulare 512 Bit Multiplikationen. Bei quadratischer Komplexität der Modulomultiplikation entspricht das etwa $1258 \cdot (512/1024)^2 \approx 315$ modularer 1024-Bit Multiplikationen. Im Vergleich dazu benötigen wir für eine DSA-Signatur nur etwa 207+20=227 1024 Bit Multiplikationen, wobei der Aufwand für die Berechnung von $s=(k^{-1}(h(m)+xr)$ mod q mit 20 1024 Bit Multiplikationen veranschlagt wurde. Für eine EC-DSA Signatur benötigen wir mit der Speicherung eines zusätzlichen Punktes nach (9) durchschnittlich 81 Punktverdoppelungen und rund 47 Punktmultiplikationen. Das heißt insgesamt benötigen wir dazu 81·10+47·15+22=1537 160 Bit Multiplikationen. Auch hier wurde die Transformation in affine Koordinaten (1 Inversion und 2 Multiplikationen) mit 20+2=22 Multiplikationen 'verbucht'. Wenn wir quadratische Komplexität der Modulo-Multiplikation unterstellen, so entspricht das nur etwa $1537 \cdot (160/1024)^2 \approx 38$ modularer 1024 Bit Multiplikationen. Die Ergebnisse sind in Tabelle 3 zusammengefaßt. Wir unterstellen, daß im EEPROM der Karte die folgenden Parameter gespeichert sind:

- RSA (ohne CRT) d,n

- RSA (mit CRT) $p,q,d_p,d_q,invq$

- DSA p,q,g,x

- EC-DSA $a,b,p,q,x,P_1(x_1,y_1),\ P_2(x_2,y_2)$

Außerdem wird hier unterstellt, daß der Kurvenparameter a und x_1 als 'Shortinteger' gewählt sind und deshalb nur 4 Byte Speicher benötigen. Die letzten beiden Spalten sind die zu erwartenden Laufzeiten auf den Prozessoren Philips 83C852 und Siemens 44C200 (siehe [29]). Die mit * gekennzeichneten Werte sind extrapoliert.

Algorithmus	Modul	EEPROM für Parameter (Byte)	mod. Mult. # (Bit)	$\approx$ mod. Mult (1024 Bit)	Philips 83C852 in ms	Siemens 44C200 in ms
RSA ohne CRT	1024	256	1239 (1024)	1239	8493*	640*
RSA mit CRT	1024	448	1258 (512)	315	2160*	163*
RSA ohne CRT	768	192	934 (768)	525	3600	271
RSA mit CRT	768	336	952 (384)	134	919*	69*
DSA-Sign	1024	296	227 (1024)	227	1557*	117*
DSA-Sign	768	232	227 (768)	127	817*	65*
DSA-Verify	1024	s.o.	22 (160) + 414 (1024)	415	2845*	214*
DSA-Verify	768	s.o.	22 (160) + 414 (768)	233	1598*	120*
EC-DSA-Sign	160	145	1537 (160)	38	260*	19*
EC-DSA-Verify	160	s.o.	3864 (160)	94	645*	49*

Tabelle 3: Signatur-Algorithmen mit Coprozessor im Vergleich

An dieser Stelle sei noch einmal darauf hingewiesen, daß diese Abschätzungen äußerst grob sind. So wurde beispielsweise der Aufwand für die Berechnung des Hashwertes einfach vernachlässigt. Außerdem wurde unterstellt, daß der Aufwand für die Modulo-Multiplikation quadratische Komplexität besitzt, was für bestimmte Chips nicht genau zutrifft. Die Ergebnisse dieser Tabelle können daher nur 'ein Gefühl' für die zu erwartende Laufzeit vermitteln. Allerdings ist bereits deutlich ersichtlich, daß die Implementierung Elliptischer Kurven selbst auf (leicht angestaubten) Chips, wie dem Philips 83C852[4] mit *sehr guter Performance* möglich ist.

Zu beachten ist schließlich ein im Vergleich zu RSA ein etwas *erhöhter (EEP)ROM-Bedarf* für die Kurvenarithmetik und ggf. die zusätzliche Implementierung des EEA zur Inversion mod p.

4.2 Implementierung auf Karten ohne Coprozessor

Da die Leistungsfähigkeit des Coprozessors (z.B. modulare 1024 Bit-Multiplikation) bei der Implementierung ohnehin nicht völlig ausgenutzt wird, liegt der Gedanke nahe, ganz auf diesen zu verzichten. Karten ohne Coprozessor sind etwa für den *halben Preis* zu haben, als vergleichbare Karten mit einer Kryptoeinheit. Das bedeutet aber, daß neben der Kurvenarithmetik, wie im vorherigen Kapitel erläutert, eine möglichst effiziente Körperarithmetik im entsprechenden 8051- bzw. 68HC05-Assembler zu implementieren ist, was wiederum etwas *mehr (EEP)ROM* beansprucht. Für diese Karten gibt es verschiedene Möglichkeiten der Implementierung, die jeweils weiterer Forschung bedürfen:

- *GF(p)-Artihmetik*, wobei $p>2^{160}$

- *Polynomarithmetik über GF(2)*, d.h. *GF(2^m)*-Arithmetik, wobei $m>160$

- *Polynomarithmetik über GF(p)*, wobei $p<256$, d.h. *GF(p^k)*-Arithmetik mit $k>20$

Daß eine solche Implementierung möglich ist, hat die Firma Certicom [4] in Zusammenarbeit mit Motorola und Schlumberger bereits gezeigt. Es wurde ein Signatursystem auf Basis Elliptischer Kurven auf der sog. *Multiflex-Karte ohne Koprozessor implementiert*. Der Chip auf dieser Karte ist ein 68HC05-SC28. Aus Tabelle 2 sehen wir, daß 12,8 kByte ROM, 240 Byte RAM und 8 kByte EEPROM zu verfügung stehen. Davon wurden ca. 90 Byte RAM und rund 4 kByte EEPROM für die Implementierung des Signatursystems benötigt. Die Erstellung der Signatur dauert ca. 600 ms, was für durchaus vergleichbar zu entsprechenden RSA-Implementierungen auf Karten mit Coprozessor ist.

4.3 Implementierung auf Java™-Karten

Da durch das Java-Interpreter-Konzept eine mögliche Implementierung etwa 100 mal langsamer ist, als eine Assembler-Programmierung auf dem gleichen Chip, ist die Implementierung Elliptischer Kurven auf diesen Karten *nicht mit akzeptabler Performance möglich*. Mit einer Änderung der Java-Card-API, d.h. die Möglichkeit den vorhandenen Co-Prozessor durch Java-Kommandos anzusteuern, oder leistungsfähigeren 16- oder 32-Bit-Chips könnte diese Möglichkeit in einigen Jahren sehrwohl interessant werden.

---

[4] Z.B. 1024 Bit RSA-Operationen sind mit diesem Chip nicht nur zu langsam, sondern durch das relativ kleine (EEP)ROM (2kB bzw. 6kB) sogar unmöglich.

5 Zusammenfassung

In dieser Arbeit wiederholten wir recht ausführlich die nötigen Grundlagen Elliptischer Kurven über Primkörpern und diskutierten Ansätze die Implementierung möglichst effizient zu gestalten. Außerdem gaben wir einen Überblick über die zur Zeit gebräuchlichen Prozessor- und Kartentypen, die potentiell für die Implentierung Elliptischer Kurven in Frage kommen und schätzten grob die zu erwartende Laufzeit auf Karten mit kryptographischem Coprozessor ab. Dies zeigte, daß selbst Chips, die heute (z.B. für RSA) nicht mehr tauglich sind, als Plattform für die Implementierung Elliptischer Kurven dienen können. Das bedeutet, daß die gleiche Sicherheit mit *preisgünstigeren Chipkarten* erreicht werden kann. Es ist sogar möglich ganz *auf den Coprozessor verzichten*. Welche der möglichen Alternativen zu bevorzugen ist muß weiter untersucht werden. Ein weiterer offener Punkt für die Zukunft ist die Integration Elliptischer Kurven in Anwendungsstandards (wie z.B. HBCI). Der Schritt des BSI, Elliptische Kurven für Signaturen im Sinne des Signatur-Gesetzes vorzuschlagen sollte nicht der letzte in diese Richtung bleiben.

Literatur

[1]   E. Brickell, D. Gordon, K. McCurley, D. Wilson: „Fast Exponentiation with Precomputation", Proceedings of EUROCRYPT '92, LNCS 658, Springer Verlag, Berlin 1993, SS. 200-207

[2]   J. Buchmann, J. Loho, J. Zayer: „An implementation of the general number field sieve", Advances in Cryptology Crypto (1993), Lecture Notes in Computer Science, 773, SS. 159-165

[3]   Bundesamt für Sicherheit in der Informationstechnik (BSI): "Maßnahmenkatalog zur digitalen Signatur, Teil 6.1. - KryptoalgorithmenV", 1997, via http://www.bsi.bund.de/aktuell/index.htm

[4]   Certicom: „Motorola and Certicom Demonstrate Elliptic Curve Digital Signatures on Smart Card", Mai 1997, via http://www.certicom.com/press/97/may2097.htm

[5]   B. Dodson and A.K. Lenstra: "NFS with four large primes: An explosive experiment", Proceedings of Crypto '95, Springer-Verlag, 1995, SS. 372-385

[6]   T. ElGamal: „A Public Key Cryptosystem and a Signature Scheme based on discrete Logarithms", Proceedings of CRYPTO '84, Springer, Berlin 1985, SS. 10-18

[7]   D. Fox, A. Röhm: "Effiziente Digitale Signatursysteme auf der Basis elliptischer Kurven", Tagungsband "Digitale Signaturen", DuD Fachbeiträge, Vieweg, 1996, SS. 201-220

[8]   W. Fulton: "Algebraic Curves", Benjamin, 1969

[9]   S. Goldwasser, J. Kilian: „Almost all primes can be quickly certified", Proceedings of the 18th Annual ACM Symposium on Theory of Computing, 1986, SS. 316-329

[10] D. Gordon: "Discrete Logarithms in GF(p) Using the Number Field Sieve", Siam Journal on Discrete Mathematics 6, 1993, SS. 124-138

[11] J. Guajardo, C. Paar: "Efficient Algorithms for Elliptic Curve Cryptosystems", erscheint in den Proceedings der CRYPTO '97, 1997

[12] D. Hühnlein: "Effiziente Exponentiation und optimale Punktdarstellung für Signatursysteme auf Basis elliptischer Kurven", Tagungsband "Digitale Signaturen",

DuD Fachbeiträge, Vieweg, 1996, SS. 221-235

[13]   D. Husemöller: „Elliptic Curves", Graduate Texts in Mathematic - 111, Springer Verlag, Berlin, 1986, ISBN 3-540-96371-5

[14]   IEEE: "IEEE P1363 Working Draft", z.B. über
ftp://stdsbbs.ieee.org/pub/p1363/predrafts

[15]   ISO 7816: "Identification cards - Integrated circuit(s) card with contacts",
"Part 1: Physical characteristics", 1987
"Part 2: Dimensions and location of the contacts", 1988
"Part 3: Electronic signals and transmission protocol", 1989
"Part 3-Amd 1: Amendment 1: Protocol type T=1, asynchronous half duplex block transmission protocoll", 1992
"Part 3-Amd 2: Amendment 2: Protocol type selection", 1994
"Part 4: Inter-industry commands for interchange ", 1995
"Part 5: Numbering system and registration procedure for application identifiers", 1994
"Part 5-Amd 1: Registration of identifiers", 1995
"Part 6: Inter-industry data elements", 1995
"Part 7: Enhanced inter-industry commands", 1995
"Part 8: Inter-industry security architecture", 1995

[16]   N. Koblitz: "Elliptic Curve Cryptosystems", Math. Comp., vol. 48, 1987, 203-209

[17]   D. E. Knuth: „The Art of Computer Programming-Vol.2 Seminumerical Algorithms", 2nd Ed., Addison-Wesley, Massachusets 1981

[18]   K. Koyama, Y. Tsuruoka: „A signed binary window method for fast computing over elliptic curves", Proceedings of CRYPTO 1992, LNCS 740, Springer-Verlag, Berlin, 1993, SS. 345-357

[19]   G-J. Lay und H.G. Zimmer: „Constructing elliptic curves with given group order over large finite fields", in L. Adleman (Ed.), ANTS-I (1994), Lecture Notes in Computer Science - 877, Berlin, Springer Verlag

[20]   A.K. Lenstra, H.W. Lenstra: "The Development of the Number Field Sieve (LNM 1554), Springer, 1993

[21]   LiDIA-Group: "Library for computational number theory",
http://www.informatik.th-darmstadt.de/TI/LiDIA/Welcome.html

[22]   C.H.Lim, P.J.Lee: „More Flexible Exponentiation with Precomputation", Pre-Proceedings of CRYPTO 1994, Springer Verlag, Berlin 1994, SS. 95-107

[23]   A.J. Menezes: „Elliptic Curve Public Key Cryptosystems", Kluwer Academic Press, Dordrecht 1993, ISBN 0-7923-9368-6

[24]   A.J. Menezes, T. Okamoto und S.A. Vanstone: "Reducing elliptic curve logarithms to logarithms in finite fields", Proceedings of 23rd Annual ACM Symposium on Theory of Computing (STOC), 1991, SS. 80-89

[25]   A.J. Menezes, P.C. van Oorschot, S.A. Vanstone: „Handbook of Applied Cryptography", CRC Press, 1996, ISBN 0-8493-8523-7

[26]   V. Miller: "Use of Elliptic Curves in Cryptology", Advances in Cryptology: Proceedings of Crypto '85, LNCS 218, Springer, 1986

[27]   Motorola: "M68HC05SC Family - At a Glance"
http://mot-sps.com/csic/SMARTCRD/sctable.htm

[28]   V. Müller, S. Paulus: „On the generation of cryptographical strong elliptic curves",
eingereicht zur Eurocrypt 1998, preprint via
http://www.informatik.th-darmstadt.de/TI/reports

[29]   D. Naccache: "Arithmetic Co-processors: State of the Art", Hand-out zu Vortrag
auf der Eurocrypt '95, Saint-Malo, 1995

[30]   National Institute of Standards and Technology (NIST): Digital Signature Standard
(DSS). Federal Information Processing Standards Publication 186 (FIPS-186), 19th
May, 1994

[31]   Philips: "Integrated Circuits and Modules for CHIP CARDS", Produktblatt, 1997

[32]   J.J. Quisquater, C. Couvreur: "Fast Decipherment Algorithm for RSA Public-Key
Cryptosystem", Electronic Letters, vol. 18, no. 21, Oct 1982, SS. 905-907

[33]   W. Rankl, W. Effing: "Handbuch der Chipkarten", Hanser-Verlag, München, 1996,
ISBN 3-446-18893-2

[34]   R. Rivest, A. Shamir, L. Adleman: „A method for obtaining Digital Signatures and
Public-Key-Cryptosystems", Communications of the ACM, v.21,n.2, Feb 1978,
SS. 120-126

[35]   J. Sauerbrey, A. Dietel: „Resource Requirement for the Application of Addition
Chains in Modulo Exponentiation", Proceedings of Eurocrypt 1992, Lecture Notes
in Computer Science - 658, Springer Verlag, Berlin 1993, SS. 174-182

[36]   O. Schirokauer, D. Weber, T. Denny: "Discrete Logarithms and the Effectiveness
of the Index Calculus Method", Proceedings of ANTS II, Springer, 1996,
SS. 337-362

[37]   C.P. Schnorr: „Efficient Identification and Signatures for Smart-Cards", Proceedings of CRYPTO '89, Springer Verlag, Berlin 1990, SS. 237-252

[38]   Schlumberger: " First-Ever Java-Based Smart Card Demonstrated by Schlumberger", Pressemitteilung, 02. April, 1997,
http://www.slb.com/ir/news/et-java0497.html

[39]   Siemens: "Security Controller IC'S",
http://www.sci.siemens.com/htdocs/catalog/ICs/smartcard/scics.html

[40]   J. Silverman: "The Arithmetic of Elliptic Curves", Graduate Texts in Mathematics
- 106, Berlin, Springer-Verlag, 1986

[41]   Telesec: "B1- Das Konzept für ein universelles Chipkartenzugangsgerät",
http://www.telesec.de/b1.htm

[42]   Telesec: "TCOS-Telesec Chipcard-Operating System",
http://www.telesec.de/tcos.htm

[43]   E. Waterhouse: "Abelian Varieties over Finite Fields", Ann. Sci. Ecole Norm.
Sup., vol. 2, 1969, SS. 521-560

[44]   E. de Win, A. Bosselaers, S. Vandenberghe, P. de Cersem J. Vandewalle: "A Fast
Software Implementation for Arithmetic Operations in GF(2n)", Proceedings of
Asiacrypt '96, Springer, 1996

Efficient Algorithms for Multiplication on Elliptic Curves

Volker Müller

Technische Universität Darmstadt
Fachbereich Informatik
Alexanderstr. 10
64283 Darmstadt, Germany
vmueller@cdc.informatik.tu-darmstadt.de

Abstract

We describe new fast algorithms for multiplying points on elliptic curves over finite fields of characteristic greater three. In contrary to the standard binary algorithm, these algorithms use representations of the multiplier with negative coefficients. Timings of the new algorithms show that they are up to 25% faster than the standard binary multiplication algorithm. This running time improvement is especially important for using elliptic curve cryptosystems on smart cards.

1 Introduction

The growing importance of public key cryptography in the last decade induced the search for optimal algorithms for fast exponentiation in various groups. Fast exponentiation is the main bottleneck for improving the speed of several cryptosystems as RSA and ElGamal. In recent years, elliptic curve public key cryptosystems are becoming more and more popular. These cryptosystems are variants of the ElGamal scheme, but they use the group of points on an elliptic curve over a finite field (for a description of such systems, see [Koblitz87], [Miller86] or [IEEE97]). Here, multiplication of a point with a large integer is the most time consuming operation of the encryption and decryption procedure. In this paper, we describe four new algorithms for this key operation, which use special properties of elliptic curves. These new algorithms lead to a running time improvement of up to 25%. Moreover, the algorithms are memory efficient, such that they can also be used for elliptic curve cryptosystem implementations on smart cards.

We start with a short introduction to elliptic curves over finite fields of characteristic greater three. It should be mentioned that the techniques of this paper can also be used for elliptic curves over fields of characteristic two.

Let $p > 3$ be a prime, and let $\mathbb{F}_q$ be the finite field with $q = p^n$ elements. An elliptic curve E over $\mathbb{F}_q$ can be defined by an equation of the form

$$y^2 \;=\; x^3 + a_4\, x + a_6 \,, \tag{1}$$

where $a_4,\ a_6 \in \mathbb{F}_q$ and $4a_4^3 + 27a_6^2 \neq 0$. The set $E(\mathbb{F}_q)$ of points on E over $\mathbb{F}_q$ is given by the set of solutions in $\mathbb{F}_q^2$ to (1) together with a "point at infinity" $\mathcal{O}$. This set $E(\mathbb{F}_q)$ forms a finite abelian (additive) group. There exist simple algebraic formulas for adding two arbitrary points in $E(\mathbb{F}_q)$ (see [Connell95]).

For the speed of elliptic curve cryptosystems, the number of elementary field operations for point addition is important. Here we are just interested in "quadratic" field operations, i.e. we do not care about operations which can be done in linear time. One important observation is then the fact that negating a point is "for free", since for any nonzero point $P = (x, y) \in E(\mathbb{F}_q)$ the negative point is given as $-P = (x, -y)$. If we count the quadratic field operations of the other basic point operations, we get the following results: Doubling a point takes one multiplication, two squarings and one inversion, adding two different points can be done with one multiplication, one squaring and one inversion in $\mathbb{F}_q$. In practice, the inversion is by far the most time consuming part of these operations (in the computer algebra library LiDIA, one inversion of a random element in a 155 bit prime field takes about the same time as 25 multiplications in this field).

Let in the remainder of this paper $m \in \mathbb{N}_{>1}$, and let $P \in E(\mathbb{F}_q)$ be a non zero point on some given elliptic curve E. In the following sections, we describe several new algorithms for computing the multiple $m \cdot P \in E(\mathbb{F}_q)$. These algorithms are designed to especially take care of the special properties of elliptic curves.

2   Left-to-Right ± 1 Addition Chains

The usual method for computing $m \cdot P$ is a variant of the binary exponentiation. It is easy to see that the running time of this algorithm depends on the bit length of m and on the number of ones in the binary decomposition of m. Morain and Olivos [Morain90] developed an extension of this binary algorithm which uses a decomposition of the form

$$m \;=\; \sum_{i=0}^{k} m_i \cdot 2^i, \quad m_i \in \{0,\, 1,\, -1\} \,. \tag{2}$$

Moreover, the number of non zero coefficients in this representation is smaller than the number of ones in the binary decomposition of m. Since for elliptic curves negating a point is for free, their algorithm is therefore faster than the usual binary method (see Section 5, where we list some timings). The algorithm works as follows: it reads the bits of the binary decomposition of m "from the right to the left" (i.e. from the low order bit to the high order bit). For each bit, the algorithm reacts according to the actual state of a given finite automaton, changes the state and multiplies on the fly. In [Morain90], two suitable finite automatons are given.

We generalize the idea of [Morain90] to describe an algorithm which uses a decomposition like (2), but reads the bits of m in the opposite direction, i.e. bits are handled "from the left to the right" (the high order bit to the low order bit). Again, the multiplication algorithm is "given" by a finite automaton.

2.1   The Basic Version

The basic idea of [Morain90] is the observation that blocks of 1's in the binary decomposition of m can be substituted by "equivalent" bit blocks, which have fewer non zero entries. For example, the computation of $15 \cdot P$ with the binary method takes 3 doublings and 3 point additions, but using the equality $15 \cdot P = 16 \cdot P - P$ it can also be done in 4 doublings and 1 addition. In this example, we have substituted a bit block $(1\,1\,1\,1)_2$ by the "equivalent" block $(1\,0\,0\,0\,-1)_2$. In general, we substitute a block $(1^a)_2$, $a \geq 2$, in the binary decomposition of m by the block $(1\,0^{a-1}\,-1)_2$.

A multiplication algorithm which uses this idea can be described by a finite automaton. The states of this automaton "store" the current situation:

state 0: The algorithm has read a 0-bit.

state 11: This state indicates that the algorithm is inside a block of 1's.

state 1: The previous bit was a 0-bit, but the current bit is a 1. We do not know whether the current 1-bit starts a block of 1's or not. Therefore we have to use "lazy evaluation" and wait for the next bit. If the next bit is 0, then we have an isolated 1 and we go back to state 0, otherwise we are in a block of 1's and we switch to state 11.

The following Figure 1 describes the actions of finite automaton A in a graph. The current bit and the corresponding operation are written at the edges of this graph.

Note that the algorithm induced by Automaton A needs two doublings and one addition, when it is in state 1 and reads a 0-bit. For this situation, it might be advantageous to precompute (and store) $2 \cdot P$ and use the equation $2 \cdot (2 \cdot H + P) = 4 \cdot H + 2 \cdot P$. This transformation is especially useful with the observation which we will describe in Theorem 1. Note further that the correctness of this algorithm follows directly from construction.

2.2   The Improved Version

We can improve the algorithm induced by Figure 1 even more with the following observation already used in [Morain90]: If there is an isolated 0 between two blocks of 1's, then we can use the substitution $(-1\,1)_2 = (0\,-1)_2$ to do the transformation

$$(1^a\,0\,1^b) \longrightarrow (1\,0^{a-1}\,-1\,1\,0^{b-1}\,-1)_2 \longrightarrow (1\,0^a\,-1\,0^{b-1}\,-1)_2 \ .$$

We change Automaton A appropriately to take care of this equation by introducing a new state 110. If we are leaving a block of 1's (i.e. we are in state 11 and we read a 0-bit), we have to delay the computation until we know the bit following the 0-bit. Therefore we go to state 110 and read the next bit. After this bit input, the algorithm can decide whether the 0-bit really is an isolated bit between two blocks of 1's or not and react correctly. We describe the corresponding finite Automaton B in Figure 2.

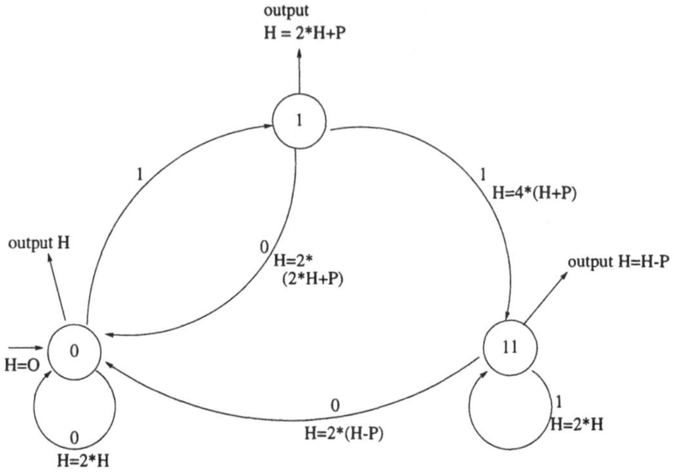

Figure 1: Finite Automaton, Version A

Note that the correctness of Automaton B follows directly from the correctness of Automaton A and the construction. Moreover, the remarks made to Automaton A remain true: it might be advantageous to replace the operations $H = 2 \cdot (2 \cdot H \pm P)$ by a precomputation and the corresponding operations $H = 4 \cdot H \pm 2 \cdot P$.

It should be observed that Automaton B does not always induce a method with fewer additions as the standard binary method. If we choose for example $m = 26 = (1\,1\,0\,1\,0)_2$, then the algorithm induced by Automaton B needs one doubling more than the standard method. Nevertheless the new algorithm is in practice very often better than the standard method, as we will see in Section 5.

3   Using a 4-adic Decomposition of the Multiplier

The ordinary binary algorithm uses a 2-adic decomposition of the multiplier m. In this section, we describe a "left-to-right" multiplication algorithm which uses a 4-adic decomposition of m. Let the 4-adic representation of m be given as

$$m \;=\; \sum_{i=0}^{s} n_i \cdot 4^i, \quad 0 \le n_i < 4,\; n_s \neq 0. \tag{3}$$

A multiplication algorithm based on (3) can process the coefficients n_i either in ascending or in descending order. Note that the processing direction is of great importance, since only for descending order $n_s, n_{s-1}, \ldots$ the algorithm can use a precomputed table of points. This claim follows directly from the equation

$$m \cdot P \;=\; 4 \cdot \Big(\ldots 4 \cdot \Big(4 n_s \cdot P + n_{s-1} \cdot P\Big) \ldots + n_1 \cdot P\Big) + n_0 \cdot P \,.$$

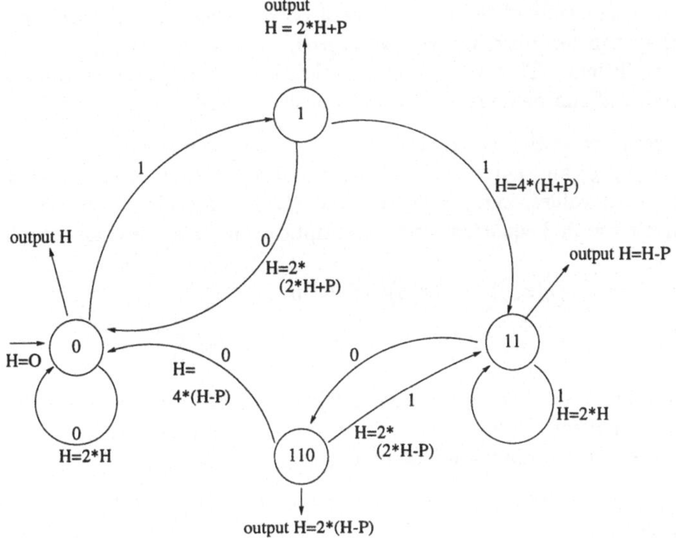

Figure 2: Finite Automaton, Version B

It is easy to see from this equality that only additions of points $r \cdot P$ for $0 \le r < 4$ are necessary, and these points can be precomputed and stored in a table.

3.1 Computing $4 \cdot H$

Another interesting point is the computation of $4 \cdot H$ for various points $H \in E(\mathbb{F}_q)$. This operation obviously is a key operation in a 4-adic multiplication algorithm. The naive algorithm would double H twice. Such an algorithm would need two inversions, two multiplications and four squarings in the given field. In this section, we describe an alternative algorithm which only needs one inversion (but more multiplications and squarings).

We use the theory of division polynomials as explained in [Connell95, page 145]. Using these polynomials, we can express multiplication of a "formal point" by a pair of rational functions. Computing $4 \cdot H$ for some given non zero point $H \in E(\mathbb{F}_q)$ then means evaluating these rational functions.

First we define some division polynomials which we will need in the alternative algorithm:

$$
\begin{aligned}
\psi_2(x,y) &= 2y, \\
\psi_3(x,y) &= \left(\left(3x^2 + 6a_4\right) \cdot x + 12a_6\right) \cdot x - a_4^2, \\
\omega_2(x,y) &= \left(\left(\left(\left(2x^2 + 10a_4\right) \cdot x + 40a_6\right) \cdot x - 10a_4^2\right) \cdot x - 8a_4a_6\right) \cdot x - 2a_4^3 - 16a_6^2, \\
\psi_4(x,y) &= \psi_2(x,y) \cdot \omega_2(x,y).
\end{aligned}
$$

Note that the coefficients of these polynomials only depend on the used elliptic curve E. Therefore they can be precomputed and stored; we neglect the cost of computing the polynomial coefficients. Thus we can assume that the evaluation of all these polynomials at a given point H can be done with 7 multiplications and 1 squaring in $\mathbb{F}_q$.

A useful observation is the fact that $4 \cdot H = \mathcal{O}$ if and only if $\psi_4(H) = 0$. Therefore the polynomials $\omega_2(x, y)$ and $\psi_2(x, y)$ should be evaluated at first. Apart from these values, we need two other values: $\lambda(H) = \psi_3(H)^3$ and $\eta(H) = \psi_2(H)^4 \cdot \omega_2(H)$. All these values can be computed with 3 squarings and 2 multiplications. Then we can compute

$$
\begin{aligned}
\phi_4(x, y) &= \big(\eta(x, y) - \lambda(x, y)\big) \cdot \psi_3(x, y)\,, \\
\omega_4(x, y) &= 2^{-1} \left(\big(-2\,\lambda(x, y) + 3\,\eta(x, y) - \omega_2^2(x, y)\big) \cdot \lambda(x, y) - \eta(x, y)^2\right).
\end{aligned}
$$

It should be mentioned that multiplication with 2^{-1} can be performed in linear time if we use the fact that for $r \in \mathbb{F}_p$ we have $2^{-1} r = (r/2)$, if r is even, and $2^{-1} r = (r + p)/2$, if r is odd (here, p is the characteristic of $\mathbb{F}_q$).

The connection to the original problem is described in [Connell95, Prop. 1.7.8, page 147]. We can deduce that

$$
4 \cdot H = \left(x(H) - \frac{\phi_4(H)}{\psi_4(H)^2},\ \frac{\omega_4(H)}{\psi_4(H)^3}\right).
$$

Therefore we can find $4 \cdot H$ by inverting $\psi_4(H)$ and using the values $\phi_4(H)$ and $\omega_4(H)$. If we count the cost for all these operations, we get the following theorem.

Theorem 1 *There exists an algorithm which computes $4 \cdot H$ in at most 14 multiplications, 7 squarings and one inversion in $\mathbb{F}_q$ for any point $H \in E(\mathbb{F}_q)$.*

In the introduction of this paper, we mentioned that in a lot of finite field implementations one inversion has about the same cost as approximately 25 multiplications. If we then compare the naive algorithm with the algorithm described in this section, we get the following result: the naive algorithm needs about 52 multiplications and 4 squarings, the new algorithm needs 39 multiplications and 7 squarings. Since in a clever implementation one squaring is up to twice as fast as a multiplication, we expect the new algorithm to be approximately 21% faster than the naive algorithm. A comparison of practical timings can be found in Section 5.

3.2　The 4-adic Multiplication Algorithm

Using (3) and the previously mentioned facts, we immediately can describe the following 4-adic multiplication algorithm.

Algorithm 1 (Multiplication of Points Using 4-adic Decompositions)

INPUT: $m \in \mathbb{N}$ and $P \in E(\mathbb{F}_q)$.
OUTPUT: $m \cdot P$.

---

(1) compute and store $T_i = i \cdot P$ for all $1 \leq i \leq 3$.
(2) compute the representation $m = \sum_{i=0}^{s} n_i \cdot 4^i, \quad 0 \leq n_i < 4, \; n_s \neq 0$.
(3) set $H = T_{n_s}$.
(4) **for** $(i = s - 1$ **downto** $0)$ **do**
(5) set $H = 4 \cdot H$.
(6) **if** $(n_i > 0)$ **then**
(7) set $H = H + T_{n_i}$.
(8) **od**
(9) **return** (H)

We can assume that for a random multiplier m about half the bits in the binary decomposition of m are 1-bits. Therefore we expect that the standard binary multiplication algorithm needs about $\log_2(m)$ point doublings and $\approx \frac{1}{2}\log_2(m)$ point additions.

The length of a 4-adic decomposition (3) of m is only half the binary length of m. Since multiplication with 4 can be done faster than two doublings, we expect the "doubling part" of Algorithm 1 to be faster than the corresponding part of the binary method. Unfortunately, Algorithm 1 needs one additional point addition for each non zero coefficient n_i in (3). If we assume that for a random integer m the coefficients n_i behave like random elements in $\mathbb{Z}/4\mathbb{Z}$, then about a fourth of these coefficients should be zero. Therefore we expect that $\approx \frac{3}{8}\log_2(m)$ point additions are necessary in step (7) of Algorithm 1. We will describe the practical behavior of this algorithm in Section 5.

The next section combines the two main ideas of this paper to reduce the number of non zero coefficients in (3).

4   Addition Chains and 4-adic Decompositions

We have already mentioned that the expected number of non zero coefficients in (3) is about one fourth of all coefficients. Fortunately, we can combine the ideas of the 4-adic multiplication Algorithm 1 with the ideas of ± 1-addition chains introduced in Section 2. Again we substitute blocks of 1's in the binary decomposition by "equivalent" blocks. In the 4-adic situation, there is the small difficulty that not every substitution gives an improvement. For example, we know that $23 = (1\,1\,3)_4 = (1\,0\,1\,1\,1)_2 = (1\,1\,0\,0\,-1)_2 = (1\,2\,-1)_4$, but this substitution does not improve the number of nonzero coefficients in the 4-adic expansion of 23. This fact aggravates the description of a finite automaton which defines the improved 4-adic multiplication algorithm.

The idea for the reduction is however simple: Assume that the actual situation is $(x\,3\,y)_4$, and we currently read the 3-coefficient. If we use the equality $3 = 4 - 1$, we get the equivalent block $((x + 1)\,0\,(y - 4))_4$. Testing all possible values for x, y, we find the situations where a substitution will increase the number of zero coefficients. These cases are stored in a finite automaton. We do not describe the automaton with a graph, but we give a list of states and corresponding actions.

The construction of this finite automaton is straightforward (the states store the last read coefficient, as in Section 2) except that input coefficients 3 are handled more careful. If the automaton reads a 3-coefficient directly after a 0-coefficient, then it goes to state **delay** to wait for the next coefficient. If we are in a block of 3-coefficients of length at least two, then this block can be exchanged, otherwise a substitution does not pay. The finite automaton starts in state **n**, if $0 \leq n = n_s < 3$, and in state **delay** if $n_s = 3$. Moreover, we initialize $H = \mathcal{O}$. Then it reads the coefficients of the 4-adic decomposition (3) of m in descending order, starting with n_{s-1}. Let $0 \leq n \leq 3$ be the last coefficient which the algorithm has read. The following list describes the actions which the automaton should perform in a given state with input n.

state 0: If $0 \leq n \leq 2$, then goto state **n**, and set $H = 4 \cdot H$. If $n = 3$, then goto state **delay**.

state 1: If $0 \leq n \leq 2$, then set $H = 4 \cdot H + P$, otherwise set $H = 4 \cdot H + 2 \cdot P$. Goto state **n**.

state 2: If $0 \leq n \leq 2$, then set $H = 4 \cdot H + 2 \cdot P$, otherwise set $H = 4 \cdot H + 3 \cdot P$. Goto state **n**.

state 3: If $1 \leq n \leq 3$, then set $H = 4 \cdot H$, otherwise set $H = 4 \cdot H - P$. If $n = 2$, then goto state **-2**, if $n = 1$, then goto state **-3**, otherwise goto state **n**.

state -2: If $0 \leq n \leq 2$, then set $H = 4 \cdot H - 2 \cdot P$, otherwise set $H = 4 \cdot H - P$. Goto state **n**.

state -3: If $0 \leq n < 3$, then set $H = 4 \cdot H - 3 \cdot P$, otherwise set $H = H - 2 \cdot P$. Goto state **n**.

state delay: If $0 \leq n \leq 1$, then set $H = 16 \cdot H + 3 \cdot P$, else set $H = 4 \cdot (4 \cdot H + P)$. If $n = 2$, then goto state **-2**, else goto state **n**.

If the automaton has read all coefficients of (3) and is currently in state **n**, the algorithm should output the result

- $H = 4 \cdot H + n \cdot P$ for n = 0, 1, 2, -2, -3,

- $H = 4 \cdot H - P$ for n = 3,

- $H = 16 \cdot H + 3 \cdot P$ for n = delay.

The correctness of this procedure follows directly by construction. Obviously, all point additions in this procedure should preferably be done as fast as possible. Therefore the points $r \cdot P$ for $2 \leq r \leq 3$ should be precomputed and stored in a table. Note again that we can derive the points $-2 \cdot P$ and $-3 \cdot P$ for free. Moreover, multiplication with 4 should be done with the new algorithm introduced in Theorem 1. In the following section, we give running times for all algorithms described in this paper. We will see that the ideas of this section improve the speed of the 4-adic multiplication algorithm again by about 5%.

5 Timings of the Algorithms

First we count the number of elementary field operations that have to be performed. We restrict our attention to the algorithms using 4-adic expansions, since the variants of the Morain/Olivos algorithm can be analyzed exactly as in [Morain90].

We assume that m behaves like a random integer. Then we can expect that about half the bits of the binary expansion of m are zero. Therefore the binary algorithm has to do $\log_2(m)$ doublings and $\approx \frac{1}{2}\log_2(m)$ additions. Moreover, we expect that about one fourth of the coefficients in a 4-adic expansion like (3) are zero. Since the length of such an expansion is only half the binary length of m, Algorithm 1 needs $\frac{1}{2}\log_2(m)$ multiplications with four and $\approx \frac{3}{8}\log_2(m)$ additions. The expected number of additions in the improved 4-adic algorithm is slightly smaller, while the length of the expansion remains the same. If we look at blocks $(x\,3\,y)_4$ and count the "good" values for x, y, then we find exactly 9 such possibilities. Therefore the probability that a 3-coefficient can be exchanged to 0 without "deletion" of another 0-coefficient is $\frac{9}{16}$. Therefore we expect the number of additions to be $\approx \frac{39}{128}\log_2(m)$. Using the result of Theorem 1, we compute the expected number of elementary field operations as follows.

Operation	Binary Method	4-adic Algorithm	Improved 4-adic Algorithm
Multiplication	$3\log_2(m)$	$\frac{31}{4}\log_2(m)+4$	$\frac{487}{64}\log_2(m)+4$
Squaring	$\frac{5}{2}\log_2(m)$	$\frac{31}{8}\log_2(m)+3$	$\frac{487}{128}\log_2(m)+3$
Inversion	$\frac{3}{2}\log_2(m)$	$\frac{7}{8}\log_2(m)+2$	$\frac{103}{128}\log_2(m)+2$

If we compare the expected number of operations of the binary method with the 4-adic algorithm, then the 4-adic version should be superior if one inversion takes longer than 10 multiplications. If we assume that squarings take the same time as a multiplication, and that one inversion takes about 25 multiplications, then we expect the 4-adic algorithm to need about 78% of the time of the binary algorithm, the improved 4-adic algorithm should even be a bit faster and take 74% of that time.

After these theoretical observations, we list practical timings. We implemented the standard binary method for multiplication, the second improved algorithm of Morain/Olivos [Morain90] and all algorithms described in this paper. The basis for these implementation is the computer algebra library LiDIA (see [LiDIA97]). All tests were done on a sparc4 machine.

In the first table, we compare the naive algorithm for computing $4 \cdot H$ with the new idea of Section 3.1. We list the average time (in milliseconds) of such an operation for the smallest prime field with the given bit length. This average time was computed by multiplying 10000 random points on 100 random elliptic curves over $\mathbb{F}_p$ with 4.

Bit length of p	Double twice	New Method	Rate
100	0.68	0.53	79%
150	0.99	0.80	81%
200	1.24	0.99	80%
250	1.40	1.12	80%
300	1.83	1.50	82%

This table shows that the running time improvements which we expected in Section 3.1 can almost be achieved in practice.

The following table lists timings for the standard binary multiplication algorithm, the second improved Algorithm of Morain/Olivos [Morain90], the new Algorithm Version A (Figure 1), the new Algorithm Version B (Figure 2), the 4-adic Algorithm 1 and the Improved 4-adic Algorithm. We chose five random elliptic curves over the smallest prime field $\mathbb{F}_p$, where p has the given bit length. For each curve, we multiply a random point P with 200 random integers $0 \leq m < p$. The table lists the average time (in milliseconds) of one such multiplication and the relative time compared to the standard binary method.

$\log_2(p)$	Std	Morain/Oli.	Version A	Version B	4-adic	Impr. 4-adic
100	175	157 (90%)	154 (89%)	148 (85%)	148 (85%)	140 (80%)
150	387	350 (90%)	343 (88%)	325 (84%)	323 (84%)	307 (79%)
200	662	594 (90%)	578 (87%)	551 (83%)	545 (82%)	517 (78%)
250	1041	933 (90%)	907 (87%)	862 (83%)	853 (82%)	808 (78%)
300	1615	1447 (90%)	1393 (86%)	1317 (82%)	1288 (80%)	1219 (75%)
350	2247	2015 (90%)	1952 (87%)	1847 (82%)	1798 (80%)	1701 (76%)
400	3053	2739 (90%)	2655 (87%)	2525 (83%)	2512 (82%)	2364 (77%)

These timings show that the new algorithms really lead to a significant running time improvement. The description of the new multiplication algorithms is very simple such that these algorithms can also be used on smart cards. Depending on the memory capacity of the card, one can achieve either an improvement of up to 18% with Algorithm Version B (no additional memory necessary), or 25% with the Improved 4-adic Algorithm (only about $4 \log_2(p)$ bit additional memory required). Since speed is an important requirement for smart card applications, the new algorithms is of great importance for smart card implementations.

Finally, it should be remarked that obviously the ideas of this paper can also be used for elliptic curves defined over finite fields of characteristic two.

References

[Connell95] I. Connell: *Elliptic Curve Handbook*, Draft July 1995, available on `ftp://math.mcgill.ca/pub/ECH1/` .

[IEEE97] *IEEE P1363 Working Draft: Public Key Cryptography*, Draft, August 6 1996, available on `ftp://stdsbbs.ieee.org/pub/p1363/` .

[Koblitz87] N. Koblitz: *Elliptic Curve Cryptosystems*, Mathematics of Computation, **48**, 1987, 203 – 209.

[LiDIA97] LiDIA – *A Library for Computational Number Theory*, available on `http://www.informatik.tu-darmstadt.de/TI/`

[Menezes93] A. Menezes: *Elliptic Curve Public Key Cryptosystems*, Kluwer Academic Publishers, 1993.

[Miller86] V.S. Miller: *Use of Elliptic Curves in Cryptography*, Advances in Cryptology - CRYPTO 85, Lecture Notes in Computer Science No. 218, 1986, 417 – 426.

[Morain90] F. Morain and J. Olivos: *Speeding up the Computations on an Elliptic Curve using Addition-Subtraction Chains*, in F. Morain, *Courbes Elliptiques et Tests de Primalité*, Doctoral Thesis, Université Lyon I, 1990.

Chipkarten als Signier- und Verifizierkomponente

Patrick Horster[1] · Klaus Keus[2]

[1]Universität Klagenfurt
Institute für Informatik - Systemsicherheit
pho@ifi.uni-klu.ac.at

[2]Bundesamt für Sicherheit in der Informationstechnik
Bonn
keus@bsi.de

Zusammenfassung

Der vorliegende Beitrag gibt zunächst einen Einblick in das Gesetz zur digitalen Signatur und seiner Zielsetzung bezüglich Zweck und Anwendungsbereich. Danach werden die erforderlichen technischen Anforderungen im Kontext der zugehörigen Signaturverordnung im Hinblick auf die technischen Komponenten allgemein und anschließend auf die Einzelkomponente Chipkarte als Signier- und Verifizierkomponente interpretiert und übertragen. Zudem werden Aspekte der erforderlichen technischen Komponentenprüfung für Chipkarten im Umfeld des Signaturgesetzes bzw. der Signaturverordnung und die damit verbundenen Problemen aufgezeigt. Abschließend werden die behandelten Aspekte nochmals kritisch betrachtet und in einem Ausblick zusammengefaßt.

1 Einleitung

Das Gesetz zur digitalen Signatur und die zugehörige Verordnung definieren keine detaillierten technischen Vorgaben, sondern lassen bewußt genügend Spielraum für zukünftige innovative technische Lösungen. Gleichwohl lassen sich implizit aus den aus der Verordnung abgeleiteten Sicherheitsanforderungen und Maßnahmen konkrete technische Lösungen im Sinne von Sicherheitsfunktionalitäten ableiten. Dies gilt insbesondere unter Berücksichtigung der zugehörigen Begründung zur Verordnung. So zeigt etwa der vom Bundesamt für Sicherheit in der Informationstechnik (BSI) in Zusammenarbeit mit Industrie, Wirtschaft und Wissenschaft erstellte Maßnahmenkatalog zu §16 (6) der Verordnung beispielhaft technische Leitvorgaben für Hersteller und Anwender auf.

Dabei kommt der Signier- bzw. Verifizierkomponente die sicherheitstechnische Zentralfunktion zu. Ihrem Einsatz wird sowohl bei der Erstellung einer Signatur als auch im Prozeß der Verifizierung einer Signatur die zentrale sicherheitskritische Bedeutung zugemessen. Während für den Masseneinsatz aufwendige und kostspielige Alternativlösungen in Form von Sicherheitsboxen (Signierboxen) eher Anwendung finden werden, wird insbesondere im endanwenderbezogenen Einzelfall eine geeignete Chipkarte zum Einsatz kommen. Als kom-

durch ihre Technik und ihre multifunktionale Einsatzbreite einerseits optimale Voraussetzungen für eine vertrauenswürdige Speicherung des erforderlichen privaten Signaturschlüssels, andererseits kann die Chipkarte als Zentralkomponente zur Erstellung der eigentlichen digitalen Signatur angesehen werden [Keus97].

Der vorliegende Beitrag behandelt das Gesetz zur digitalen Signatur und betrachtet die Chipkarte als die technische Lösung, die einerseits eine angemessen sichere digitale Signatur ermöglicht, und andererseits kostenmäßig vertretbar ist. Der Schwerpunkt dieses Beitrags liegt auf Anforderungen des Gesetzes an die Signierkomponente und interpretiert diese im Hinblick auf eine technische Realisierung durch die Chipkarte. Grundlegende Anforderungen an den praktischen Einsatz der Chipkarte im Anwendungsspektrum der digitalen Signatur runden den Artikel ab.

2 Das Gesetz zur digitalen Signatur

Am 18. April 1997 wurde der Entwurf des aus 11 Artikeln bestehenden „Informations- und Kommunikationsdienste Gesetz" (IuKDG) in der ersten Lesung im Deutschen Bundestag beraten und am 1. August 1997 in Kraft gesetzt [IuKD97]. Dieses Gesetz beinhaltet in Artikel 3 das Gesetz zur digitalen Signatur. Eine begleitende Verordnung zum Gesetz zur digitalen Signatur wurde im Oktober 1997 nach langer Diskussion abschließend behandelt und am 1. November in Kraft gesetzt [SigV97]. Das Gesetz soll Rahmenbedingungen schaffen, bei deren Einhaltung eine digitale Signatur als mindestens gleichwertig sicher zu einer eigenhändigen Unterschrift angesehen werden kann. Bei entsprechender gesetzmäßiger Umsetzung wird dies zu weitreichenden Konsequenzen führen, deren Relevanz nur in Teilbereichen (etwa Electronic Commerce, Electronic Banking, Digitale Verwaltungsakte, Internetanwendungen) erahnt werden kann.

Um die an eine digitale Signatur gestellten Anforderungen erfüllen zu können, muß sichergestellt werden, daß Fälschungen weitgehend ausgeschlossen oder zumindest erkennbar sind. Dies betrifft nicht nur die eigentliche digitale Signatur, sondern insbesondere die zu signierenden Daten bzw. die damit verbundenen Informationen.

In der heutigen, zunehmenden Nutzung der modernen Informations- und Kommunikationstechnik werden elektronische Daten übermittelt oder gespeichert, die nur unter großem Aufwand vor unbefugter und unbemerkter Einsichtnahme (Verlust der Vertraulichkeit) und unbefugter oder unbeabsichtigter Veränderung der Daten (Verlust der Integrität) gesichert werden können. Das Internet mit seinen Möglichkeiten und Grenzen kann dabei als Paradebeispiel angesehen werden. Darüber hinaus gewinnt die authentische elektronische Kommunikation zwischen Partnern zunehmend an Bedeutung. Hierbei ist es erforderlich, daß eine Nachricht einem Urheber eindeutig zugeordnet und diese Zuordnung nachweislich belegt werden kann.

Damit dieses Sicherheitsziel, insbesondere auch in Verbindung mit zukünftigen rechtsverbindlichen Willenserklärungen in elektronischer Form oder bei der elektronischen Archivierung beweiserheblicher Daten und Informationen ermöglicht werden kann, muß ein gesamtheitliches System zur Verfügung stehen, das durch geeignete Realisierung den Urheber und die Unverfälschtheit der signierten Daten zuverlässig erkennen und nachweislich belegen läßt.

Das Signaturgesetz ist so beschaffen, daß nur die Ziele vorgegeben werden. Da es keine Vorgaben bezüglich der inhaltlichen und technischen Ausgestaltung macht, werden die Anforderungen an die technischen Komponenten betreffend ihres Einsatzzweckes in einer zugehörigen Verordnung (sichergestellt durch § 16 des Gesetzes) näher spezifiziert [SigV97]. Aber selbst in der zugehörigen Verordnung sind die Vorgaben für die Beteiligten und die jeweiligen Lösungsansätze bewußt so offen gehalten, daß unterschiedliche Lösungen zulässig sind, um innovativen Entwicklungen Spielraum zu lassen. Zeitgleich mit der Verabschiedung der Verordnung hat das BSI Orientierungshilfen zu einem Maßnahmenkatalog vorgelegt, die in einem komplexen Projekt gemeinsam mit Experten aus Wirtschaft, Wissenschaft, Industrie und Behörden erstellt wurden [MKat97]. Basierend auf dieser Grundlage wird von der zukünftigen Regulierungsbehörde (die Nachfolgebehörde des heutigen Bundesamtes für Post- und Telekommunikation - BAPT) als zuständige staatliche Steuerungsbehörde ein abstrakter Maßnahmenkatalog erstellt, der dann durch exemplarische Vorgaben allen Beteiligten zur Verfügung steht. In diesem Maßnahmenkatalog (zu § 12 und § 16 der Verordnung) werden insbesondere Vorgaben hinsichtlich Technik, Organisation und Personal behandelt.

3 Technische Komponenten

Durch digitale Signaturen kann, in Verbindung mit einer kollisionsresistenten Hashfunktion und Zertifikaten, die Urheberschaft und der Nachweis der Integrität signierter Daten ermöglicht werden [HoKr96]. Folgt man dem Signaturgesetz, so kann eine digitale Signatur als digitales „Siegel" aufgefaßt werden, wobei das Siegel auf der Basis der Nutzung des privaten Schlüssels eines Benutzers erstellt werden kann.

Digitale Sigaturen beruhen auf asymmetrischen (Public-Key-) Kryptosystemen. In Abhängigkeit des zur Anwendung kommenden Kryptosystems verfügt dabei jeder Benutzer über systembedingte Schlüsselparameter. Insbesondere besitzt aber jeder Benutzer einen privaten Signaturschlüssel (geheime Schlüsselkomponente) und einen zugehörigen öffentlichen Signaturschlüssel (öffentliche Schlüsselkomponente). Der private Signaturschlüssel kann zur Signaturbildung und der öffentliche Signaturschlüssel zur Verifikation einer Signatur dienen.

Die Schlüsselparameter müssen durch einen vertrauenswürdigen Prozeß generiert und durch einen vertrauenswürdigen Dritten einer bestimmten Person nachweislich zugeordnet werden. Diese Zuordnung wird durch ein sogenanntes Zertifikat bestätigt, das durch einen vertrauenswürdigen Dritten (Zertifizierungsstelle) ausgestellt wird.

In folgender Abbildung ist das Prinzip einer digitalen Signatur wiedergegeben. Zunächst berechnet der Sender A den zur Nachricht m gehörenden Hashwert h=H(m), hierzu verwendet er eine öffentlich bekannte Hashfunktion H. Die Signatur s zur Nachricht m wird mittels einer Signierfunktion S aus dem Hashwert h und dem geheimen Signaturschlüssel GSA des Senders A gebildet: s=S(h,GSA). Die Nachricht m wird gemeinsam mit der Signatur s und dem Zertifikat (zum öffentlichen Schlüssel OSA des Senders A) an den Empfänger übermittelt. Dabei kann das Zertifikat auch zur Signaturbildung verwendet werden.

Der Empfänger überprüft zunächst die Gültigkeit des öffentlichen Schlüssels OSA, indem er beispielsweise die Korrektheit des zugehörigen Zertifikats unter Verwendung des öffentlichen Schlüssels der ausstellenden Zertifizierungsstelle verifiziert. Anschließend wird der Hashwert h=H(m) berechnet, um mittels des Prädikats V(h,s,OSA), das die Werte „true" und „false" annehmen kann, die Integrität und Authentizität der Nachricht m zu prüfen.

Das Ziel von Signaturverfahren ist es, zu garantieren, daß V(H(m),s,OSA)=true genau dann gilt, wenn s=S(H(m),GSA) ist. Die konkreten Realisierungen der Funktionen S und V weichen bei den unterschiedlichen Verfahren (z.B. RSA-Verfahren [RiSA78], ElGamal-Verfahren [ELGa85] und Digital Signature Standard [NIST94]) stark voneinander ab. Grundlagen, Realisierungen, rechtliche Aspekte und Anwendungen digitaler Signaturen werden in [Hors96] ausführlich betrachtet.

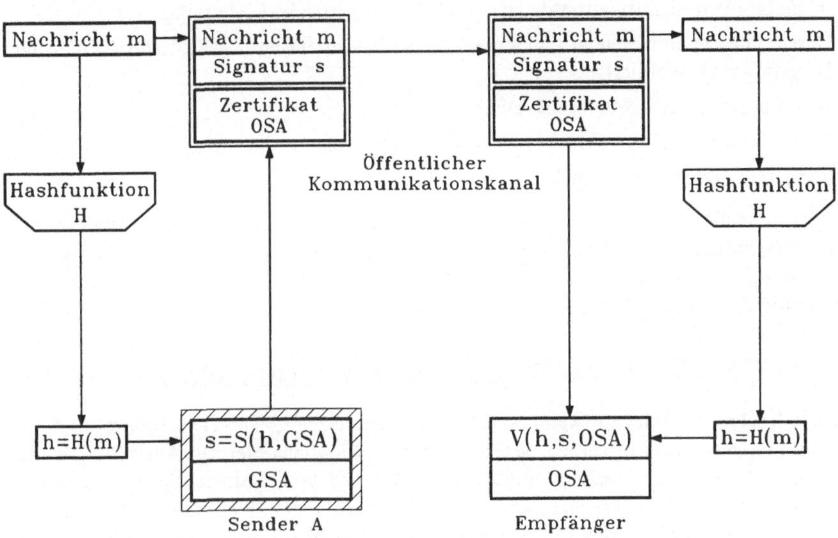

Abb. 1: Prinzip einer digitalen Signatur mit Hashfunktion und Zertifikat

Alle erforderlichen Prozesse, angefangen von der Erstellung der Schlüsselparameter und der Zuordnung der Schlüssel zu einem persönlichen Zertifikat über die Speicherung des privaten Signaturschlüssels bis zur Umsetzung und Anwendung eines Signier- und Verifiziervorgangs sind aufgrund ihrer Bedeutung extrem sicherheitskritisch und erfordern angemessene vertrauensbildende Maßnahmen [HoWo97].

Folglich werden entsprechende Sicherheitsanforderungen an alle Beteiligten gestellt. Dies sind insbesondere Maßnahmen an die Regulierungsbehörde als Wurzelinstanz, an die Betreiber von Zertifizierungsstellen und Anforderungen an zwingend erforderliche Infrastrukturen. Die Wurzelinstanz hat dabei im wesentlichen die Kontroll- und Steuerungsaufgabe, dient als „oberste Zertifizierungsstelle" und erteilt die Genehmigungen für die zukünftigen, privatwirtschaftlich organisierten Zertifizierungsstellen (vgl. insbesondere § 12 der Signaturverordnung). Da es sich um ein komplexes Gesamtsystem handelt, berücksichtigen die Sicherheitsanforderungen außerdem technische, organisatorische, personelle und materielle Aspekte.

Alle eingesetzten technischen Komponenten müssen einen angemessenen Sicherheitsgrad nachweisen. Eine eher rahmenmäßige Festlegung bezüglich Inhalt und Prüfgrundlage gibt die Verordnung (§§ 16 und 17) vor.

Nach Konzeption und Analyse des Gesamtschemas ergibt sich nachfolgende Mindestausstattung für ein Gesamtsystem hinsichtlich der Komponenten zur

- Schlüsselerzeugung (z.B. Chipkarten, PCMCIA-Karten, Sicherheitsboxen),
- Erstellung von Zertifikaten,
- Speicherung des privaten Signaturschlüssels,
- Erzeugung und Prüfung von digitalen Signaturen (z.B. Chipkarte oder Sicherheitsbox),
- Erfassung von Identifikationsdaten (z.B. PIN, biometrische Merkmale),
- Darstellung zu signierender und signierter Daten – in diesem Zusammenhang hat sich der Begriff Viewer etabliert,
- Abrufung von Zertifikaten (z.B. DFÜ-Anwendung),
- Überprüfung der Korrektheit von Zertifikaten (Verzeichnisdienste),
- Realisierung von Zeitstempeln.

Bei der Umsetzung fordert die Verordnung in Teilbereichen eine Unterscheidung zwischen einer Anwendung im privaten Umfeld und einer geschäftsmäßig zur Verfügung gestellten Anwendung. Dies betrifft insbesondere den erwähnten Viewer - hier sind noch keine allgemeingültigen Lösungen in Sicht.

4 Die Chipkarte als Signier- und Verifizierkomponente

Für die Realisierung der im Rahmen des Signaturgesetzes geforderten Signier- und Verifizierkomponente stehen verschiedene technologische Möglichkeiten zur Auswahl. Die technische Ausprägung kann von der Einsatzumgebung und vom beabsichtigtem Anwendungszweck beeinflußt sein. Chipkarten können als Grundbestandteil eines Gesamtsystems zur Erzeugung und zur Verifikation digitaler Signaturen angesehen werden und stellen eine spezielle Ausgestaltung der Signierkomponente dar. Ihr Anwendungsbereich ist in Bezug auf Performance, Funktionsumfang und Preis eher am unteren Ende des möglichen Spektrums unterschiedlicher Komponenten angesiedelt. Chipkarten bieten gegenüber anderen möglichen Signierkomponenten den Vorteil, daß sie aufgrund ihrer Leistungsfähigkeit und kompakten Bauart erlauben, persönliche Informationen und geheime Daten sicher und mobil bereitzustellen.

In kommerziellen Umgebungen wird sicherlich eher auf andere Komponenten, etwa auf Sicherheitsboxen zurückgegriffen, da diese erhebliche Mehrfunktionalität bieten können und weitergehende Anforderungen, etwa Mehrfachsignaturen und schnellere Bearbeitungszyklen, abdecken können. Aufgrund ihres Preises sind Sicherheitsboxen allerdings für normale Endanwender weniger angemessen. Andere Lösungen, etwa in Form von PCMCIA- oder PC-Karten, sind dagegen als Alternativlösungen nicht ausgeschlossen.

Als Favorit ist aber die Chipkarte anzusehen, die im Einsatzgebiet der digitalen Signatur vielfältige Aufgaben haben kann, von denen nachfolgend einige beispielhaft aufgeführt sind:

- Schlüsselerzeugung, wobei ein Zufallszahlengeneratoren von Bedeutung sein kann.
- Speicherung eines Zertifikats oder mehrerer Zertifikate, wobei dies entscheidend von der Zertifikatsgröße in Relation zur Speicherkapazität der Chipkarte abhängt.
- Träger des privaten Signaturschlüssels.
- Träger des öffentlichen Signaturschlüssels (zumindest) einer Zertifizierungsstelle.

- Authentifikation des Benutzers und Speicherung der Authentisierungsdaten.

- Speicherung der verwendeten mathematischen Verfahren und Algorithmen, eventuell Berechnung des Hashwertes, wobei dies entscheidend von der Speicherkapazität der Chipkarte und ihrer Übertragungsrate abhängt.

- Berechnung digitaler Signaturen (aus einem Hashwert).

- Verifikation digitaler Signaturen (mittels eines Hashwerts).

Da die Chipkarte nur eine Komponente im Gesamtsystem ist, ist es wichtig, ihre Abgrenzung bzw. Einbindung im Gesamtsystem zu betrachten. Eine wichtige Schnittstelle ist dabei ihre Anbindung zur „Komponente zur Erfassung von Identifikationsdaten". Beispiele sind PCMCIA-Kartenleser, POS-Kartenterminals, PCs mit externem Kartenterminal oder integriertem Kartenleser.

Bei den in der Anwendung der digitalen Signatur einzusetzenden Chipkarten handelt es sich ausschließlich um Prozessorchipkarten, mindestens ausgestattet mit EEPROM (Electrical Erasable Programmable Read Only Memory), RAM (Random Access Memory), ROM (Read Only Memory), CPU (Central Processor Unit), einer Schnittstellensteuerung sowie einem kryptographischen Coprozessor (Crypto Controller).

Aus den unterschiedlichen Einsatzgebieten lassen sich vielfältige Bedrohungen ableiten, seien es Bedrohungen technischer Art oder Bedrohungen durch Personen, wobei sowohl Innentäter als auch Außentäter in Betracht zu ziehen sind. Ausgerichtet an den Sicherheitszielen Vertraulichkeit: Schutz gegen unbefugte Kenntnisnahme von Daten, Integrität: Schutz gegen unbefugte oder unbeabsichtigte Veränderung von Daten sowie Verfügbarkeit: Schutz gegen unbefugtes Vorenthalten von Daten oder Diensten, seien hier nur einige sicherheitsrelevante Aspekte aufgeführt, die bei der Konzeption entsprechend zu berücksichtigen sind:

- Unautorisierte Benutzung der Signaturkomponente,

- Diebstahl bzw. Austausch der Signaturkomponente,

- Verwechslung der Chipkarte,

- Duplizierung der Signaturkomponente,

- Unautorisierte oder unbeabsichtigte Erstellung einer Signatur,

- Verwendung eines trivialen Paßwortes bzw. einer unzureichend langen PIN, fehlende Verfallsdaten,

- Unautorisierter Zugriff auf Authentisierungsinformationen,

- Generierung kryptographisch schwacher oder doppelter Schlüsselparameter,

- Implementierte Hashfunktion oder Signieralgorithmus ist kompromittiert,

- Kenntnisnahme und Weitergabe des privaten Signaturschlüssels,

- Fehlerhafte Berechnung oder Weitergabe einer Signatur,

- Fehlerhafte Ausführung einer Signaturprüfung,

- Veränderungen an Teilnehmerzertifikaten,

- Ungeeignete Entsorgung ausgesonderter Signaturkomponenten und damit erhöhtes Risiko bezüglich Reverse Engineering,

- Ungesicherte Übertragung sensitiver Daten (Informationen),

- Unautorisiertes oder generelles Nachladen zusätzlicher Software, die möglicherweise das Ausspähen vertraulicher Daten zum Ziel hat,

- Unvollständige Protokollierung, Ausfall der Protokollierung sicherheitsrelevanter Vorkommnisse oder Veränderungen an Protokollaufzeichnungen,

- Manipulation an sicherheitstechnischen Komponenten; Angriffe über geänderte Betriebsparameter, Auswertung interner Signale oder Strukturanalysen,

- Beschädigung durch Umwelteinflüsse sowie unerkannte Speicherfehler.

Sicherheitsrelevante Maßnahmen für Chipkarten können unterteilt werden in Basisschutzmaßnahmen und erweiterte Sicherheitsmaßnahmen. Diese beinhalten Maßnahmen auf der Karte selber und organisatorische Maßnahmen im Umfeld der Herstellung und Anwendung. Auf der Karte umfaßt dies beispielsweise die Durchführung der I/O-Kontrollen über alle Schnittstellen, die Sicherstellung der Interferenzfreiheiten einzelner Anwendungen und den Verzicht auf Trace- und Debug-Funktionen. Organisatorische Maßnahmen können dazu beitragen, daß ein Angreifer keine Kenntnis von (technischen) Informationen zur Chipkarte erhält, sie können aber auch dazu dienen, den gewinnbringenden Diebstahl von Chipkarten zu erschweren.

Aus dem Signaturgesetz und der zugehörigen Verordnung lassen sich für Chipkarten im wesentlichen die folgenden Sicherheitsanforderungen und Empfehlungen ableiten:

- Hardware: Authentisierungsdaten und der private Signaturschlüssel des berechtigten Benutzers dürfen aus der Chipkarte nicht mit vertretbarem Aufwand auslesbar sein, und die Neutralität der Signierkomponente muß garantiert sein.

- Schlüsselerzeugung in der Chipkarte: Die absichtliche Erzeugung von Schlüsseldubletten muß ausgeschlossen sein und die Implementierung der Schlüsselgenerierung muß garantieren, daß mit an Sicherheit grenzender Wahrscheinlichkeit keine sicherheitsrelevanten Schlüsselparameter doppelt erzeugt werden. Zudem darf keine Beziehung zu anderen Schlüsseln ableitbar sein. Der Startwert für die jeweilige Schlüsselgenerierung muß für jede Signaturkomponente individuell sein und intern berechnet werden. Daten oder Zustände, die in die Startwert-Berechnung eingehen, dürfen nicht auslesbar sein. Aus dem öffentlichen Signaturschlüssel darf insbesondere der private Signaturschlüssel nicht berechnet werden können.

- Initialisierung und Personalisierung: Zertifizierungsstellen müssen sich davon überzeugen können, ob eine vorliegende Chipkarte als Signierkomponente geeignet ist. Werden Schlüsselkomponenten außerhalb einer Chipkarte erzeugt, so müssen Mechanismen bereitgestellt werden, die die Geheimhaltung der privaten Signaturschlüssel bei der Personalisierung gewährleisten. Das Nachladen von Software, die das Auslesen oder Ändern von Authentisierungsdaten, privaten oder öffentlichen Schlüsselparametern ermöglichen können, ist zu unterbinden. Während der (durch vertrauenswürdiges Personal in einer vertrauenswürdigen Umgebung) durchgeführten Personalisierung können die öffentlichen Signaturschlüssel der Wurzel und der Zertifizierungsstelle in die Chipkarte geladen werden. Darüber hinaus können gegebenenfalls auch die zugehörigen Zertifikate geladen werden. Falls bei der Personalisierung eine (Initial-) PIN oder ein Paßwort voreingestellt wird, sollte die Chipkarte vor der erstmaligen Benutzung durch ihren Inhaber erzwingen, daß diese Voreinstellung durch den Chipkarten-Inhaber zwangsweise geändert werden muß. Es sind Regelungen hinsichtlich der zuverlässigen Übergabe der Schlüssel- und

Authentisierungsparameter sowie eines sicheren Auslieferungsverfahrens für die Signierkomponente zu treffen.

- Vernichtung der Chipkarte: Ist das Zertifikat abgelaufen oder wird der private Signaturschlüssel nicht mehr benötigt, so ist die Chipkarte zuverlässig unbrauchbar zu machen.

- Identifikation und Authentisierung: Jeder Benutzer muß sich vor dem Signiervorgang gegenüber der Chipkarte eindeutig identifizieren und authentisieren. Die Authentisierungsinformationen müssen so gespeichert sein, daß sie nicht auslesbar sind und nur autorisiert verändert werden können. Wird die Chipkarte in Verbindung mit technischen Komponenten (z.B. Chipkartenlesegeräte, PIN-Pads) eingesetzt, die gewerbsmäßig Dritten zur Nutzung angeboten werden, so muß die Chipkarte die Echtheit der Komponente prüfen. Darüber hinaus muß sie feststellen können, ob sicherheitstechnische Veränderungen an der Komponente stattgefunden haben. Echtheit und sicherheitstechnische Veränderungen müssen für den Benutzer erkennbar angezeigt werden.

- Zugriffskontrolle: Der private Signaturschlüssel darf ausschließlich in der Chipkarte gespeichert werden und unterliegt dort einer Zugriffskontrolle. Auf diesen Signaturschlüssel darf lediglich der dafür vorgesehene Prozeß zum Zwecke der Signaturbildung zugreifen. Private Signaturschlüssel dürfen nicht aus der Chipkarte gelesen werden können, insbesondere darf der private Signaturschlüssel nicht duplizierbar sein. Die Erzeugung von digitalen Signaturen darf nur nach erfolgreicher Authentifikation des berechtigten Benutzers möglich sein. Bei multifunktionalen Chipkarten muß sichergestellt sein, daß diese keinen Zugriff auf Authentisierungsdaten oder vertrauliche Schlüsselparameter erhalten. Die Chipkarte muß sicherstellen, daß die zur Signaturbildung und Verifikation eingesetzte Software, gespeicherte Zertifikate, öffentliche Signaturschlüssel und alle geheimzuhaltenden Schlüsselparameter (z.B. Transportschlüssel) einer geeigneten Zugriffskontrolle unterliegen, so daß lediglich autorisierte Personen oder Prozesse einen Zugriff erhalten. Damit sichergestellt ist, daß der Signiervorgang mit Wissen und Wollen des Teilnehmers erfolgt, sollte unmittelbar vor der Ausführung von Signiervorgängen eine Authentifikation des Chipkarteninhabers erfolgen. Dabei kann der Signiervorgang das Signieren mehrerer Datensätze umfassen. Als Grundzustand sollte die Chipkarte allerdings vor jedem Signiervorgang eine explizite Authentifikation erfordern. Nach dem Aufruf anderer Applikationen ist grundsätzlich eine erneute Authentifikation für Signiervorgänge erforderlich.

- Beweissicherung und Protokollauswertung: In der Chipkarte sollten Informationen über Sicherheitsprobleme (z.B. mehrfache falsche PIN-Eingabe) gespeichert werden, wobei nur berechtigte Benutzer ein Einsichtsrecht auf die zur Beweissicherung gespeicherten Daten erhalten können. In der Chipkarte sollten sicherheitsrelevante und sicherheitskritische Aktionen sowie die Durchführung der letzten Signiervorgänge protokolliert werden. Hierzu könnten Datum, Terminalkennung, Hashwert und Dateiname des signierten Dokuments, Informationen zu Wartungsarbeiten, Manipulationsversuche oder identifizierte Fehlerzustände gespeichert werden. Die Protokolldaten sollten von autorisierten Benutzern gelesen werden können und gegen unberechtigte Veränderung geschützt sein.

- Wiederaufbereitung: Multifunktionale Chipkarten müssen sicherstellen, daß Speicherbereiche, die die Signaturanwendung verwenden, vor der Weiterbenutzung durch andere Anwendungen gelöscht werden.

- Unverfälschtheit: Sicherheitstechnische Veränderungen an der Chipkarte und am Chip sollten für den Benutzer erkennbar sein.

- Übertragungssicherheit: Die Datenübertragung zwischen Chipkarte und Anwendungsbereich sollte vor Manipulationen und Störungen gesichert werden.

- Das administrative Umfeld: Chipkarten für Signaturanwendungen (im Sinne des Signaturgesetzes) müssen so ausgeführt werden, daß Zertifizierungsstellen sich davon überzeugen können, daß sie im Hinblick auf Schlüsselerzeugung und persönliche (Identifizierung und) Authentifizierung geeignet sind.

Damit den aufgeführten Bedrohungen entsprechend begegnet werden kann und zugleich die Sicherheitsanforderungen abgedeckt werden, sind nachfolgende Maßnahmen umzusetzen. So lassen sich bzgl. der Chipkarten-Hardware insbesondere folgende (mehr oder weniger) sinnvolle Maßnahmen ergreifen:

- Verwendung von Spezialkontrollern,

- Eingeschränkter Zugang zu Test- und Entwicklungsequipment,

- Geheimhaltung von Maskenlayouts,

- Wirksames Abschalten von Testmöglichkeiten,

- Einbau von Dummy-Strukturen im Chiplayout,

- Kapselung des Chips in einem Spezialgehäuse,

- Scrambling interner Busse,

- Sicherstellung einer einheitlichen Stromaufnahme,

- Anbringen von Schutzschichten über dem Chip,

- Spezielle Speicher-Strukturen,

- Fortlaufende Seriennummer des Chips,

- Integration eines Passivierungsschicht-Sensors,

- Integration einer Power-On-Erkennung,

- Integration einer Spannungs- und Frequenzüberwachung,

- Einschränkung der Gültigkeitsdauer.

Werden Schlüsselparameter in der Chipkarte erzeugt, so sollten ausschließlich geeignete und geprüfte Verfahren verwendet werden. Dies kann den Einsatz eines Zufallszahlengenerators einschließen. Im Rahmen der Initialisierung und Personalisierung ist darauf zu achten, daß ein sicheres Nachladen von Applikationen ermöglicht wird. Es muß aber verhindert werden, daß durch Nachladen von Applikationen und Software das Auslesen oder Verändern von schutzbedürftigen Daten (Schlüssel, Paßwörter) möglich wird.

Durch physikalische Maßnahmen am Kartenkörper der Chipkarte lassen sich bestimmte Basisbedrohungen ausschalten. Beispiele für diese Maßnahmen sind:

- Normgerechte physikalische Stabilität,

- Anbringen von Name, Bild und Unterschrift des Karteninhabers,

- Anbringen von Hinweisen im Verlustfall, für Karteninhaber und Finder,

- Individuelles Merkmal zur Verhinderung des Chipaustausches.

Außerdem sollten postaktive Maßnahmen vorgesehen werden. Dies betrifft nicht nur die Vernichtung der Chipkarte als Ganzes und die Vernichtung aller gespeicherten sensiblen Daten, sondern auch die Behandlung bei Verletzung der zugrundeliegenden Sicherheitsstrategie, die beispielsweise sicherheitstechnische Veränderungen detektiert und für Auswertungen bereitstellt.

Neben den Hardware-Anforderungen beinhalten die aufgeführten Betrachtungen konkrete Maßnahmen in Bezug auf die eingesetzte Software, dies betrifft insbesondere das Betriebssystem. Einzelheiten bezüglich der Funktionalität können vergleichsweise den relevanten Funktionalitätsklassen der ITSEC entnommen werden.

5 Grundlagen der technischen Komponentenprüfung

Die Vertrauenswürdigkeit des Gesamtsystems einer Zertifizierungsstelle in einem geprüften und bestätigten Sicherheitskonzept berücksichtigt die (bekannten) technischen, materiellen, personellen und organisatorischen Aspekte. Dabei ist insbesondere der Nachweis der Vertrauenswürdigkeit der relevanten technischen Komponenten zu erbringen. Ermöglicht das Signaturgesetz noch eine hinreichende Prüfung nach dem Stand der Technik (§ 14 (4)) mit einer Bestätigung durch eine durch die Regulierungsbehörde anerkannten Stelle, so sind die Forderungen zur Komponentenprüfung in § 17 der Signaturverordnung fest definiert. Als wesentliche Grundlage zur Komponentenprüfung werden die harmonisierten europäischen Sicherheitskriterien „Kriterien für die Bewertung der Sicherheit von Systemen der Informationstechnik - ITSEC" in Verbindung mit der zugehörigen „Evaluationsmethodologie - ITSEM" festgelegt [ITSE91, ITSE94].

Als Träger der höchst vertraulichen Information, dem privaten Signaturschlüssel, steht die Komponente „Chipkarte" im besonderen Blickpunkt. Chipkarten enthalten aufgrund ihrer Konstruktion geheime Informationen, die in der Regel sowohl in ihrer Hardware als auch in ihrer Software gespeichert sind [Issel97, RaEf95]. Betrachtet man den speziellen Aufbau von Chipkarten, den Anwendungsbereich im Umfeld der digitalen Signatur und berücksichtigt zudem die einzelnen Phasen in ihrem Lebenszyklus, so ergeben sich unterschiedliche Sicherheitsprobleme und Bedrohungsszenarien.

Ausgehend von den Bedrohungen lassen sich entsprechende Anforderungen und Maßnahmen festlegen. Die Angaben des BSI zum Maßnahmenkatalog gemäß SigV [MKat97] geben hierzu detaillierte Hinweise und Empfehlungen. Konsequenterweise ergeben sich Anforderungen an die Prüfgrundlage und an die Prüftechnik, und zwangsläufig müssen neben Software-Prüfungen entsprechende Hardware-Prüfungen durchgeführt werden [WoFo97]. Da derzeit noch keine diesbezügliche Norm für Chipkarten besteht, gibt es folglich auch noch keine standardisierte Prüfgrundlage oder Prüftechnik für die hier relevanten Sicherheits- und Konformitätsprüfungen. In der zuständigen Arbeitsgruppe des Deutschen Instituts für Normung - DIN (Sektorkomitee NI 17) sind allerdings entsprechenden Standards in Bearbeitung.

Die Software-Prüfung, etwa im Umfeld des eingesetzten Betriebssystems, läßt sich (mit dem bekannten Aufwand) auf der Basis der ITSEC durchführen. Hierzu lassen sich entsprechende Maßnahmen (Sicherheitsfunktionen und Mechanismen) festlegen. Daneben werden weitere allgemeine Sicherheitsmaßnahmen empfohlen, die sich auf die einzelnen Phasen des Lebenszyklus beziehen. Hier sind insbesondere die Maßnahmen bei der Herstellung, der Initialisierung und dem Versand (die durch die ITSEC nur geringfügig und daher nicht befriedigend abgedeckt sind) sowie Sicherheitsmerkmale des Kartenkörpers und anwendungsrelevante Sicherheitsmechanismen zu nennen. Mechanismen der Chipkarten-Software, etwa Algorithmen und Betriebssystemkomponenten, können im Rahmen der ITSEC-Prüfung abgedeckt werden. Dabei steht noch zur Diskussion, ob die zugrunde liegenden Algorithmen und die Speicherung geheimer Schlüssel in Form von Software erbracht werden kann, oder ob diese nicht angemessener Maßen und bedingt durch die Forderung der Signaturverordnung in Form von Hardware (was nach heutigem technischen Stand durchaus möglich wäre) realisiert werden sollte. Die Vorgabe „Mechanismenstärke hoch" läßt, nach ITSEC-Interpretation des Begriffs „hoch", beide Varianten zu. Denkbar ist auch eine verteilte Lösung im Sinne „verteilter geheimer Schlüssel".

Zudem ist offen, ob die Algorithmen zur Erstellung des erforderlichen Hashwertes (sowohl bei der Signaturgenerierung als auch bei der Signaturverifikation) außerhalb der Chipkarte zur Anwendung kommen sollen. Die Signaturverordnung läßt diese Lösung implizit zu. Sollten die passiven und aktiven hardwaremäßigen Sicherheitsmechanismen der Chipkarte gemäß ITSEC zu prüfen sein, so bedarf es einer ausführlichen Interpretation dieser Kriterien. Selbst im europäischen Vergleich sind solche Hardware-Prüfungen auf der Basis der ITSEC nicht unumstritten.

Grundsätzlich lassen sich solche Prüfungen als Black- und White-Box-Tests durchführen. Im wesentlichen handelt es sich bei der hier notwendigen Prüfung - analog zur allgemeinen Evaluierung nach ITSEC - einerseits um den Nachvollzug (Konformitätstest) von Herstellerangaben; in der Terminologie der ITSEC entspricht dies einer Korrektheitsprüfung. Anderseits handelt es sich um Penetrationstests, die mit und ohne Insiderwissen durchgeführt werden können; hier werden insbesondere Wirksamkeitsprüfungen durchgeführt.

Obwohl Hardware-Angriffe nachweislich erfolgreich durchgeführt worden sind (mit und ohne Insiderwissen), lassen sich entsprechende Abwehrmechanismen nur bedingt realisieren. Allerdings ist dies auch eine Frage der Angemessenheit. Analyseverfahren wie Timing-Attacks, der bekannte Bellcore-Angriff, der Angriff mittels differentieller Fehler-Analyse oder Kombinationen solcher (hybriden) Angriffe sind eher einer extremen professionellen Technik zuzuordnen und nur unter Einsatz hochtechnologischer Ausstattung möglich [Hora97, WoFo97]. Unter Verwendung der ITSEC als Prüfkriterien würden derart attackierbare Mechanismen aber dennoch der Mechanismenstärke „hoch" entsprechen und damit den Anforderungen der Signaturverordnung genügen.

Erste Ansätze bzgl. einer Evaluierungsgrundlage auf der Basis der zukünftig die ITSEC ersetzenden Kriterien, die sogenannten Common Criteria [CoCr97], liegen mit dem Protection Profile „Common Criteria for IT Security Evaluation Protection Profile: Smartcard Integrated Circuit Protection Profile" vor [CCPP97]. Hierzu ist allerdings zu bemerken, daß die dort beschriebene Evaluierungsstufe „nur" EAL4 abdeckt. Diese entspricht im wesentlichen der Stufe E3 der ITSEC, die Signaturverordnung fordert aber eine Stufe E4 in Kombination mit der Mechanismenstärke „hoch"!

6 Schlußfolgerungen und Ausblick

Erfolgreiche Angriffe auf die Hardware von Chipkarten können nach dem aktuellen Stand der Technik nicht absolut ausgeschlossen werden, obwohl ständig Gegenmaßnahmen erarbeitet und beim Reengineering (zumindest teilweise) berücksichtigt werden. Es muß allerdings auch angemerkt werden, daß die meisten Angriffe theoretischer Art sind und nur sehr schwer in praxisrelevante Attacken umgesetzt werden können. Dies läßt aber weder den Schluß zu, daß Chipkarten absolut (praktisch) sicher gemacht werden können, noch sollte der Gedanke aufkommen, daß eine Chipkarte schutzlos jeder denkbaren Attacke ausgesetzt ist.

Als Signier- und Verifizierkomponente im Umfeld des Signaturgesetzes ist es aber zwingend erforderlich, daß die zur Anwendung kommenden Chipkarten als Träger privater Signaturschlüssel und als Medium zur Ausführung sicherheitsrelevanter Algorithmen, insbesondere der Signaturgenerierung, als vertrauenswürdige Komponente akzeptiert und eingesetzt werden kann. Da die ITSEC nicht die Gesamtbandbreite der dafür erforderlichen Untersuchungen abdecken, sollten auch andere, den Stand der Technik berücksichtigende und erprobte (Spezial-) Kriterien einsetzt werden. Dies ist von besonderer Bedeutung, denn die ITSEC sind, bedingt durch ihre Historie, ursprünglich nicht für Hardware-Komponenten ausgelegt.

Darüber hinaus muß das zugrunde liegende Kriterienspektrum um die Common Criteria erweitert werden. Hier bietet es sich an, die Struktur der Common Criteria zu nutzen und entsprechende „Protection Profiles" für Chipkarten und für die übrigen Komponenten im Anwendungsgebiet digitaler Signaturen zu erstellen. Von Bedeutung ist darüber hinaus, daß aus den expliziten Sicherheitsmechanismen für den Herstellungsprozeß als auch aus den impliziten Anforderungen der zugrunde liegenden Prüfkriterien Vorgaben für den Entwicklungsund Herstellungsprozeß ermittelt werden sollten.

Obwohl die Signaturverordnung zuläßt, daß die Berechnung eines Hashwerts außerhalb der zur Anwendung kommenden Chipkarte stattfinden kann, bedeutet dies doch eine gewisse Einschränkung der Sicherheit. Unter bestimmten Voraussetzungen betrifft dies die Erzeugung einer digitalen Signatur oder die Verifikation einer gegebenen digitalen Signatur. Dies gilt insbesondere dann, wenn dem Gesamtsystem nicht das erforderliche Vertrauen entgegengebracht werden kann. Der Fall kann bereits dann eintreten, wenn auf einem PC lediglich ein (ungeprüftes) Betriebssystem und die entsprechende Signaturanwendung laufen. Die Alternative, Hashwerte auf der Chipkarte zu ermitteln, kann aus Gründen der Performance derzeit nicht realisiert werden. Zukünftige Techniken und Technologien werden hier vielleicht einen Ausweg aufzeigen.

Nach dem heutigen Kenntnisstand können Systeme zur Realisierung digitaler Signaturen mit der Hilfe von Chipkarten relativ leicht und angemessen sicher verwirklicht werden. Dies gilt hauptsächlich dann, wenn es sich um Anwendungen in geschlossenen Benutzergruppen, sogenannten isolierten Domänen [HoWo97] handelt. Die Propheten einer neuen digitalen Signaturwelt werden aber mit großer Wahrscheinlichkeit den Realitäten noch oft ins Auge sehen müssen, bis sie die komplexen Strukturen des gesamten erforderlichen Umfeldes so eingeordnet haben, daß flächendeckende sichere und kompatible Lösungen ermöglicht werden können. Außerdem werden beispielsweise in Electronic-Commerce-Anwendungen zusätzliche Signaturschemata (etwa blinde und empfängerspezifische digitale Signaturen [PeMH96]) zum Einsatz kommen.

Das Signaturgesetz soll in vielen Bereichen die Grundlage für eine zukünftige rechtliche Gleichstellung von eigenhändiger Unterschrift und digitaler Signatur bilden. Die Signaturverordnung definiert dafür einen generischen Rahmen, und die Maßnahmenkataloge enthalten Beispiele zur inhaltlichen Ausgestaltung konkreter Realisierungen. Signaturgesetz, Signaturverordnung und Maßnahmenkataloge können somit einen wertvollen Beitrag für die Realisierung rechtsrelevanter digitaler Signaturen liefern. Gleichwohl werden in einer ersten

Phase nicht alle Anforderungen in die Praxis umgesetzt werden können. In konkreten Projekten wird sich nach einer ersten Bewährungsprobe aber schnell zeigen, welche Aspekte hilfreich und welche störend und damit zu revidieren sind. Dies betrifft speziell das Problem der Interoperabilität zwischen isolierten Anwendungsbereichen im nationalen und im internationalen Umfeld. Zuvor ist allerdings noch eine Sensibilisierung der zukünftigen Nutzer und insbesondere eine entsprechende Aufklärungsarbeit erforderlich.

Literatur

[CoCr97] Common Criteria, Version 2.0, Dezember 1997, www.csrc.nist.gov/cc/ccv1x/ wip_list.htm

[CCPP97] Common Criteria for IT Security Evaluation Protection Profile: Smartcard Integrated Circuit Protection Profile, Issue October 1997, Registered at the French Certification Body under the number PP/9704.

[ElGa85] ElGamal, T.: A Public Key Cryptosystem and a Signature Scheme Based on Discrete Logarithms, IEEE Transactions on Information Theory 31 (1985) 469-472.

[Hora97] Horak, M.: Sicherheit von Chipkarten, Tagungsband ChipCard 97, Computas (1997).

[Hors96] Horster, P. (Hrsg.): Digitale Signaturen, DuD-Fachbeiträge, Vieweg (1996).

[HoKr96] Horster, P., Kraaibeek, P.: Grundüberlegungen zu digitalen Signaturen, Tagungsband Digitale Signaturen, Vieweg (1996) 1-14.

[HoWo97] Horster, P., Wohlmacher, P.: Sicherheitsinfrastrukturen als Quelle des Vertrauenes, Tagungsband Sicherheitsinfrastrukturen in Wirtschaft und Verwaltung - SiS-WV 97, Computas (1997).

[Isse97] Isselhorst, H.: Chipkarten - ausreichend sicher?, KES 2 (1997) 21-27.

[ITSE91] Kriterien für die Bewertung der Sicherheit von Systemen der Informationstechnik (ITSEC), Kommission der Europäischen Gemeinschaft, EGKS-EWG-EAG (1991) ISBN 92-826-3003-X.

[ITSE94] Information Technology Security Evaluation Manual (ITSEM), Kommission der Europäischen Gemeinschaft, ECSC-EEC-EAEC (1993) ISBN 92-826-70 87-2.

[IuKD97] Gesetz zur Regelung der Rahmenbedingungen für Informations- und Kommunikationsdienste (Informations- und Kommunikationsdienste-Gesetz - IuKDG), Bundesgesetzblatt 1869, Teil I G5702 (1997) 1869-1880.

[Keus97] Keus, K.: Chipkarten im Umfeld des Signaturgesetzes, Tagungsband ChipCard 97, Computas (1997).

[MKat97] Angaben des BSI zum Maßnahmenkatalog gemäß SigV §§ 12(2) und 16(6), Ausgabe Version 1.0 vom 18.11.1997, www.bsi.bund.de

[NIST94] National Institute of Standards and Technology: Federal Information Processing Standard FIPS PUB 186, Digital Signature Standard (1994).

[PeMH96] Petersen, H., Michels, M., Horster, P.: Taxonomie digitaler Signaturkonzepte, Tagungsband Digitale Signaturen, Vieweg (1996) 63-79.

[RaEf95] Rankl, W., Effing, W.: Handbuch der Chipkarten, München, Wien: Carl Hanser-Verlag, 1995.

[RiSA78] Rivest, R. L., Shamir, A., Adleman, L.: A Method for Obtaining Digital Signatures and Public Key Cryptosystems, Communications of the ACM (1978) 120-126.

[SigV97] Verordnung zur digitalen Signatur (Signaturverordnung - SigV), Stand: 8. Oktober 1997, www.iid.de/aktuelles/index.html/#iukdg

[WoFo97] Wohlmacher, P., Fox, D.: Hardwaresicherheit von Smartcards, DuD - Datenschutz und Datensicherheit 21 (1997) 260-265.

Digitale Signatur-Anwendung nach Signaturgesetz und Signaturverordnung

Bruno Struif

GMD – Forschungszentrum Informationstechnik
Rheinstr. 75, D-64295 Darmstadt
struif@darmstadt.gmd.de

Zusammenfassung

Am 22.07.1997 wurde das Gesetz zur digitalen Signatur (Signaturgesetz – SigG) verabschiedet. Das BSI (Bundesamt für Sicherheit in der Informationstechnik) als vom BMI mit der Ausarbeitung der Maßnahmenkataloge beauftragte Instanz stellte im Einvernehmen mit der Regulierungsbehörde (oberste nationale Zertifizierungsstelle) beim DIN den Antrag auf Erstellung einer Spezifikation für Chipkarten mit digitaler Signatur-Anwendung entsprechend SigG und SigV (Signaturverordnung). In diesem Beitrag wird die vom AK DIN NI-17.4 ausgearbeitete Spezifikation (Stand 0.75 vom 10.12.97) vorgestellt.

1 Einleitung

Die DIN-Spezifikation [DIN-SIG] berücksichtigt die Vorgaben aus

- Signaturgesetz
- Signaturverordnung und
- Maßnahmenkatalog

und verwendet als Grundlage für die Spezifikation der Schnittstelle zur Chipkarte die relevanten ISO/IEC-Standards, insbesondere

- ISO/IEC 7816-4, -8 [ISO7816-4, -8] und
- ISO/IEC 9796-2 [ISO9796-2].

Aus der Sicht der DIN-Spezifikation gibt es eine Hierarchie von Anforderungen, wie sie in Abb. 1 dargestellt ist.

Gesetzliche Regelung	§ Signaturgesetz SigG
Ausführungsbestimmungen	Signaturverordnung SigV
Technische Anforderungen	Maßnahmenkataloge
Technische Spezifikationen	Technische Spezifikationen für Chipkarten, TrustCenter, ...

Abb. 1: Einordnung der DIN-Spezifikation

2 SigG-Anwendung

Die nachfolgende Abbildung zeigt den Ablauf der Signatur-Anwendung in der Chipkarte.

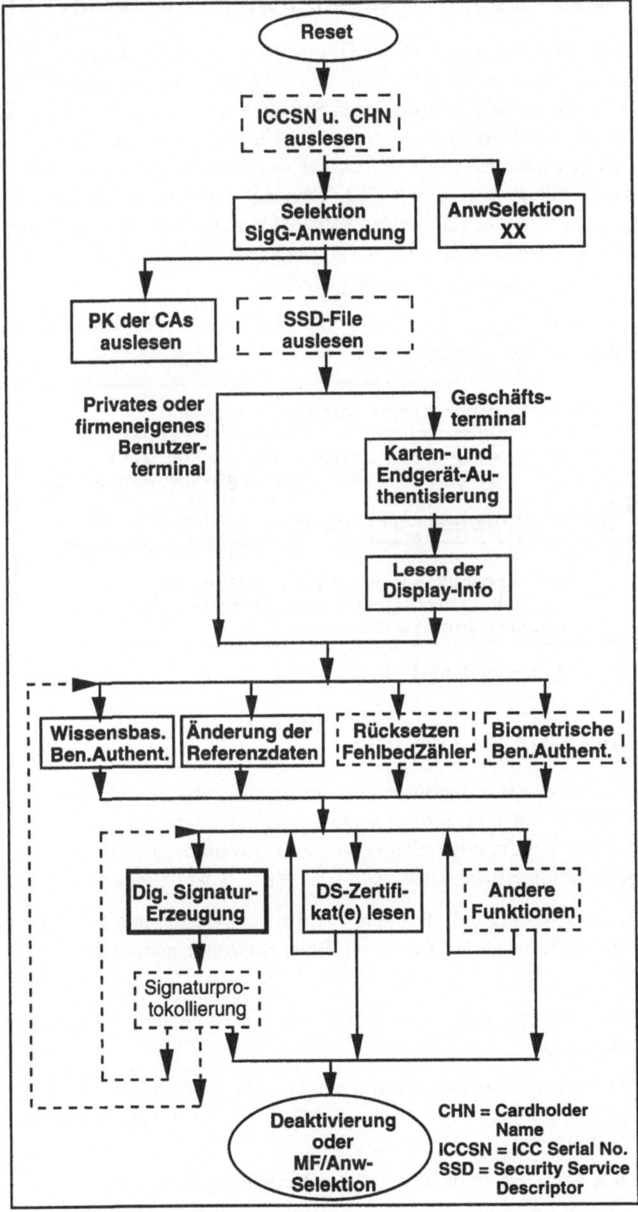

Abb. 2: Genereller Ablauf der SigG-Anwendung in einer Chipkarte

Wegen des großen Umfangs der Spezifikation können im nachfolgenden nur die wichtigsten Aspekte dargestellt werden.

2.1 Reset, Lesen globaler Datenobjekte, Anwendungsselektion

Nach erfolgtem Reset ist die Chipkarte im Grundzustand. In diesem erlaubt sie das Lesen von ElementaryFiles, die direkt unter dem MasterFile (Wurzel des Chipkarten-Filesystems) aufgehängt sind und z.B. globale Informationen enthalten, und natürlich die Selektion von Anwendungen. Als globale Datenobjekte sollen die Chipkarten-Serien-Nummer (ICCSN) und der Name des Karten-Inhabers in der Form, wie er üblicherweise auf dem Karten-Cover zu finden ist, bereitgestellt werden. Die ICCSN kann z.B. für Karten-Sperrlisten (ICC revocation list) verwendet werden, der Cardholder Name ist vor allem für Plug-in-Karten (Chipkarten im Kleinformat wie sie z.B. in Handys benutzt werden) wichtig, so daß auch hier die Information zur Verfügung steht, wem die Karte gehört.

Zur Anwendungsselektion werden weltweit eindeutige Application Identifier verwendet (siehe [ISO7816-5]). Das Anwendungskennzeichen (AID) der SigG-Anwendung zeigt Abb. 3.

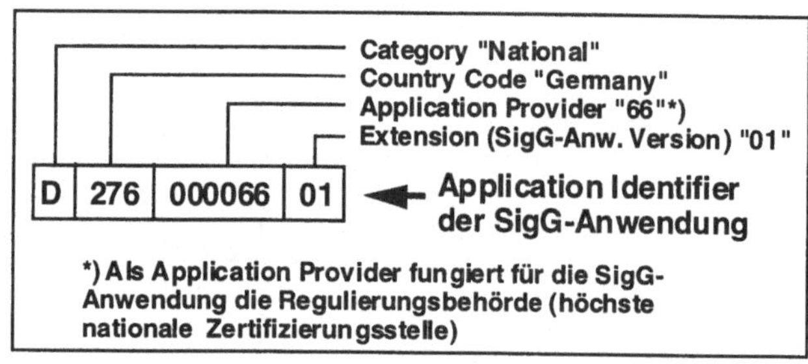

Abb. 3: Anwendungskennzeichen (AID) der SigG-Anwendung

2.2 Benutzer-Authentisierung

Der Karteninhaber muß entweder durch eine wissensbasierte Identifikation (PIN- oder Paßwort-Eingabe) oder durch eine biometrische Identifikation nachweisen, daß er der rechtmäßige Benutzer der betreffenden Chipkarte ist. Das wissensbasierte und das biometrische Benutzer-Authentisierungsverfahren können koexistent in der Chipkarte realisiert sein.

Chipkarten können je nach Konfigurierung bei der Personalisierung bzw. der Fähigkeit des jeweiligen Betriebssystems im Bezug auf die Benutzer-Authentisierung zwei Verhaltensweisen zeigen:

1.   Die Karte verlangt vor jeder Ausführung der digitalen Signatur eine Benutzer-Authentisierung.

2.   Die Karte verlangt nur eine einmalige Benutzer-Authentisierung, d.h. aus der Sicht der Chipkarte können dann im Prinzip beliebig viele Signaturen nacheinander angefertigt werden.

Bei der 2. Variante kann jedoch in Anwendungs-Umgebungen, wo

-    vor jeder digitalen Signatur oder

- vor einer digitalen Signatur nach n digitalen Signaturen oder

- vor einer digitalen Signatur nach Ablauf einer definierbaren Zeitspanne x nach der letzten digitalen Signatur

erneut die Benutzer-Authentisierung durchlaufen werden soll, vom Karteninhaber über eine entsprechende Bedienerführung oder durch Konfigurierung des Anwendungssystems die Einstellung der gewünschten Variante veranlaßt werden. Aus Systemsicht bedeutet dies, daß sofort nach Eintreten des entsprechenden Ereignisses (Signatur erzeugt, n Signaturen erzeugt oder Zeit abgelaufen) das Anwendungssystem entweder

- ein Reset zur Chipkarte oder

- einen Rücksprung zum übergeordneten DF (d.h. in der Regel zum MF) durch Selektion des übergeordneten DFs

veranlaßt, so daß bei erneutem Signierwunsch des Benutzers eine erneute Anwendungsselektion erfolgt und anschließend wieder die Benutzer-Authentisierung angefordert wird.

Der Maßnahmenkatalog fordert bei wissensbasierter Benutzerauthentisierung eine Mindestlänge der PIN bzw. des Paßwortes von 6 Stellen. Auch wird gefordert, daß die PIN bzw. das Paßwort änderbar sein muß.

Die in der Chipkarte abgelegten Daten zur Benutzer-Authentisierung (PIN, Paßwort, biometrische Daten) werden entsprechend ISO/IEC 7816-4 [ISO7816-4] als Reference Data bezeichnet, die mit den von außen angelieferten Verifikationsdaten in der Chipkarte verglichen werden müssen (siehe Abb. 4).

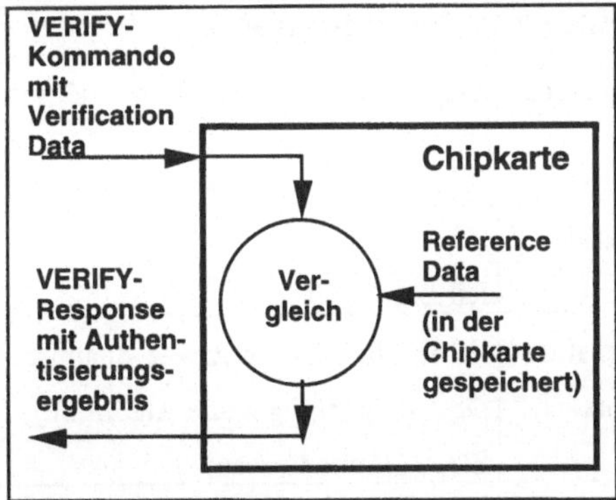

Abb. 4: Vergleich von Verifikationsdaten und Referenzdaten

Bei wissensbasierter Benutzer-Authentisierung werden die Benutzereingabedaten formatiert und im Datenteil des VERIFY-Kommandos zur Chipkarte übertragen, die einen Vergleich auf Übereinstimmung mit den Referenzdaten vornimmt.

Bei biometrischer Benutzer-Authentisierung müssen die biometrischen Meßdaten zunächst einer Merkmalsextraktion unterworfen, dann formatiert und schließlich im Datenteil des VERIFY-Kommandos zur Chipkarte übertragen werden, die mit einem Merkmals-Vergleichs-Algorithmus die Verifikationsdaten auf hinreichende Übereinstimmung mit den Referenzdaten zu prüfen hat (siehe Abb. 5). An der Realisierung von Vergleichsalgorithmen in der Chipkarte wird gearbeitet, d.h. derzeit ist dieses Verfahren selbst für biometrische Verfahren mit geringerer Komplexität noch nicht verfügbar.

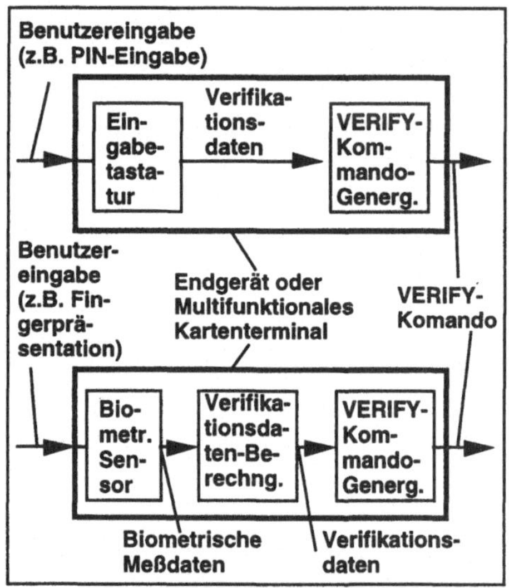

Abb. 5: Aufbereitung und Formatierung der Verifikationsdaten

2.3 Erzeugung einer digitalen Signatur

Derzeit sind die in Tab.1 angegebenen Verfahren als Hash-Funktionen und Signatur-Algorithmen vorgesehen.

Hash-Funktion	SHA-1
	RIPEMD-160
Digital Signature-Algorithmus	RSA (Algorithm of Rivest, Shamir, Adleman)
	DSA (Digital Signature Algorithm)
	ELC (Elliptic Curves)

Tab. 1: Hash-Funktionen und Digital Signature-Algorithmen

Hash-Funktionen lassen sich schon in Chipkarten implementieren. Bei längeren Texten ist es jedoch aus Performance-Gründen besser, das Hashen im PC oder Terminal vorzunehmen. Grundsätzlich sieht die DIN-Spezifikation jedoch drei Varianten in Bezug auf die Arbeitsteilung vor (siehe Abb. 6).

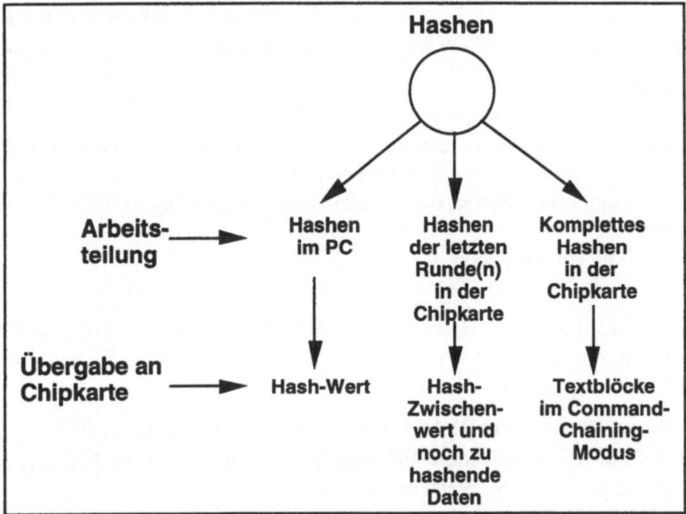

Abb. 6: Mögliche Arbeitsteilung beim Hashen zwischen Endgerät und Chipkarte

Ein Signatur-Prozeß umfaßt folgende Arbeitsschritte:

- Hashen der Nachricht
- Formatieren des „Digital Signature Inputs (DSI)"
- Berechnen der digitalen Signatur

Abb. 7 zeigt dieser Arbeitsschritte in ihrer Abfolge.

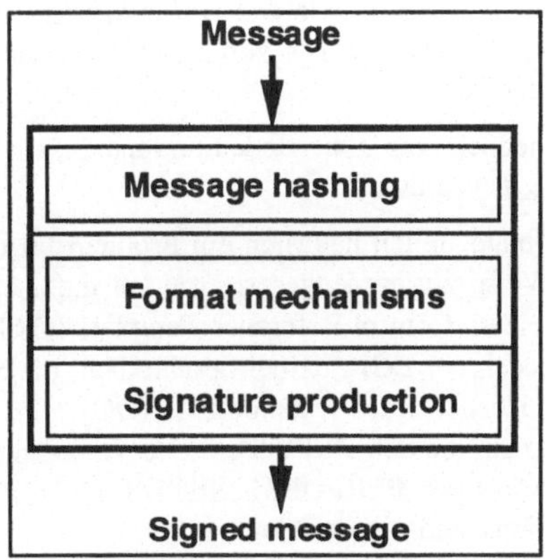

Abb. 7: Signatur-Prozeß

Bei Signatur-Algorithmen mit DSI Recovery wie z.B. RSA ist die Formatierung des Digital Signature Inputs von besonderer Bedeutung.

Als SigG-Grundformat für die DSI-Formatierung bei Nutzung von RSA wird die Konvention basierend auf ISO/IEC 9796-2 [ISO9796-2] wie folgt verwendet:

- Header: 2 bits (=01)
- More-data bit = 1 (Mn not empty)
- Padding field : zero, one or more bits equal to 0 followed by a single bit set to 1 (recommended: 12 bits 0 followed by a single bit equal to 1)
- Data field: random no. inserted by the card (length depending on RSA key length, min length 8 bytes)
- Hash field: kh bits of hash-code
- Trailer: 1 byte: ´BC´

Die Chipkarte padded also den Hash-Wert mit einer Zufallszahl, verändert somit den DSI und erzwingt dadurch, daß selbst bei gleichem Hash-Wert die digitale Signatur immer einen anderen Wert hat.

Im Internet werden überwiegend RSA-basierte Signaturen verwendet, deren Signatur-Format sich an PKCS #1 [PKCS1] orientiert. Für digitale Signaturen sind die Blocktypen 0 und 1 vorgesehen. Blocktyp 1 hat z.B. folgenden Aufbau:

- Eingangsbyte: ´00´
- Blocktypanzeige: ´01´
- Padding-String: ´FF ...FF´
- Separator: ´00´
- DigestInfo: digestAlgorithm und digest (als ASN.1-Sequenz codiert)

Bei diesem Signatur-Format wird der Chipkarte ebenfalls normalerweise nur der Hash-Wert (Digest) übergeben, d.h. die übrigen Teile des DSI werden von der Chipkarte ergänzt (die meisten Chipkarten erlauben jedoch auch die komplette Übergabe des DSI von außen). Das bereits 1993 festgelegte PKCS #1-Format hat den Nachteil, daß jedesmal die gleiche Signatur erzeugt wird, wenn der Hash-Wert identisch ist, da kein veränderliches Datenelement (Signaturzähler, Zufallszahl) von der Chipkarte in den DSI integriert wird. Das PKCS #1-Format ist daher anfällig gegen einen bestimmten Typ von Angriff, der unter der Bezeichnung Bellcore-Attack bekannt geworden ist.

Da Internet-Anwendungen jedoch eine hohe Bedeutung haben, sind beide Signatur-Formate (ISO/IEC 9796-2 und PKCS #1) zulässig.

Die verschiedenen DSI-Formate stellen natürlich ein Problem für die Chipkarte dar, da der DSI auf unterschiedliche Weise zusammengesetzt werden muß. Mit Hilfe einer sog. Headerlist [ISO7816-8], die in einem Control Reference Template (CRT) für Digitale Signaturen abgelegt werden kann, läßt sich der DSI flexibel beschreiben: die Sequenz der Tag-Length-Werte (Tag-Length ist der Header eines Datenobjekts (DO)) gibt die Sequenz der Datenobjekte bzw. Datenelemente an, die den DSI bilden. Die Verwendung der CRTs im Umfeld des ISO/IEC 7816-8-Kommandos PERFORM SECURITY OPERATION (PSO) zeigt Abb. 8. Als Security Operations sind u.a. definiert

- HASH
- COMPUTE DIGITAL SIGNATURE
- VERIFY DIGITAL SIGNATURE
- VERIFY CERTIFICATE.

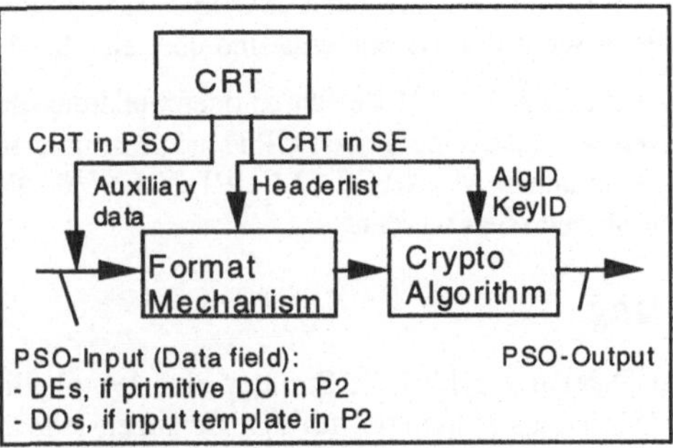

Abb. 8: Verwendung von Control Reference Templates (CRT)

Wendet man die ISO-DSI-Beschreibungsmethode an, dann ergeben sich daraus ein CRT für DSI nach ISO/IEC 9796-2 (Abb. 9) und ein CRT für DSI nach PKCS # 1 (Abb. 10). Der Tag ´4D´ für das Headerlist DO legt fest, daß nur der Value-Teil der in der Headerlist aufgeführten Datenobjekte konkateniert werden soll.

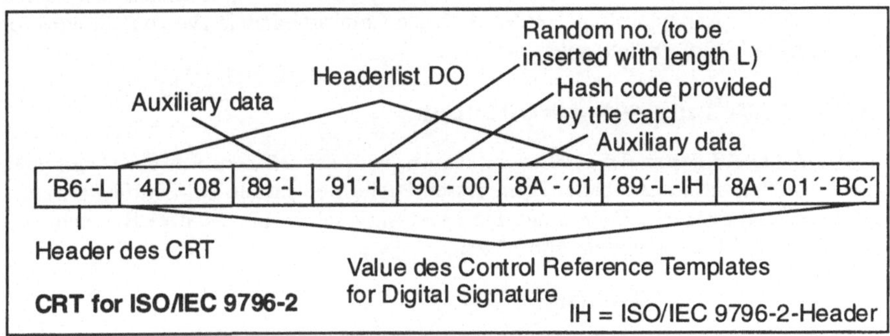

Abb. 9: CRT für DSI entsprechend ISO/IEC 9796-2

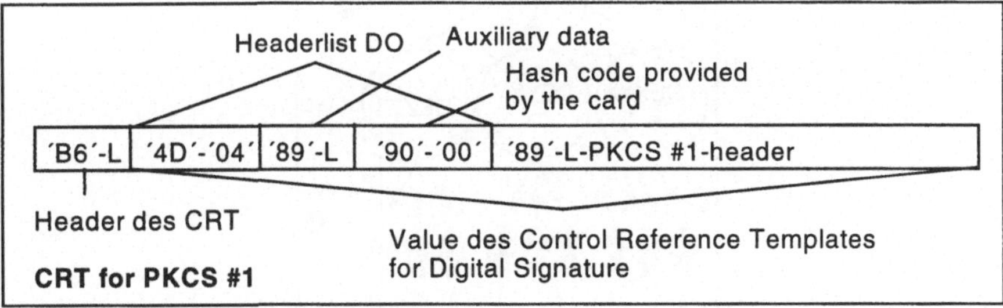

Abb. 10: CRT für DSI entsprechend PKCS #1

Bei Bedarf kann auch in einem CRT ein Key Identifier (KID) spezifiziert werden. In dem über die KID referenzierten Key Record ist dann üblicherweise auch der zugehörige Algorithmus angegeben. Falls nicht, besteht auch im CRT die Möglichkeit, einen Algorithm Identifier zu spezifizieren. CRTs sind enstprechend ISO/IEC 7816-8 Bestandteil von Security Environments (SE). Solche Security Environments sind über eine SE # referenzierbar.

Untersützt eine Chipkarte beide RSA-DSI-Verfahren oder sind koexistent RSA und DSA implementiert, dann müssen die unterschiedlichen CRTs auch in unterschiedlichen SEs abgelegt werden. Mit dem Kommando MANAGE SECURITY ENVIRONMENT, Option SET, kann dann das benötigte SE selektiert werden.

2.4 Protokollierung

Optional kann in einer Datei mit zyklischer Struktur eine Protokollierung von Signaturen vorgenommen werden. Sie ist als Hilfsmittel für den Benutzer, jedoch nicht im Sinne einer juristischen Beweissicherung zu sehen. Ein Record-Inhalt besteht aus

- Datum,
- Terminalkennung,
- Dokumenten-Kennung und
- Signaturzähler.

Der Stand des Signaturzählers kann dem letzten gespeicherten Record entnommen werden oder von der Chipkarte abgefragt werden, falls die Chipkarte diesen Typ von „Progression Value" (siehe ISO/IEC 7816-8) unterstützt.

2.5 Bereitstellung von Zertifikaten

Um eine digitale Signatur (DS) prüfen zu können, wird das DS-Zertifikat des Unterschreibenden benötigt, das in gesicherter Form dessen Public Key enthält. DS-Zertifikate werden von Zertifizierungsstellen (ZS) erzeugt. SigG und SigV sehen eine 2-stufige Hierarchie der Zertifizierungsstellen vor, wie sie Abb. 11 zeigt.

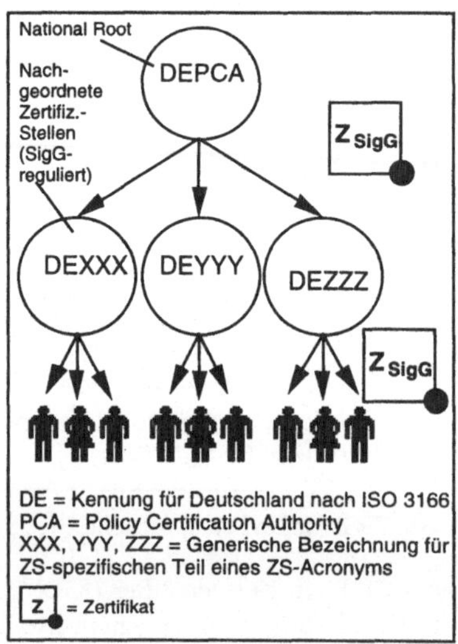

Abb. 11: Zertifizierungsstellen und Zertifikate

Der Aufbau der Zertifikate ist nicht Gegenstand der DIN-Spezifikation, wohl aber, wie sie aus der Chipkarte ausgelesen werden können, falls sie dort abgelegt sind.

2.6 Bereitstellung von öffentlichen ZS-Schlüsseln

Zertifikate tragen eine digitale Signatur der Zertifizierungsstelle, die das Zertifikat ausgegeben hat. Die prüfende Instanz benötigt daher den öffentlichen Schlüssel der Zertifizierungsstelle. Die Chipkarte kann z.B. den Public Key der höchsten ZS direkt oder in einem „ZS-Selbstzertifikat" bereitstellen, so daß er der Software, die ein DS-Zertifikat prüft, zur Verfügung steht.

2.7 Erstellung von digitalen Signaturen an Geschäftsterminals

In SigG/SigV wird unterschieden, ob eine digitale Signatur an einem Geschäftsterminal, das von Dienstleistungserbringern Dritten zur Verfügung gestellt wird (z.B. öffentliches E-mail-System), oder aber an einem privaten PC bzw. einem PC im Büro ausgeführt werden soll. Für den Fall des Geschäftsterminals wird erhöhte Sicherheit gefordert, die in einer gegenseitigen Authentisierung von Chipkarte und Terminal besteht. Damit der Karteninhaber erkennen kann, ob die gegenseitige Authentisierung erfolgreich durchlaufen wurde, soll eine kartenspezifische Display-Message angezeigt werden, die nur von der Karte freigegeben wird, wenn die gegenseitige Authentisierung positiv verlaufen ist.

Literatur

[DIN-SIG]   DIN NI-17.4: Spezifikation der Schnittstelle zu Chipkarten mit digitaler Signatur-Anwendung/Funktion nach SigG/SigV. Version 0.75 vom 10.12.97

[SigG]   Gesetz zur digitalen Signatur (Signaturgesetz - SigG) vom 22.07.1997 (BGBl. I S. 1870, 1872), verkündet als Artikel 3 des „Gesetzes zur Regelung der Rahmenbedingungen für Informations- und Kommunikationsdienste (Informations- und Kommunikationsdienste-Gesetz - IuKDG)"

[SigV]   Verordnung zur digitalen Signatur (Signaturverordnung - SigV) vom 22.10.1997 (BGBl. I S. 2498)

[ISO7816-3] ISO/IEC 7816-3: 1997 (2nd edition) Information technology - Identification cards - Integrated circuit(s) cards with contacts - Part 3: Electronic signals and transmission protocols

[ISO7816-4] ISO/IEC 7816-4: 1995 Information technology - Identification cards - Integrated circuit(s) cards with contacts - Part 4: Interindustry commands for interchange - Part 4 / AM1: Impact of secure messaging on the structures of APDU messages

[ISO7816-5] ISO/IEC 7816-5: 1995 Information technology - Identification cards - Integrated circuit(s) cards with contacts - Part 5: Numbering system and registration procedure for application identifiers

[ISO7816-6] ISO/IEC 7816-6: 1995 Information technology - Identification cards - Integrated circuit(s) cards with contacts - Part 6: Interindustry data elements

[ISO7816-8] ISO/IEC 7816-8: Final CD 1997 Information technology - Identification cards - Integrated circuit(s) cards with contacts - Part 8: Security related interindustry commands

[ISO9796-2] ISO/IEC 9796-2: 1997 Information technology - Security techniques - Digital signature schemes giving message recovery - Part 2: Mechanisms using a hash-function

[ISO10118] ISO/IEC 10118: 1997 Information technology - Security techniques - Hash functions - Part 3: Dedicated hash-functions

[PKCS1] PKCS #1: RSA Encryption Standard - Version 1.5, Nov. 1993

Personalisierung von Chipkarten nach Signaturgesetz und Signaturverordnung

Gisela Meister

Giesecke & Devrient
Postfach 80 07 29, D-81607 München
GiMei@compuserve.com, Gisela.Meister@gdm.de

Zusammenfassung

Am 22.07.1997 wurde das Gesetz zur digitalen Signatur (Signaturgesetz – SigG) verabschiedet. In diesem Beitrag wird basierend auf der vom AK DIN NI-17.4 im Auftrag des BSI und der Regulierungsbehörde ausgearbeiteten Spezifikation (Stand 0.75 vom 10.12.97) und aufbauend auf dem von beiden Behörden gemeinsam vorgestellten Maßnahmenkatalog für digitale Signaturen (Stand 1.0 vom 18.11.97) ein Konzept zur Personalisierung von Chipkarten vorgestellt.

1 Einleitung

Bei der Personalisierung von Chipkarten, die gemäß Signaturgesetz und Signaturverordnung (SigG und SigV) eingesetzt werden können, spielt insbesondere einerseits die Instanz eine Rolle, die die Schlüsselgenerierung von Teilnehmerschlüsseln vornimmt und andererseits ist das Medium entscheidend, das die Schlüsselgenerierung durchführt. Aus Sicht der Chipkarte sind daher zunächst folgende Ansätze zu unterscheiden:

- Die Schlüsselgenerierung für das Signaturschlüsselpaar eines Teilnehmers mit geheimen und öffentlichen Anteil erfolgt außerhalb seiner Karte unter Kontrolle eines Trustcenters, z.B. der zugeordneten Zertifizierungsstelle (ZS).

- Die Schlüsselgenerierung für das Signaturschlüsselpaar erfolgt in der Teilnehmer-Karte.

Bei beiden Ansätzen ist es entscheidend, die Kommunikation zwischen Kartenhalter und Zertifizierungsstelle abzusichern.

Im ersten Fall wird der Schlüssel bei seiner Erstellung unter Kontrolle der ZS erzeugt. Der geheime Schlüssel wird unter Wahrung der Integrität in die Karte geladen. Der öffentliche Schlüssel wird in ein sogenanntes Public-Key-Zertifikat gebracht, das den Namen des Kartenhalters oder ein Pseudonym in eindeutiger Weise mit seinem öffentlichen Schlüssel, dem Public-Key (PK) verknüpft und durch die Unterschrift der Zertifizierungsstelle beglaubigt.

Im zweiten Fall kann der Kartenhalter entweder der Teilnehmer oder aber der Kartenherausgeber sein. Wenn z.B. der Kartenteilnehmer sein eigenes Schlüsselpaar erzeugt, wird der öffentliche Schlüsselteil unter Wahrung der Integrität zur Zertifizierungsstelle geschickt und nach Protokoll verknüpft mit weiteren Datenelementen wie z.B. dem Namen des Kartenhalters und unter Umständen der eindeutigen Kartennummer (ICCSN).

Beide Ansätze werden im folgenden detailliert beschrieben, wobei hier ausschließlich asymmetrische Verfahren für die Kommunikationsabsicherung benutzt werden. Diese Verfahren spielen zunehmend eine Rolle, da ihr Einsatz nicht auf geschlossene Systeme beschränkt ist, sondern auch beim Einsatz in offenen Systemen wie dem Internet ihre Anwendung findet.

Für die Spezifikation einer Personalisierungsschnittstelle, wie sie das Bundesamt für Sicherheit in der Informationstechnik (BSI) beim Deutschen Institut für Normung (DIN) beantragt hat, sollten beide Ansätze berücksichtigt werden. Das BSI hat in seinen Maßnahmenkatalogen exemplarisch zum zweiten Ansatz einer möglichen Personalisierung eines Schlüsselhalter-Mediums, eines *Personal Security Environment* (PSE), Stellung bezogen. Hier wird dieser Ansatz detailliert für das individuelle und flexible Medium, die PSE in ihrer Ausprägung als Chipkarte, spezifiziert, wobei auf Chipkarten-Standards und auf Standards für kryptographische Verfahren Bezug genommen wird.

Im folgenden Abschnitt werden zunächst grundsätzliche Dienstleistungen einer Zertifizierungsstelle erläutert.

2 Zentrale und dezentrale Zertifizierungsstellen

Die Basisfunktion einer Zertifizierungsstelle ist die Erstellung und Verteilung von sogenannten Public-Key-Zertifikaten für den digitalen Signaturdienst an registrierte Teilnehmer.

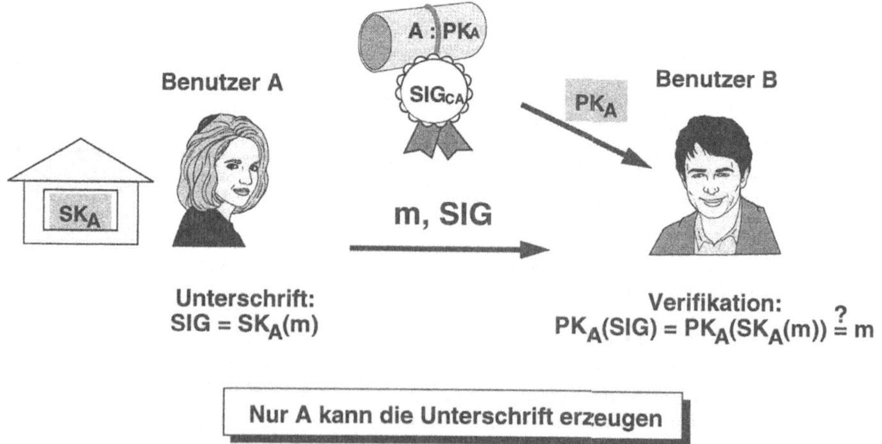

Abb.1: Zertifikatsbasierte Public-Key (PK)-Systeme, generelles Modell

Ein Public-Key-(PK)-Zertifikat ist allgemein als eine Datenstruktur definiert, die vorher digital von der zuständigen Instanz, der *Zertifizierungsstelle*, signiert wurde, um den im Zertifikat enthaltenen öffentlichen Schlüssel mit dem diesbezüglichen Namen (*distinguished name*) des Teilnehmers zu verbinden. Dabei kann ein PK-Zertifikat auch noch zusätzliche Informationen enthalten.

Eine Zertifizierungsstelle nach SigG und SigV ist nach dem Maßnahmenkatalog [RBPT], 2.1 ein Trustcenter mit der folgenden Aufgabe:

Verfügt der Teilnehmer nicht über ein selbst generiertes Schlüsselpaar, ist von der ZS ein Schlüsselpaar für diesen Teilnehmer zu generieren. Dieses Schlüsselpaar besteht aus einem privaten und einem öffentlichen Schlüssel. Der private Schlüssel wird vom Teilnehmer für die Bildung digitaler Signaturen, der öffentliche Schlüssel für die Verifikation der Signaturen benötigt. Dabei ist elementar, daß die privaten Schlüssel nach der Übergabe an den Teilnehmer in der ZS vernichtet werden, und daß jedes Schlüsselpaar nur einmal vorkommt. Um eine authentische Zuordnung der vom Teilnehmer erzeugten digitalen Signaturen ermöglichen, muß die Zuordnung des Schlüsselpaares zu diesem Teilnehmer ebenfalls in authentischer Weise erfolgen. Diese Aufgabe wird vom Schlüsselerzeugungsdienst (KG) wahrgenommen.

Die folgende Abbildung aus dem Maßnahmenkatalog erläutert das Zusammenspiel der einzelnen Dienste einer ZS.

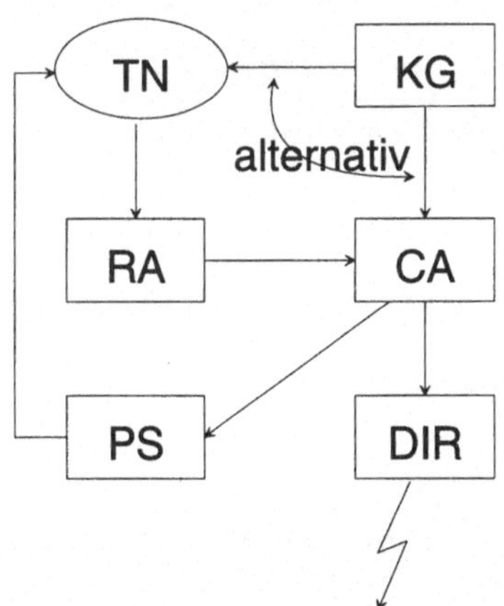

Abb.2: Instanzen einer ZS nach [ISO14516-2] bzw.[RBPT]

Zunächst wird vom Teilnehmer (TN) selbst oder in der Zertifizierungsstelle ein Schlüsselpaar generiert (KG Instanz). Der Teilnehmer läßt sich bei der Registrierungsstelle (RA) registrieren und identifizieren und beantragt ein Zertifikat. Dieses Zertifikat wird von der Zertifikats-Ausstellungs-Instanz (CA) erstellt und sowohl an das Personalisierungssystem als auch an den Verzeichnisdienst (DIR) übermittelt. Das Personalisierungssystem (PS) überträgt die für den Teilnehmer relevanten Daten, die noch nicht auf der Signierkomponente vorhanden sind (also ggf. auch das Schlüsselpaar), auf die Signierkomponente, welche nun dem Teilnehmer übergeben werden kann. Der Verzeichnisdienst ist über öffentliche Kommunikationseinrichtungen erreichbar.

Die nach SigG vorgesehene Zertifizierungshierarchie ist zweistufig. Diese bezieht sich allerdings ausschließlich auf Zertifikate für den Signaturdienst. Die Regulierungsbehörde übernimmt dabei die Rolle der Wurzelinstanz und zertifiziert öffentliche Signaturschlüssel der akkreditierten Zertifizierungsstellen mit jeweils angeschlossener Zertikats-Ausstellungs-Instanz., der Certification Authority (CA). Diese wiederum zertifizieren die öffentlichen Signaturschlüssel registrierter Teilnehmer.

Für Zertifikate, die für Authentisierungsdienste, eingesetzt werden, ist nicht notwendig eine zweistufige Hierarchie der CAs erforderlich.. D.h. es bzw. können für diesen Dienst auch andere CAs, insbesondere nicht notwendig genehmigte CAs nach SigG und SigV eingesetzt werden.

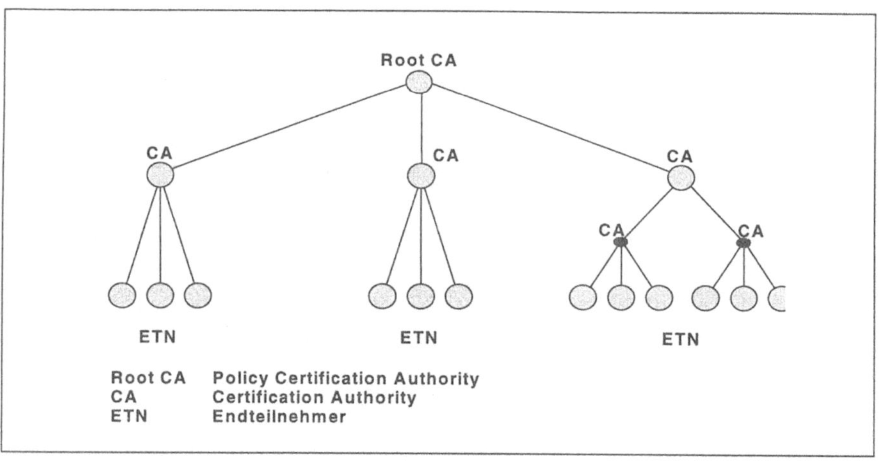

Abb.3: Zertifikats-Hierarchie, generelles Modell

Aus Sicht der ZS sind grundsätzlich folgende Modelle zu unterscheiden (siehe Maßnahmenkatalog, 5.3):

- **Zentrales Modell**

  In diesem vollständig zentral orientiertem Modell besitzt die ZS keine Außenstellen, die Teilnehmerschlüssel werden durch die ZS erzeugt.

  Die für den Betrieb einer Zertifizierungsstelle notwendigen Dienstleistungen werden alle von einer Institution innerhalb eines eng begrenzten Gebietes (i.d.R. innerhalb eines Gebäudekomplexes) realisiert.

- **Dezentrales Modell**

  In diesem möglichst verteilt operierenden Modell betreibt die ZS Außenstellen, welche die Aufgaben Registrierung (und Identifizierung) und Personalisierung wahrnehmen. Die Schlüsselerzeugung erfolgt zudem durch die Teilnehmer. Alle anderen Dienste werden weiterhin zentral abgewickelt.

Mischformen der Lösungen 1 und 2 sind in verschiedenen Variationen vorstellbar.

3 Personalisierungsmodelle für Zertifizierungsstellen

Im nächsten Abschnitt wird unabhängig von der Organisation der ZS der Ablauf der Schlüsselgenerierung dargestellt.

3.1 Exemplarische Ablaufbeschreibung zur Personalisierung von PK-Schlüsseln, dezentrales Modell

Der Ablauf des gesamten Personalisierungsprozesses inklusive Schlüsselerzeugung in der Chipkarte wird im folgenden beschrieben.

Dabei kann sich jede CA einer Zertifizierungsstelle (ZS) von der Integrität der Chipkarte (abgekürzt als ICC bezeichnet) überzeugen. Die Schlüsselerzeugung des Teilnehmer-Schlüssel-Paares wird in der ICC selbst vorgenommen und im Fall, daß diese durch den Teilnehmer angestoßen wird, erfolgt sie erst nach Aushändigung der Chipkarte (siehe 3.1.1).

Ebenfalls möglich nach SigG und SigV ist der Fall, das der Kartenherausgeber als Kartenhalter die Schlüsselgenerierung in der Karte anstößt, siehe 3.1.2.

Die CA kann aus dem vom Kartenhalter vorgelegten öffentlichen Schlüssel, jedoch ohne Vorlage der Chipkarte, feststellen, ob die Chipkarte das Schlüsselpaar tatsächlich selbst erzeugt hat.

Eine wichtige zusätzliche Nutzungsmöglichkeit eines zur Authentisierung verwendeten asymmetrischen Schlüsselpaares besteht in der Geräteauthentisierung, z.B. zwischen Terminal und Karte. Diese ist nach SigV für den Einsatz in einer öffentlichen Umgebung notwendig. Wird zusätzlich ein von der Root-Instanz ausgestelltes Zertifikat für die entsprechende CA verwendet, so kann der Authentisierungsschlüssel auch für die Authentisierung zweier Geräte verwendet werden, die nicht notwendig von derselben CA personalisiert worden sind.

Bei einem vereinfachten Modell, bei dem nur eine ZS pro Kartenhersteller und Teilnehmer eine Rolle spielt, kann man anstatt asymmetrischer Authentisierungsschlüssel auch symmetrische Authentisierungsschlüssel einsetzen.. Diese werden dann jeweils für den Issuer-Schlüssel und für den individuellen Chipkarten-Schlüssel (ICCSN) von einem sogenannten Master-Schlüssel (unter Einfluß der ZS) abgeleitet.

Vorausgesetzte Aktionen: Vorinitialisierungsphase beim Kartenhersteller/Herausgeber

1. Erzeugung eines Issuer-Public-Key-Paares (SK.I, PK.I) beim Issuer , Einbringung von PK.I und / oder gegebenenfalls zusätzlich Einbringung eines Card Verifiable (CV) Issuer Zertifikates für Authentisierung von der Root- Instanz durch den Kartenhersteller/ Issuer in die Karte

2. Im Betriebssystem der Chipkarte verankerter Lifecycle mit den Nutzungsphasen [Ph0] –[Ph4] und den jeweils entsprechenden für diese Phasen gültigen Befehlssätzen

---

Hinweis:

Die Signatur-Anwendung nach SigG /SigV [DINS] wurde bereits vom Chipkartenhersteller/ Issuer in die Chipkarte ohne Personalisierungsdaten eingebracht.

Die Einbringung persönlicher Daten und gegebenenfalls der PIN erfolgt erst durch die Zertifizierungsstelle.

Die Einbringung der Benutzerschlüssel erfolgt in 3.1.1 beim TN , in 3.1.2. beim Kartenhersteller bzw. Herausgeber (Issuer).

3.1.1 Personalisierung von PK-Schlüsseln durch den Teilnehmer

Die Schlüsselgenerierung für das Signaturschlüsselpaar erfolgt in der Teilnehmer-Karte unter eigener Verantwortung und gegebenenfalls unter Einflußnahme des Teilnehmers z.B. durch Registrierung zufälliger Anschläge der Tastatur bei der Erzeugung der Startwerte für die Primzahlengenerierung.

Vor der Einbringung des Schlüssels durch den Teilnehmer (TN) erfolgt die Initialisierungsphase beim Kartenhersteller, Phase [Ph0], und die Vorpersonalisierungsphase bei der CA , Phase [Ph1].

Phase [Ph0] (Initialisierungsphase beim Issuer)

Erläuterung:

Während dieser Lifecycle-Phase wird ein Authentisierungsmechanismus etabliert, der die Authentisierung der Chipkarte ermöglicht. Dadurch kann sich die CA vergewissern, daß nur von ihr anerkannte ICCs in den Personalisierungsprozeß gelangen. Die Chipkarte läßt sich nur durch die berechtigte CA personalisieren. Diese Phase kann in einem Arbeitsgang mit der Vorinitialisierungsphase durchgeführt werden.

> **Ablaufbeschreibung [Ph0] Initialisierungsphase beim Issuer :**
>
> 1. Erzeugung eines Chipkarten-spezifischen Public-Key-Paares (SK.ICC.AUT, PK.ICC.AUT) zur Authentisierung, (Referenzierung des öffentlichen Schlüssels z.B. durch Angabe der eindeutigen Chipkartennummer, der ICCSN)
>
> 2. Transport der Chipkarte zur CA

Anmerkung:

Es ist nur eine Zuordnung zwischen Authentisierungsschlüssel und Chipkarte erfolgt, die Zuordnung zum Benutzer erfolgt erst in der CA.

Phase [Ph1] (Vorpersonalisierungsphase bei der Zertifizierungsstelle)

Erläuterung:

Während dieser Lifecycle-Phase wird die Zuordnung zwischen Chipkarte und Teilnehmer vorgenommen sowie der Authentisierungsmechanismus zwischen Teilnehmer und Chipkarte etabliert.

> **Ablaufbeschreibung [Ph1] Vorpersonalisierungsphase bei der CA:**
>
> 1. Identifikation der Chipkarte in derCA mittels ICCSN und PK.ICC. Gegenseitige Authentisierung von ICC und CA
>
> 2. Laden der Personalisierungsdaten in die Chipkarte

> 3. Registrierung der Zuordnung (ICC / TN) mittels (PK.ICC:AUT/ TN.ID) in der CA
>
> 4. Aktivierung des PIN-Systems (oder gleichwertiges Verfahren)
>
> 5. Integerer Transport der Chipkarte zum TN (z. B. über ICC-Ausgabe)
>
> 6. Vertraulicher Transport der PIN zum TN (z. B. PIN-Brief)

Phase [Ph2] (Schlüsselerzeugungsphase beim TN)

Erläuterung:

Während dieser Lifecycle-Phase wird das Signaturschlüsselpaar durch den berechtigten Teilnehmer erzeugt. Der öffentliche Verifikationsschlüssel des Teilnehmers wird von der Chipkarte signiert. Die CA kann sich mittels dieses signierten Datensatzes davon überzeugen, daß das Teilnehmer-Signaturschlüsselpaar in der dem Teilnehmer zugeordneten Chipkarte erzeugt wurde z.B. durch zusätzlichen Input der eindeutigen Chipkarten-Identifikationsnummer ICCSN (siehe [DINS].

> **Ablaufbeschreibung [Ph2] Schlüsselerzeugungsphase beim TN:**
>
> 1. Authentisierung des TNs gegenüber Chipkarte mittels PIN
>
> 2. Erzeugung des Public-Key-Paares (SK.TN.SIG, PK.TN.SIG) in der Chipkarte
>
> 3. Erzeugung einer Signatur über PK.TN.SIG und ggf. ICCSN mit SK.ICC.AUT

Phase [Ph3] (Personalisierungsphase)

Erläuterung:

Während dieser Lifecycle-Phase wird nach erfolgtem Erhalt des Zertifikates durch die CA durch Einbringen des Zertifikats in die Chipkarte die Regelphase freigegeben.

> **Ablaufbeschreibung [Ph3] Personalisierungsphase:**
>
> 1. Transport des signierten PK.TN.SIG zur CA
>
> 2. Verifikation der Signatur über PK.TN mittels PK.ICC in der CA (durch diese Verifikation ist sichergestellt, daß der TN-Schlüssels in der Chipkarte erzeugt wurde, die dem Teilnehmer zugeordnet wurde)
>
> 3. Erzeugung des Zertifikats ZertTN über PK.TN
>
> 4. Transport des Zertifikats ZertTN zum Teilnehmer
>
> 5. Laden des Zertifikats ZertTN in die Chipkarte
>
> 6. Konsistenzprüfung des Zertifikats anhand der in der Chipkarte bereits vorhandenen Daten
>
> 7. Verifikation des Zertifikats mittels des Verifikationsschlüssels (öffentlicher Zertifizierungsschlüssel) der CA evtl. inklusive des vollständigen Zertifikatspfades

Phase [Ph4] (Regelphase)

Erläuterung:

Während dieser Lifecycle-Phase ist das Erstellen und Verifizieren von Signaturen für den berechtigten Teilnehmer möglich.

Ablaufbeschreibung [Ph4] Regelphase:

1. Authentisierung des Teilnehmers gegenüber der Chipkarte

2. Signaturerzeugung und/oder Signaturverifikation (inkl. Zertifikate)

3.1.2 Personalisierung von PK-Schlüsseln durch den Kartenhersteller

Erläuterung:

Die Schlüsselgenerierung für das Signaturschlüsselpaar erfolgt in der Teilnehmer-Karte unter Verantwortung des Kartenherstellers oder Kartenherausgebers. Daher wird [Ph0] von 3.1.1 ersetzt durch [Ph0.1]. Die Authentizität der vom Issuer gelieferten Chipkarten wird von der CA verifiziert. Die Chipkarte wird vollständig von der CA personalisiert. Daher entfällt die Personalisierungsphase beim Teilnehmer.

Ablaufbeschreibung [Ph0.1] Initialisierungsphase beim Issuer :

1. Erzeugung eines Public-Key-Paares (SK.ICC.AUT, PK.ICC.AUT) zur Authentisierung in der Chipkarte und eines PK–Paares zur Signaturerstellung/Verifikation (SK.ICC.SIG, PK.ICC.SIG) beim Issuer, unter Angabe der Verwendung des geheimen Schlüssels und des öffentlichen Schlüssels z.B. durch Angabe der ICCSN und des Verwendungszwecks (AUT bzw. SIG)

2. Laden des Issuer- Zertifikates für PK.I in die Chipkarte, Laden des PK.CA unter Name der CA, ID.CA, in die Karte

3. Transport der Chipkarte zur CA

Ablaufbeschreibung [Ph1.1] Personalisierungsphase bei der CA:

1. Identifikation der Chipkarte in der CA mittels ICCSN und PK.ICC. Gegenseitige Authentisierung von ICC und CA

2. Laden der Personalisierungsdaten in die Chipkarte

3. Registrierung der Zuordnung (ICC / TN) mittels (PK.ICC / TN.ID) in der CA

4. Erzeugung des Zertifikats ZertTN über PK.TN und Einbringen in die Karte

5. Aktivierung des PIN-Systems (oder gleichwertiges Verfahren)

6. Integerer Transport der Chipkarte zum TN (z. B. über ICC-Ausgabe)

7. Vertraulicher Transport der PIN zum TN (z. B. PIN-Brief)

Hinweis:

[Ph0] und [Ph2] werden durch [Ph0.1] ersetzt.

[Ph1] und [Ph3] werden durch [Ph1.1] ersetzt.

[Ph4] wird wie unter 3.1.1 durchgeführt.

3.2 Exemplarische Ablaufbeschreibung zur Personalisierung von PK-Schlüsseln, zentrales Model

Erläuterung:

Die Schlüsselgenerierung für die PK Schlüssel der Teilnehmer erfolgt in der ZS. Daher ist nur die Authentizität der Chipkarten vom Hersteller von der ZS in der Phase [Ph0] zu verifizieren. Anschließend wird [Ph1.2] durchgeführt. Die Chipkarte wird vollständig von der CA personalisiert. Daher entfällt die Personalisierungsphase beim Teilnehmer.

Ablaufbeschreibung [Ph1.2] Personalisierungsphase bei der ZS:

1. Identifikation der Chipkarte in der CA mittels ICCSN und PK.ICC. Gegenseitige Authentisierung von ICC und CA

2. Erzeugung des Public-Key-Paares (SK.TN, PK.TN)in der ZS

3. Laden der Personalisierungsdaten in die Chipkarte

4. Registrierung der Zuordnung (ICC / TN) mittels (ICC.TN / TN.ID) in der CA

5. Erzeugung des Zertifikats ZertTN über PK.TN und Einbringen in die Karte

6. Aktivierung des PIN-Systems (oder gleichwertiges Verfahren)

7. Integerer Transport der Chipkarte zum TN (z. B. über ICC-Ausgabe)

8. Vertraulicher Transport der PIN zum TN (z. B. PIN-Brief)

Hinweis:

[Ph0] wird wie unter 3.1 durchgeführt

[Ph1], [Ph2]und [Ph3] werden durch Phase [Ph1.2] ersetzt.

[Ph4] verläuft wie unter 3.1 beschrieben.

Die Abfolge der Kommandos in der Karte erfolgt nach den Chipkarten-Standards ISO/IEC 7816-4 und DIS 7816-8. (siehe [DINS]) und wird Gegenstand der erwähnten DIN-Spezifikation sein.

Alle Phasen sind durch Kommandosequenzen zur Karte realisierbar, bei denen durch Kommandos wie VERIFY, GENERATE_PK_PAIR, VERIFY_CERTIFICATE, INTERNAL_AUTHENTICATE und EXTERNAL_AUTHENTICATE eine Rolle spielen. Auf eine detailliertere Beschreibung wird an dieser Stelle verzichtet.

Literatur

[ISO14516-2] Guidelines for the use and management of Trusted Third Parties - Part 2: Technical aspects, ISO/IEC Draft 1997

[RBPT] Regulierungsbehörde für Telekommunikation und Post nach Angaben des Bundesamtes für Sicherheit in der Informationstechnik, Maßnahmenkatalog für digitale Signaturen - auf Grundlage von SigG und SigV –18.11.1997, Version 1.0

[SigG] Gesetz zur digitalen Signatur (Signaturgesetz - SigG) vom 22.07.1997 (BGBl. I S. 1870, 1872), verkündet als Artikel 3 des "Gesetzes zur Regelung der Rahmenbedingungen für Informations- und Kommunikationsdienste (Informations- und Kommunikationsdienste-Gesetz - IuKDG)

[SigV] Verordnung zur digitalen Signatur (Signaturverordnung - SigV) vom 22.10.1997 (BGBl. I S. 2498)

[DINS] Spezifikation der Schnittstelle zu Chipkarten mit Digitaler Signatur-Anwendung/Funktion nach SigG und SigV, DIN NI-17.4, Stand 0.75 vom 10.12.97

[ISO7816-4] ISO/IEC 7816-4: 1995, Information technology - Identification cards - Integrated circuit(s) cards with contacts - Part 4: Interindustry commands for interchange, Part 4 / AM1: Impact of secure messaging on the structures of APDU messages

[ISO7816-5] ISO/IEC 7816-5: 1995, Information technology - Identification cards - Integrated circuit(s) cards with contacts - Part 5: Numbering system and registration procedure for application identifiers

[ISO7816-6] ISO/IEC 7816-6: 1995, Information technology - Identification cards - Integrated circuit(s) cards with contacts - Part 6: Interindustry data elements

[ISO7816-8] ISO/IEC 7816-8: Final CD 1997, Information technology - Identification cards - Integrated circuit(s) cards with contacts - Part 8: Security related interindustry commands

Digitale Signatur per Chipkarte zur maximalen Sicherheit im Internet

Ernst-Michael Hamann

IBM Deutschland Entwicklung GmbH
Schönaicher Str. 220, D-71032 Böblingen
mhamann@de.ibm.com

Zusammenfassung

Die Öffnung des Internet's für geschäftliche Anwendungen setzt eine eindeutige Identifizierung der Kommunikationspartner und die Unveränderbarkeit der ausgetauschten Dokumente voraus. Mit Hilfe von kryptographischen Verfahren, wie z.B. der RSA Public Key Algorithmus, können Daten durch eine digitale Signatur (gleich 'elektronische Unterschrift') sicher gekennzeichnet werden. Voraussetzung für ein solches Verfahren ist die eindeutige Zuordnung von kryptographischen Schlüsseln. Für die Vergabe von eindeutigen Schlüsseln pro Benutzer und Anwendung wurden öffentliche Zertifizierungsstellen 'Certification Authorities' im Netzwerk eingerichtet, von denen ein persönliches Zertifikat für den öffentlichen Schlüssel erworben werden kann. Die sichere Generierung, Speicherung und Verwendung dieser Schlüsselpaare ist eine wichtige Bedingung für die eindeutige und rechtlich verbindliche Kennzeichnung der im Internet versendeten Dokumente.

Dieser Artikel schildert die Verwendung der 'IBM Signaturkarte' ('IBM Signature Card') im Internet als sicherer 'Krypto-Token' basierend auf internationalen Standards für Chipkarten und Kryptoadaptoren. Der Benutzer kann sich mit Hilfe der Signaturkarte von einer beliebigen Workstation mit Chipkartenleser im Intra- oder Internet eindeutig identifizieren wobei die Signaturkarte hohe Sicherheitsanforderungen erfüllt. Mit Hilfe der im PKCS#11 Standard definierten 'Cryptoki' Anwendungsprogrammschnittstelle wird die Chipkarte zu einem sehr variable einsetzbaren mobilen 'Krypto-Rechner', welcher auch zur Speicherung aller vom Netzwerkbenutzer benötigten Datenobjekte dienen kann (z.B. ID-Files, URL Adressen).

Abschließend wird an Hand von Beispielen der Einsatz dieser Karte beim Aktienhandel, im Bankenbereich ('Secure Logon') und bei der Kontrolle von Netzwerkcomputern (NC) erklärt. Zusammenfassend werden die Fähigkeiten der jetzigen Kartentechnologie und die zukünftigen Erweiterungsmöglichkeiten aufgezeigt.

1 Einleitung

Die Verwendung kryptographischer Verfahren zur sicheren Identifizierung der Kommunikationspartner und für die digitale Signatierung und Verschlüsselung von Daten sensitiver Anwendungen ist im Internet bereits weit verbreitet. Dabei werden meistens die verwendeten Schlüsselpaare in der Workstation auf den gängigen Speichermedien wie Diskette oder Festplatte mehr oder weniger gesichert gespeichert und zur Ausführung der Verschlüsselungsmethoden in den Arbeitsspeicher geladen. Besonders bei vernetzten Workstations kann dabei die Geheimhaltung der Schlüssel nicht garantiert werden.

Die Geheimhaltung und exklusive Verwendung der Schlüssel durch nur eine Person ist aber eine notwendige Voraussetzung für den geschäftlichen Datenaustausch über öffentliche Netze, als Ersatz für die papierbasierende Kommunikation mit handschriftlichen Signaturen. Nur so kann, zum Beispiel ein über das Internet erfolgter 'elektronischer' Auftrag zum Kauf von Aktien, im Streitfall eine rechtliche Anerkennung finden.

Durch die Verwendung eines kyptographischen Adaptors ('Krypto-Token') im Arbeitsplatzrechner kann die sichere Aufbewahrung der kritischen Schlüssel und die gesicherte Ausführung der Verfahren garantiert werden. Solch ein Token sollte möglichst leicht zu erstellen, handlich, und in mehreren Rechnerumgebungen einfach zu verwenden sein.

Zwei Arten von Krypto-Token mit den genannten Eigenschaften kommen in Betracht:

- ein Krypto-Adaptor in PC-Karten (PCMCIA) Format

- eine Chipkarte mit integriertem Kryptokoprozessor

Beide Token unterscheiden sich wesentlich in der Leistungsfähigkeit, der Handhabung, der notwendigen Workstationinfrastruktur und natürlich den Kosten. Idealerweise wird die Leistungsfähigkeit der Chipkarten im Laufe der nächsten Jahre in die Nähe der von PC-Karten gelangen, so daß dann die heute noch bestehenden Einschränkungen beim Einsatz von Chipkarten entfallen. Wegen der limitierten Leistungs- und Speicherfähigkeiten der Chipkarte muß man sich heute auf die wirklich kritischen Anwendungen mit den sensitivsten Daten und Schlüssel beschränken. Alle weiteren notwendigen kryptographischen Verfahren, wie die Hash-Erstellung und die Verschlüsselung größerer Datenmengen, werden in der leistungsfähigeren Workstation oder in einem Krypto-Adaptor ausgeführt. Bei 'privaten' Schlüsseln sollte unbedingt garantiert sein, daß der Schlüsselwert niemandem bekannt gemacht wird, möglichst nicht einmal dem Besitzer des Tokens. Dies wird entweder dadurch erreicht, daß die Schüsselpaare direkt auf dem hermetisch abgeschotteten Token generiert werden und nur der 'öffentliche' Schlüssel nach außen gegeben wird, oder daß das Schlüsselpaar in einer sicheren, absolut virenfreien Umgebung innerhalb des Rechners generiert wird. Im letzteren Fall muß sichergestellt werden, daß während der Generierung keine Kopie des 'privaten' Schlüssels erstellt werden kann und daß die Übertragung des privaten Schlüssels zum Token verschlüsselt erfolgt.

Da beide Tokentypen in verschieden Anwendungsbereichen Vorteile haben und auch nebeneinander in der gleichen Workstation verwendet werden können, sollten diese über eine gleichartige Anwendungsschnittstelle ansprechbar sein. Dazu eignet sich das von den RSA Laboratorien im *PKCS#11 Standard* beschriebene *'Cryptoki' API*, welches eine sehr flexible Schnittstelle für die gebräuchlichsten kyptographischen Methoden bietet und in der aktuellen Version 2.01 für Chipkarten erweitert wurde. Darüber hinaus können auch benutzerspezifische Datenobjekte inklusive der Benutzerzertifikate auf dem Token über dieses API durch ein vom Benutzer zu definierendes geheimes 'Passwort' geschützt abgelegt und darauf zugegriffen werden.

Die physikalischen Eigenschaften für Chipkarten wurden in den letzten Jahren in den CEN und ISO Standards genormt und sind die Voraussetzung für die Verbreitung der notwendigen weltweiten Infrastruktur. Für den Zugriff auf Chipkarten und Kartenterminals wurden in den letzten Monaten große Fortschritte mit der Veröffentlichung der *PC/SC Spezifikation* und der

Bereitstellung erster Implementierungen für auf Windows basierenden PC's erzielt. Die Normung von Netzwerkcomputern auf Basis der *NCRP Definition* und deren Erweiterung für Chipkarten durch die *'OpenCard Frame Work'* Spezifikation, definiert für in Java geschriebene Anwendungen im Internet eine einheitliche Schnittstelle für Chipkarten.

2 Die IBM Signaturkarte (Signature Card)

Die 'IBM Signaturkarte' basierend auf dem *IBM Kartenbetriebsystem 'MultiFunction Card (MFC) Version 4'* ist ein Krypto-Token mit den folgenden wesentlichen Eigenschaften:

- Kompatibilität zu den wesentlichen internationalen Chipkartenstandards:
- *ISO/IEC 7816* - ICC Integrated Circuit(s) Cards
- *PC/SC* (für auf Windows basierende Workstations)
- *NCRP* und *OpenCard Framework* (Netzwerkcomputer / Java)
- sicherer Speicher für private Schlüssel und Zertifikate
- erstellen von RSA Signaturen auf der Karte in sicherer Umgebung
- Standard Cryptoki Schnittstelle (*PKCS#11 Standard*)
- laden und verwenden der Schlüssel und Zertifikate über Internet Browser (z.B. *Netscape*)
- Speicherungsmöglichkeiten für weitere private oder öffentliche Datenobjekte
- Optional:
- Generierung von Schlüsselpaaren auf der Karte
- sicherer Import extern generierter Schlüssel
- weitere kryptographische Verfahren (wie z.B. DSA)

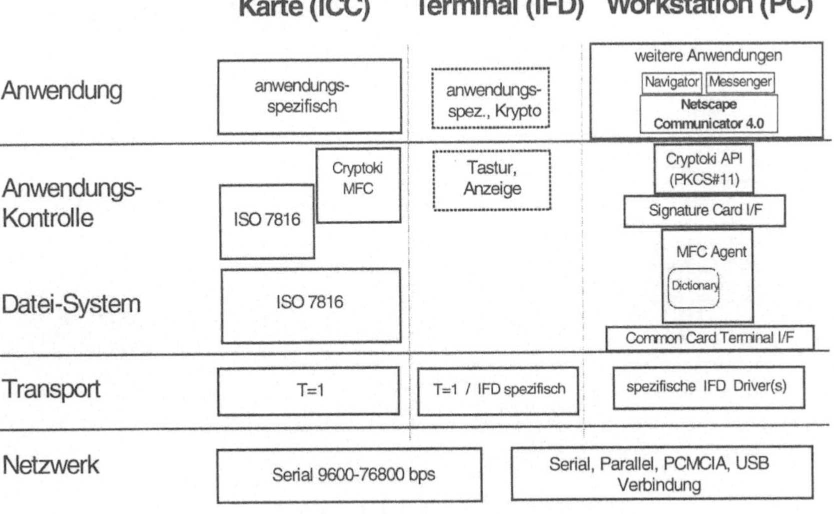

Abb. 1: Schichtenmodell der IBM Signaturkarte

Für die Unterstützung der im PKCS#11 Standard definierten 'Cryptoki' Schnittstelle wurde das multifunktionale Kartenbetriebsystem IBM MFC 4 spezifisch erweitert und besondere DF/EF-Strukturen für die geschützte Aufbewahrung der Schlüssel, Zertifikate und anderer Datenobjekte im EEPROM definiert. Über das 'Cryproki' API lassen sich nun alle notwendigen Funktionen von der Anwendung aufrufen, um die verschiedenen Objekte anzulegen, zu manipulieren oder die unterstützten kryptographischen Verfahren zu verwenden.

Die verschiedenen Kommunikationsschichten für die Chipkarten (ICC), die Kartenterminals (IFD) und die Workstation (PC) als Voraussetzung für den Einsatz der Signaturkarte sind in Abbildung 1 dargestellt.

Die Workstationanwendung muß die spezifischen Eigenschaften der Chipkarten nicht kennen und kann sich deshalb der im PKCS#11 Standard definierten Cryptoki-Funktionen bedienen, um z.B. eine digitale Signatur mit Hilfe der Signaturkarte zu erzeugen. Die Schichten auf der Workstationseite entsprechen in etwa dem *OpenCard Framework*. Die speziellen Fähigkeiten der Chipkarte werden mit Hilfe des 'MFC Agenten' abgebildet. Die Dateistrukturen auf der Chipkarte sind im 'Dictionary' definiert, welches auf Kundenwunsch mit Hilfe des *IBM Smart Card ToolKit's* beliebig erweitert werden kann. Beide Kontrollelemente können dynamisch bei der Erkennung der Kartenart geladen werden. Die Einbindung der Kartenterminals kann bei Windows NT / Windows 95 Systemen auf der PC/SC Spezifikation Teil 5 ('IFD Resource Manager' statt 'Common Card Terminal Interface') basieren.

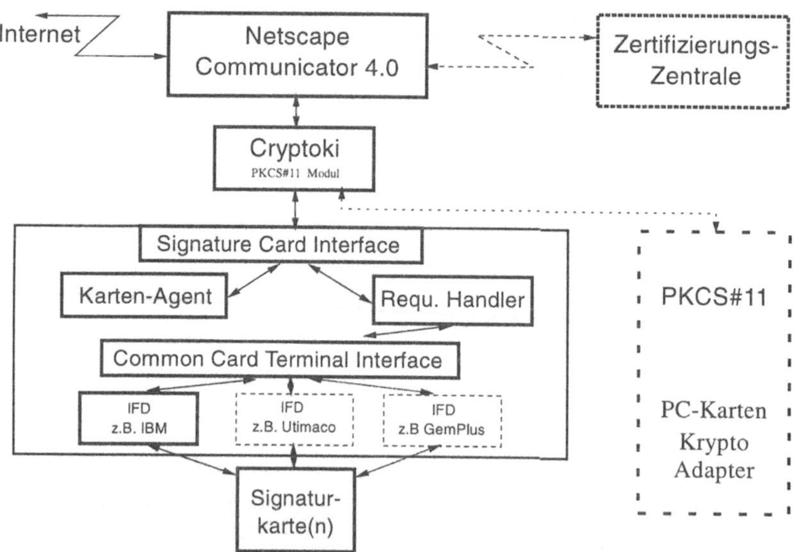

Abb. 2: Komponenten einer Cryptoki-Signaturkarten Konfiguration

Wie in der Abbildung 2 dargestellt, kann das auf der 'Signatur Card API' aufbauende PKCS#11 Cryptoki Modul mehrere unterschiedliche Krypto-Token gleichzeitig bedienen und diese auch verschiedenen Anwendungen zur Verfügung stellen. Das 'Signature Card Interface' ist in der Lage mehrere Signaturkarten in verschiedenen Kartenterminals gleichzeitig den Anwendungen zur Verfügung zu stellen. Die Verwendung eines Karten-Agenten pro Chipkartentyp erlaubt es Karten mit unterschiedlichen Kartenbetriebssystemen ohne Veränderung der Anwendungsschnittstellen zu unterstützen.

Für die im PKCS#11 Standard definierten Objekte stehen zwei Datenbereichen im EEPROM der Karte zur Verfügung (siehe Abbildung 3). Im öffentlichen Bereich 'Ctrl_public' werden die nicht schützenswerten Objekte wie das allgemein zugängliche Zertifikat abgelegt. Im privaten Bereich 'Ctrl_privat', welcher durch ein Benutzer-Paßwort geschützt ist, werden alle geheimen Objekte, wie die nicht-öffentlichen Schlüssel und private Information, abgelegt. Diese beiden Datenbereiche sind so organisiert, daß keine Datenverluste bei Stromausfall (z.B. durch ein Herausziehen der Karte) während einer Schreiboperation auftreten kann. Die Bereiche sind in gleich große Segmente von je 64 Bytes Länge aufgeteilt, wobei das erste Byte jedes Segments Kontrollzwecken dient. Je nach Größe wird ein Objekt in einem oder mehreren dieser Segmente gespeichert.

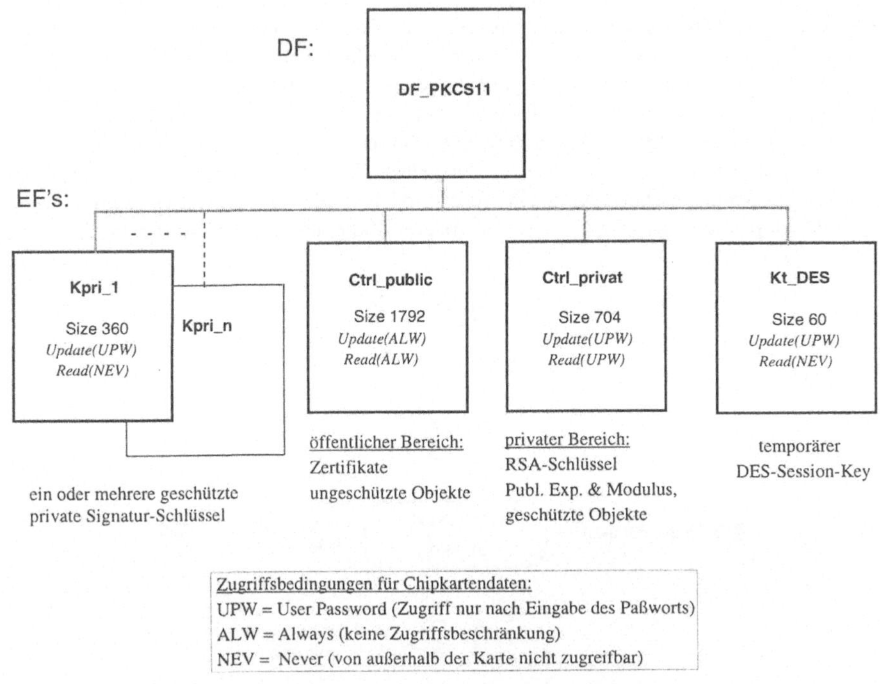

Abb.3: Organisation der Signaturkarte

Falls die Signaturkarte nicht bereits bei der Auslieferung mit Schlüsselpaaren und Zertifikat versehen wurde, kann bei der erstmaligen Benutzung der IBM Signaturkarte im Netscape Communicator über die auf PKCS#11 basierende 'Cryptographic Module' Schnittstelle ein Schlüsselpaar mit den Schlüssellänge 512, 768 oder 1024 Bits generiert werden. Der private Schlüssel wird dann sicher auf der Karte als Schlüsselobjekt abgelegt. Anschließend wird zusammen mit den notwendigen Daten des Schlüsselinhabers der öffentliche Schlüssel über das

Netz in einer SSL Session an eine Zertifizierungsstelle (z.B. von der Firma 'VeriSign') geschickt und dafür ein Zertifikat beantragt. Nach der Überprüfung der Antragsdaten wird von dieser Stelle ein Zertifikat generiert und an die Workstation zurückgesandt, um dort auf der Karte als Zertifikatsobjekt abgelegt zu werden. Danach ist die Karte über die 'Cryptoki' Schnittstelle für alle Netzwerkanwendungen für die Generierung von Signaturen mit einem der sicher auf der Karte gespeicherten privaten Schlüssel einsetzbar. Nun kann mit Hilfe der Signaturkarte im Netscape Messenger eine e-Mail gemäß dem S/MIME Sicherheitsstandard signiert werden oder die Fähigkeiten der Signaturkarte dazu verwendet werden, um Zugriff zu gesicherten Servern im Internet über den Netscape Navigator zu erlangen.

Die Kontrolle der Signaturkartenfunktionen erfolgt anolog zu den im PKCS#11 Standard beschrieben Funktionen. Nach dem Einschieben der Karte in das Kartenterminal wird dem Cryptoki Modul die Kontrollinformation der Karte übermittelt (wie z.B. die 16-stellige Kartenseriennummer und der Kartentyp). Dannach kann ein 'Open Session' für die Karte erfolgen, wonach alle 'Read Only' Objekte, falls vorhanden, ausgelesen werden können. In der Abbildung 4 werden die möglichen Cryptoki 'Read / Write' Sessionarten gezeigt. Außer der 'Public Session' kann mit Hilfe des korrekten 'User Paßworts' eine 'Private Session' aufgebaut werden, dann können alle User Funktionen ausgeführt werden (wie z.B. das Ändern des User-Paßwortes) und die geschützten Objekte gelesen und geschrieben werden. Die 'Security Officer (SO)' Session dient Sonderfunktionen zur Pflege der Karte wie z.B. das Zurücksetzen des User-Paßwortes. Das für die SO Session notwendige karten-individuelle SO-Paßwort wird beim Initialisieren der Karte festgelegt und kann entweder dem Kartenbesitzer in einem Umschlag geschützt mitgeteilt werden, oder es läßt sich nur mit Hilfe einer speziellen SO-Chipkarte vom Sicherheitsbeauftragten des Kartenherausgebers ermitteln.

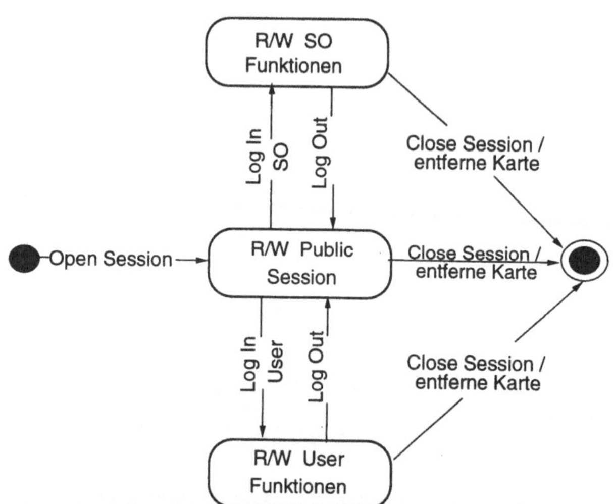

Abb. 4: Signaturkarten Sessiontypen (nach PKCS#11)

Neben den über die 'Cryptoki' Schnittstelle aufzurufenen Funktionen lassen sich auf der Signaturkarte noch Daten für weitere Anwendungen (z.B. eine Bonusfunktion des Kartenherausgebers) in ISO 7816 Filestrukturen definieren. Diese können dann mit allen Funktionen des umfangreichen IBM MultiFunction Card Betriebsystems lokal und entfernt anwendungsspezifisch genutzt werden, wie im OpenCard Framework definiert.

Die Signaturkarte kann als Basis für kundenspezifische 'Identifikationskarten', also eine Art digitaler Ausweis, dienen. Dafür kann die Karte mit Hilfe des *IBM Smart Card ToolKit's* beliebig für weitere Anwendungen erweitert und die Karten mit Bild und anderen persönlichen Merkmalen kontrolliert durch den Kunden hergestellt werden. Die Kontrolle der Karten kann dann über die 'Security Officer' Funktionen der Cryptoki Schnittstelle und mit dem *IBM Smart Card Management System* erfolgen.

3 Anwendungsbeispiele

Der Einsatz der Signaturkarte zur Erreichung hoher Sicherheit bei der Kommunikation über das Internet (einschließlich Intranets und Extranets) läßt sich mit den folgenden Anwendungsbeispielen verdeutlichen:

3.1 Aktienkauf im Internet

Unter der Annahme, daß der Kunde mit seinem Zertifikat für diesen Service beim Broker registriert und autorisiert wurde, kann dieser im Internet Aktien kaufen. Hierzu wählt er über seinen Browser die Web-Seite des Brokers und erstellt einen Kaufauftrag. Bevor der Auftrag an den Internet Broker Server geschickt werden kann, wird der Besteller von der Brokeranwendung aufgefordert, seine Signaturkarte in das Kartenterminal zu schieben und diese durch die Eingabe seines geheimen Paßworts freizugeben. Dann wird die digitale Signatur für den Kaufauftrag auf der Signaturkarte mit dem dort sicher abgelegten privaten Schlüssel generiert und zusammen mit dem Auftrag an den Internet Broker Server gesandt. Dort kann auf Grund des Zertifikats die Unterschrift geprüft und der Auftragseingang gleich bestätigt werden. Bei Bedarf kann die Bestellung durch Verwendung der Standard SSL Internet Services des Browsers verschlüsselt übertragen werden. Manipulationen des Auftrags innerhalb des Netzwerks können durch die Signatur ausgeschlossen werden.

Diese Anwendung wurde beispielhaft für die Verifikation der 'OpenCard Framework' Komponenten in Java implementiert. Über die OpenCard Framework Webpage kann die 'Internet Stock Broker Demo' als Referenzimplementation bestellt werden.

3.2 Zugangskontrolle für Systemoperatoren und Kundenbeauftragte

Bei kritischen Anwendungen speziell in großen Firmennetzen ist es wichtig, den berechtigten Personen nicht nur durch ein einfaches manuell einzugebenes Paßwort Zutrittsberechtigung zu erteilen. Des weiteren müssen Systemadministratoren meistens mit mehreren Systemen gleichzeitig kommunizieren, wofür ein einmaliges Logon mit hohen Sicherheitsgarantien eine unablässige Anforderung ist. Mit Hilfe der Signaturkarte lassen sich dafür bei Verwendung gängiger Netzwerk-Browser die höchsten Sicherheitskriterien erfüllen. Wie im nächsten Beispiel 'Netzwerkcomputer' beschrieben, wird beim Logon die Berechtigung des Kartenbesitzers überprüft. Durch geeignete Schnittstellen wird anschließend regelmäßig die Benutzung des Terminals überprüft, um bei Abwesenheit des Kartenbesitzers die Anwendungen

automatisch zu schließen. Das gleiche geschieht, wenn die Signaturkarte aus dem Kartenterminal entfernt wird.

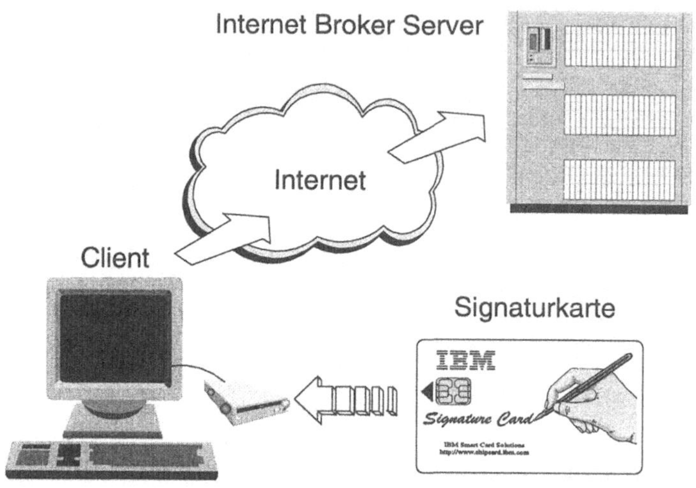

Abb. 5: Aktienkauf im Internet

Abhängig von den Sicherheitsanforderungen der Systeme und Anwendungen werden längere
Schlüssel (bis zu 1024 Bits) oder andere kryptographische Verfahren auf der Karte verwendet. Die Zertifikate für die Schlüssel haben eine kürzere Gültigkeitsdauer und müssen deshalb öfters ausgetauscht werden. Um die Administration der Signaturkarte im Netzwerk zu
vereinfachen, können spezielle, auf kryptographisch geschützten Nachrichten basierende
Schnittstellen implementiert werden. Diese Nachrichten können vom 'Superadministrator'
der Firma erzeugt und auf einem speziellen Server sicher abgelegt werden. Beim Logon eines
Benutzers mit der Karte wird dann als erstes im speziellen Administrator-Server geprüft, ob
eine Administratornachricht für diese Karte vorliegt. Falls dies der Fall ist, wird diese in einem speziell gesicherten Modus zur Karte übertragen und ausgeführt. So lassen sich zum
Beispiel am normalen Arbeitsplatz im Netzwerk das Benutzerpaßwort auf der Karte zurücksetzen und die Gültigkeit der Zertifikate verändern.

3.3 Netzwerkcomputer Zugangskontrolle

Netzwerkcomputer (NC) besitzen keine lokalen Plattenspeicher, so daß alle Programme und
Daten vom Server im Netzwerk geladen werden müssen. Das gilt auch für die größten Teile
des Betriebsystems, welches entweder das Java OS oder eine virtuelle Java Maschine (JVM)
als Basis für die Anwendungsprogramme zur Verfügung stellt. In dieser Umgebung werden
Chipkarten über die im 'OpenCard Framework' definierten Schnittstellen angesprochen. Die
Überprüfung der von dem Server zu ladenden Programmteile geschieht mit Hilfe einer digitalen Signatur jedes Programmteils, damit eine sichere virenfreie Maschine gewährleistet
werden kann. Die 'public keys' der berechtigten Server können im NC auf einer maschi-
neninternen, bei der Installation festintegrierten Signaturkarte, zusammen mit einem individuellen 'private key' des Netzwerkcomputers gespeichert werden. Wenn der NC einge-

schaltet wird, holt sich dieser von der Signaturkarte die Netzwerkadresse (URL) des zugeordneten Servers und kann dann mit Hilfe des gespeicherten 'public keys' die von dort geladene Software prüfen. Der Server kann dabei den Netzwerkcomputer auffordern, eine mit Zufallsparametern generierte Testnachricht mit dem privaten Schlüssel des NC's zu signieren und zurückzusenden. Mit dem zertifizierten öffentlichen Schlüssel des NC's kann diese Signatur auf Gültigkeit überprüft werden. Falls Netzwerkdienste pro NC individuell abgerechnet werden sollten, kann dafür die eindeutig durch die digitale Signatur erfolgte Identifizierung des NC's als Referenz benutzt werden.

Nachdem das Basisbetriebsystem und die Standardanwendungen in den NC geladen wurden kann der NC im 'Kioskbetrieb' z.B. zum Surfen im Internet benutzt werden. Dabei kann die Berechnung der Netzwerkservices benutzerunabhängig über die NC-interne Signaturkarten-ID erfolgen. Dies ist besonders dann notwendig, wenn NC, Netzwerk und Server von verschiedenen Firmen betrieben werden.

Um den NC für einen Benutzer individuell einzurichten, kann eine benutzerspezifische Signaturkarte in ein im NC integriertes Chipkartenterminal eingeschoben werden. Nachdem der Benutzer seine Signaturkarte durch Eingabe seines geheimen Paßwortes freigegeben hat liest der NC die Netzwerkadresse (URL) des Benutzerservers und beginnt die Kommunikation mit der Identifizierung des Benutzers über die Generierung einer Signatur für eine vom Server generierte Zufallsnachricht mit dem geheimen 'private key'. Wenn dies erfolgreich vom Server verifiziert wurde, werden dem Benutzer seine persönlichen Anwendungsprofile zur Verfügung gestellt. Dies idealerweise in einer Form wie er sie bei der letzten Benutzung verlassen hat. Bei Inanspruchnahme weiterer Dienstleistungen im Netz, wie im vorher beschriebenen Beispiel 'Aktienkauf' geschildert, wird man sich mit Hilfe der Signaturkarte ausweisen. Die Netzwerkkosten können auch mit Hilfe der durch die Karte erfolgten Identifikation abgerechnet werden.

Auf der Signaturkarte können für weitere Benutzeranwendungen Daten als PKCS#11 Datenobjekte abgelegt werden und über die 'Cryptoki' Schnittstelle zugegriffen werden. Falls eine Anwendung auf eine spezielle zusätzliche Karte zugreifen möchte, z.B. eine Karte mit einer integrierten Geldbörse, kann die Signaturkarte zeitweise durch eine anwendungsspezifische, dem 'OpenCard Framework' entsprechende, weitere Karte ersetzt werden.

4 Ausblick

Die IBM Signaturkarte basiert auf den heute verfügbaren Hochleistungschips mit integriertem Kryptokoprozessor für Chipkarten. Diese Technologie erlaubt es, Schlüsselpaare mit bis zu 1024 Bit Länge innerhalb der Karten zu generieren und eine digitale Signatur bei einem bereits generierten SHA-1 Hash, mit einem 1024 Bit langen Schlüssel in weniger als 300 mSek zu erstellen. Für die Verwendung längerer Schlüssel (z.B. 2048 Bit) auf der Karte können die extern generierten Schlüssel sicher über die PKCS#11 Schnittstelle importiert werden. Die geringe Übertragungsgeschwindigkeit zwischen Kartenterminal und Karte und auch die auf der Karte verwendeten langsamen EEPROM Speicher erlauben es nicht, größere Datenmengen auf der Karte in akzeptabler Zeit zu verarbeiten. Deshalb sollte die Verschlüsselung und Entschlüsselung von Daten außerhalb im Prozessor der Workstations oder bei kritischen Anwendungen in speziellen Kryptoadaptoren ausgeführt werden. Der mit zur Zeit maximal 16 kBytes geringe Speicher für Schlüssel-, Zertifikats- und Datenobjekte reicht nicht aus, um viele Zertifikate abzulegen, da jedes einen Speicherplatz von etwa 2.000 Bytes belegt.

Um dem wachsenden Bedarf für spezielle Schlüsselpaare pro Anwendung gerecht zu werden, ist es notwendig, entweder mehrere Signaturkarten mit Zertifikat pro Person auszustellen oder nur die Schlüssel ohne die Zertifikate auf der Karte abzulegen, mit einem Verweis auf die dazugehörenden Zertifikate auf anderen Datenträgern (wie z.B. einer URL Netzadresse). Um die Schlüssel zu verwalten ist es dann notwendig, diese für den Benutzer in einem 'Schlüsselbund' auf einem oder mehren Krypto-Token zu organisieren, wobei das für eine Anwendung benötigte Schlüsselpaar mehr oder weniger automatisch durch das Anwendungsprogramm ausgewählt werden kann.

Der Einsatz der Signaturkarte setzt das Vorhandensein von Chipkartenlesern an jeder internetfähigen Workstation voraus. Die Stabilisierung der Standards und der zunehmende Einsatz von Chipkarten für den bargeldlosen Zahlungsverkehr fördert die Verbreitung diese Geräte. Idealerweise wird der Chipkartenleser ein Teil der Basisausstattung jeder Workstation, speziell des Netzwerkcomputers, wobei mit der Einführung des Universal Serial Bus (USB) als Schnittstelle zwischen PC und Leser auch die Übertragungsgeschwindigkeit erhöht werden kann.

Auf der Basis einer Chipkarte mit Cryptoki Schnittstelle, wie die IBM Signaturkarte, können sehr flexibel die verschiedensten Sicherheitsanforderungen im Internet realisiert werden. Diese Schnittstelle ist darauf vorbereitet, Erweiterungen der Funktionen und Mechanismen auf der Signaturkarte innerhalb der Standarddefinition über geeignete Internet-Browser den Anwendungen zur Verfügung zu stellen.

Die Standardgremien werden die Definition der kryptographischen Eigenschaften von Chipkarten auf allen Systemplattformen kontinuierlich den neuesten Anforderungen und technologischen Fähigkeiten anpassen.

Im Zusammenhang mit der Standardisierung der Verfahren zur digitalen Signatur (siehe auch das deutsche Signaturgesetz SigG) sollte die Verwendung des Cryptoki Standards als Basis für eine international anerkannte Schnittstelle für Chipkarten in Betracht gezogen werden.

Literatur

[ISO7816]   ISO/IEC 7816-1, 2, 3, 4 International Standard for Integrated circuit(s) cards with contacts, Geneva, Switzerland 1994

[MFC 4]   IBM MultiFunction Card with Public Key: IBM Corporation (http://www.chipcard.ibm.com)

[NCRP]   Network Computer Reference Profile (siehe OpenCard)

[Netscape]   Netscape Communicator 4.0: Netscape Communications Corporation, USA (http://www.netscape.com)

[OpenCard]   Network Computer Reference Profile (NCRP) 'OpenCard': Apple, IBM Corporation, Netscape, Oracle, Sun (http://www.opencard.org)

[PC/SC]   PC/SC Draft Rev. 0.9 (1996/97) CP8 Transac, Gemplus, Hewlett-Packard, IBM Corporation, Microsoft, Schlumberger, Siemens Nixdorf Informationssysteme, Sun: Interoperability Specification for ICCs and Personal Computer Systems (http://www.smartcardsys.com/overview.html)

[PKA-RSA]   L. Rivest, A. Shamir, L. M. Adleman: "A method for obtaining digital signatures and public-key cryptosystems", Communications of the ACM, 21(2), Seite 120-126, Februar 1978

[PKCS#11]   PKCS#11: Cryptographic Token Interface Standard 'Cryptoki', RSA Laboratories, Vers. 2.0 draft, July 2, 1997 (http://www.rsa.com/rsalabs/pubs/PKCS)

[ToolKit]   IBM Smart Card ToolKit: IBM Corporation (http://www.chipcard.ibm.com)

Studierendenausweise auf Chipkartenbasis

Stefan Ptascheck

Ruhr-Universität Bochum
Dezernat für Informations- und Kommunikationsdienste
Stefan.Ptascheck@uv.ruhr-uni-bochum.de

Zusammenfassung

Seit der Immatrikulation zum Sommersemester 1997 können die Studierenden der Ruhr-Universität Bochum im Rahmen eines Pilotprojektes des Landes Nordrhein-Westfalen einen neuen chipkarten-basierten Studierendenausweis erhalten. 6000 Studierende haben dieses Angebot bereits angenommen. Mit diesem Ausweis stehen ihnen die Rückmeldung in Selbstbedienung, Bibliotheksfunktionen und elektronische Zahlmöglichkeiten an multifunktionalen Selbstbedienungsterminals zur Verfügung.

Der Studierendenausweis wird von der Universität selbst herausgegeben und basiert nicht auf einer White-Card. Die eingesetzte Krypto-Prozessorchipkarte verfügt über die Fähigkeit zur Erstellung digitaler Signaturen und speichert eine Reihe von Daten der Studierenden. Diese werden durch ein umfangreiches Sicherheitskonzept vor unbefugtem Zugriff geschützt. Die optisch lesbare Darstellung einiger relevanter Daten ist um ein wiederbeschreibbares Schriftfeld für die Gültigkeit ergänzt.

Der vorliegende Beitrag beschreibt den Verlauf des Pilotprojektes und durchleuchtet die gefällten Entscheidungen. Eine Darstellung der Erfahrungen und eine Kostenabschätzung erlauben einen Ausblick auf weitere Entwicklungen im Bereich der Hochschule.

1 Einleitung

An der Ruhr-Universität Bochum wird zur Zeit das Pilotprojekt „Studierendenausweise auf Chipkartenbasis" durchgeführt. Vor ca. 3 Jahren wurde mit den ersten Vorüberlegungen begonnen und seit ca. einem Jahr sind die ersten Ausweis im Umlauf. Das Pilotprojekt wird von der Verwaltung der Ruhr-Universität Bochum durchgeführt, die Federführung hat die Stabstelle für Informations- und Kommunikationstechnik in der Verwaltung.

1.1 Zielsetzung

Das Pilotprojekt ist mit drei miteinander verwobenen Zielen gestartet (siehe Abb.1):

Die ersten beiden Punkte sind relativ einleuchtend, beim letzten ist eine Verbindung zum Thema Chipkarte nicht sofort offensichtlich. Deshalb wird dieser Punkt im folgenden genauer betrachtet.

Der Begriff der „virtuellen Universität" geistert seit geraumer Zeit durch die Hochschulland-schaft. Das Pilotprojekt an der Fern-Universität Hagen scheint sehr erfolgreich zu verlaufen. Leider wird nur sehr selten darüber nachgedacht, wie man sich „Zutritt" zu dieser Universität verschafft. Für fast alle „wichtigen" Schritte wird dann doch der so gar nicht virtuelle Papierweg gewählt.

Abb. 1: Ziele des Pilotprojekts

Ein neuer Studierendenausweis muß der Schlüsselbund für die virtuelle Universität sein. Die Fähigkeit zur Erstellung digitaler Signaturen und die eindeutige Identifikation über Netze sind wesentliche Bestandteile davon.

Die Lösungswege zum Erreichen der beiden anderen Ziele bauen auf diesen Fähigkeiten auf. Sie werden später noch genauer aufzeigt werden.

1.2 Geplante Einsatzgebiete

Wie die Ziele mit einem Chipkartenausweis erreicht werden sollen, ist an den Anwendungsgebieten dieses Ausweises (siehe Abb. 2) erkennbar.

Die Chipkarte muß selbstverständlich die herkömmlichen Funktionen des konventionellen Ausweises abdecken. Mit dem aufgebrachten Foto und der optisch lesbaren Gültigkeit wird sie den Anforderungen eines Ausweises für die Universität, den Nahverkehr und die sonstigen „Rabattgeber" Kino und Theater gerecht. Die Bedeutung als Identifikationsmedium in Netzen, für die Zutrittskontrolle, ja auch als Medium für die Erstellung digitaler Signaturen wurde bereits dargelegt.

Die Einführung von Selbstbedienung ist eines der zentralen Mittel um die beiden Ziele „Komfort" und „Verwaltungsvereinfachung" zu erreichen. Die Selbstbedienung für die Rückmeldung und den Bescheinigungsdruck war der erste und wichtigste Punkt auf der Liste der umzusetzenden Funktionalitäten. Hierdurch konnte die persönliche Rückmeldung erheblich vereinfacht werden, eine „Entlastung" für beide Seiten – Studierende und Verwaltung.

Abb. 2: Geplante Einsatzgebiete für den Chipkartenausweis

Die Anmeldung zu Prüfungen macht aus organisatorischen Gründen etwas Probleme, aber aufgeschoben ist ja nicht aufgehoben. Auch das Entrichten von Klein- und Kleinstgebühren ist schon lange ein Dorn im Auge. Hier soll ein weiterer Einsatzschwerpunkt des neuen Ausweises liegen. Als Optionen sind hier die Bibliotheksgebühren und das Kopieren zu nennen. Leider hat das Studentenwerk es vorgezogen, eine andere Technik einzusetzen.

1.3 Selbstbedienung = Mehr Komfort?

Inwieweit die Einführung von Selbstbedienung zu einem Mehr an Komfort führen soll, ist nicht offensichtlich. Insbesondere wenn man das Beispiel der Einführung von SB-Tankstellen im Kopf hat. Mit Sicherheit waren diese nicht komfortabler als diejenigen mit Tankwart, dafür waren sie aber billiger. *Also doch wieder nur ein Mittel zur Kostensenkung ?*

Daß dies nicht unser Ziel war, soll das Beispiel der Rückmeldung in Selbstbedienung erläutern.

Die Studierenden melden sich mit Ihrer Chipkarte an einem „Rückmeldeterminal" zurück. Die nun wegfallende „persönliche Rückmeldung" im Studierendensekretariat entfällt und entlastet dieses. Somit bleibt mehr Zeit für die Beratung der Studierenden, die sonst während der Rückmeldezeiten nur sehr spärlich ausfällt. Dies führt zu zufriedeneren Kunden (Studenten) und weniger gestreßtem Personal, aber nicht zu einer Kosteneinsparung. Die Studierenden werden unabhängig von den Öffnungszeiten des Sekretariats (ein PC braucht keine Mittagspause). Die sonst üblichen Schlangen entfallen, da es erheblich mehr Rückmeldeterminals als „Schalter" im Sekretariat gibt. Somit kommt auch auf diese Art und Weise die Selbstbedienung den Studierenden zu Gute. Und auch hier ist von Kosteneinsparung nicht die Rede.

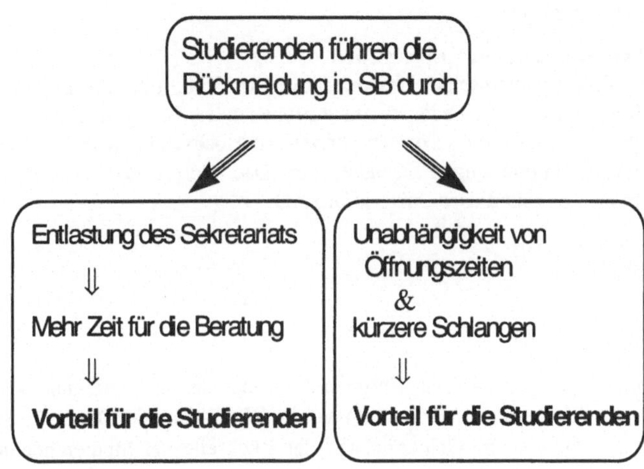

Abb. 3: Rückmeldung in Selbstbedienung

1.4 Eckdaten des Pilotprojektes

⇒ **Pilotprojekt für NRW**
Das Ministerium für Wissenschaft und Forschung des Landes Nordrhein-Westfalen hat das Projekt an der Ruhr-Universität Bochum zum Pilotprojekt erklärt und entsprechend unterstützt.

⇒ **Start im Sommersemester 97**
Seit dem Sommersemster '97 können die Studierenden den neuen Ausweis bekommen.

⇒ **Angebot an Erstsemester**
Die Teilnahme am Pilotprojekt ist freiwillig. Dies war unter anderem eine Forderung der Datenschutzbeauftragten. Diese Freiwilligkeit kann sich bei einem Vollausbau als problematisch herausstellen, aber in der Anfangsphase mußten die Systeme sowieso mit beiden Ausweisen arbeiten können. Zumal: *Wie sollten wir 36000 Ausweise auf einen Schlag umtauschen?*
Aus organisatorischen Gründen haben wir dies zunächst nur den Erstsemestern (während der Einschreibung) anbieten können.

⇒ **"Intelligente Karte"**
Um den gesetzten Zielen gerecht zu werden, ist für die eingesetzte Karte eine gewisse „Intelligenz" Pflicht. Sie mußte den folgenden Anforderungen gerecht werden:

a) Überprüfung einer **PIN** zur Identifikation des Inhabers.

b) Erstellung einer **digitalen Signatur** zur Absicherung (rechts-)verbindlicher Vorgänge.

c) Führung sicherer **Geldbörsen/Wertmarkenzähler** zur Vereinfachung von Bezahlvorgängen.

⇒ **Zusammenarbeit mit UniversCard**

In der Arbeitsgemeinschaft UniversCard haben sich, unter der Leitung von Prof. Dr. P. Horster, die verschiedenste Institutionen zusammengefunden um grundlegende Aspekte der Einführung von Chipkarten im Hochschulbereich zu erörtern und entsprechende Empfehlungen auszusprechen. Das Pilotprojekt an der RUB versteht sich als ein Pilotprojekt dieser UniversCard AG.

2 Umsetzung

2.1 Kartensystem

Als maßgeblicher Schritt war eine Entscheidung über das einzusetzende Kartensystem zu fällen. Zur Auswahl standen ein Aufsatteln auf die White-Card der Banken – auf der Vorderseite eine kontoungebundene Geldkarte, auf der Rückseite der Studierendenausweis – und die Herausgabe einer eigenen „Hochschulkarte" auf Basis einer Prozessorchipkarte mit Krypto-Kontroller.

Die White-Card ist vom Ansatz her eine Geldkarte. Die Zahlfunktion steht hier eindeutig im Vordergrund und die damit umsetzbaren Verwaltungsfunktionen müssen deutlich zurückstecken.

Die Hochschulkarte ist eine Karte der Hochschule für die Hochschule. Ihre primäre Aufgabe ist die Unterstützung der Verwaltungsprozesse und dazu ist sie maßgeschneidert. Eine Zahlfunktion kann damit – wenn überhaupt – nur innerhalb der Hochschule realisiert werden.

Eine detaillierte Betrachtung von Pro und Contra der White-Card und der Hochschulkarte liefert eine solide Entscheidungsgrundlage.

⇒ **Geldkarte**

Für die Geldkarte sprechen eine Reihe organisatorischer Gründe. Die Hochschule braucht sich weder um das Clearing, noch um die Beschaffung und Auswahl der eigentlichen Karte zu kümmern, das ist Aufgabe der Bank. Auch die universitätsübergreifende Zahlfunktion kann die Karte als Plus verbuchen.

Leider konnten sich die Banken nicht dazu durchringen einen Chip einzusetzen, der zur Erstellung von digitalen Signaturen fähig ist. Weiterhin ist auch der zur Verfügung stehende Speicherplatz nicht gerade üppig. Als besonders problematisch stellt sich die rechtliche Situation dar. Die Bank ist und bleibt Eigentümer der Karte und hat damit auch das Recht die Karte einzuziehen.

Der Vergleich mit einem Personalausweis, der McDonalds gehört, ist zwar etwas überspitzt und sarkastisch, trifft aber doch den Kern des Problems.

⇒ **Hochschulkarte**

Viele der Vor- und Nachteile der Hochschulkarte ergeben sich als direktes Analogon der entsprechenden Aspekte der Geldkarte. Einer der Vorteile soll hier dennoch hervorgehoben werden. Die Hochschule braucht eine gewisse technologische Freiheit und sich hier in die technologische Abhängigkeit von den Banken zu bringen, wäre fatal.

Vorwiegend die folgenden Argumente haben in der Ruhr-Universität Bochum zu einer Entscheidung für die Hochschul- und gegen die Geldkarte geführt:

⇒ die rechtlichen Bedenken,

⇒ die Verfügbarkeit der digitalen Signatur,

⇒ die technologische Abhängigkeit

⇒ und nicht zuletzt die Empfehlung der AG UniversCard.

2.2 Der eingesetzte Chip

Die Entscheidung für eine „Eigenlösung" brachte die Notwendigkeit der Auswahl eines einzusetzenden Chips mit sich.

Aus Kostengründen kamen nur Chips in Frage, die mit Standardbetriebssystem lieferbar waren. Die Implementierung eines eigenen ROM-Codes schien ob der geringen Stückzahl in der Anfangsphase nicht sehr sinnvoll. Im Rennen waren der Chip SLE 44 CR 80S von Siemens mit dem Betriebssystem STARCos von Gieseke&Devrient und die GPK 2000 von Gemplus mit dem Chip von SGS Thomson und dem Betriebssystem GPCos.

Der Siemens Chip bestach durch seinen relativ großen Speicher von 8 KByte – in der Designphase hätte ich mir den so manches mal gewünscht. Leider war er damals nicht in Kombination mit einer sog. ThermoChromic-Folie lieferbar. Wofür diese TC-Folie überhaupt eingesetzt wird, wird noch dargelegt.

Da der Siemens Chip nicht rechtzeitig auf dem Markt verfügbar war, ist die Wahl auf die GPK 2000 von Gemplus gefallen.

2.3 Karten-Layout

Wir haben das in Abb. 4 dargestellte Layout gewählt.

Die Besonderheit der Karte wird auf der Rückseite deutlich. Die ThermoChromic-Folie ist mit einem thermischen Verfahren beschreib- und wieder löschbar. So ist es möglich die Gültigkeit des Ausweises jedes Semester zu verlängern und auch eine etwaige Berechtigung zur freien Fahrt im Nahverkehr optisch zu kennzeichnen. Dieses Feature war uns sehr wichtig. Denn nur so ist es ohne großen technischen (mechanischen) Aufwand möglich, den Ausweis automatisiert optisch zu aktualisieren. Bei Aufklebemarken wäre ein Benutzereingriff nötig gewesen. Zudem paßt nach unserer Vorstellung zu einer modernen Karte einfach kein System von Aufklebe-Gültigkeitsmarken, auch wenn diese Marken Hologramme sind.

2.4 Auf dem Chip gespeicherte Daten

Die Menge der auf dem Chip zu speichernden Daten ist nicht ganz einfach festzulegen. Speichert man nur eine ID (z.B. die Matrikelnummer) auf dem Chip, so muß in der Regel jedes Gerät, das mit der Karte etwas Sinnvolles anfangen soll, an das Hintergrundsystem online angeschlossen sein. Es muß also eine vollständige Vernetzung aller Geräte vorliegen. Speichert man hingegen „alle" Daten der Studierenden auf der Karte, so kann im Prinzip (fast) jedes Gerät auch offline betrieben werden. Ein Netzanschluß ist nur für solche Anwendungen notwendig, die einen Zugriff auf Sperrlisten erfordern.

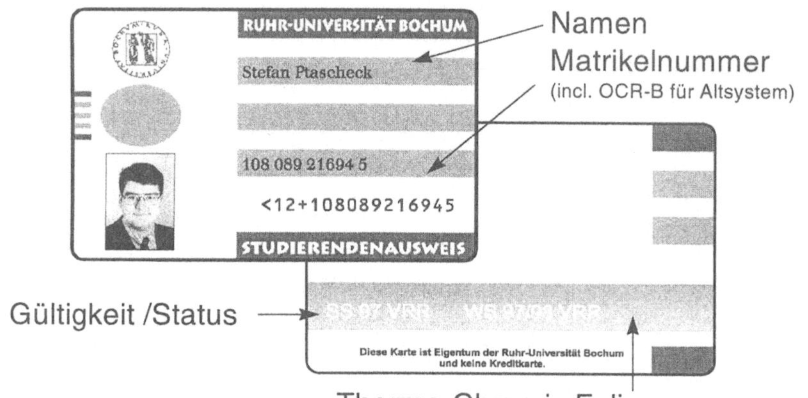

Abb. 4: Layout der Studierendenchipkarte

Je mehr Daten jedoch auf der Karte gespeichert werden, desto „interessanter" wird es diese zu manipulieren. Man stelle sich vor, die aktuellen Prüfungsergebnisse sind auf der Karte gespeichert und werden für die Zulassung zu einer Veranstaltung ausgelesen. Mit einer gefälschten Karte könnte man sich so Zugang zu dieser Veranstaltung verschaffen. Zudem ist es nur schwer sicherzustellen, daß die Daten auf der Karte stets aktuell sind.

Ein weiteres Problem ist die mit dem steigendem Datenumfang enorm anwachsende Zugriffsschutzproblematik.

All dies habe ich versucht in Abb. 5 darzustellen. Ergebnis einer solcher Überlegung war eine „*Sowenig wie möglich, soviel wie nötig*" Strategie.

Auf der Karte sind verhältnismäßig „harmlose" Informationen gespeichert:

⇒   **Gültigkeit & Studierendenstatus** (*Ordentlicher Student, Gast- oder Zweithörer*)

⇒   **Matrikelnummer**

⇒   **Name, Vorname** und **Geburtsdatum**

⇒   **Adresse**

⇒   **Hochschulsemester, Studiengänge** und **-fächer**

⇒   **Netz-Zugang** (*ID und Passwort*)

⇒   **Bibliotheksstatus**

⇒   **Wertmarkenzähler** (*im separatem Dedicated File*)

⇒   **Signatur-Information** (RSA-Schlüsselpaar, 768-Bit, CRT)

⇒   relevante Teile des **Zertifikats**

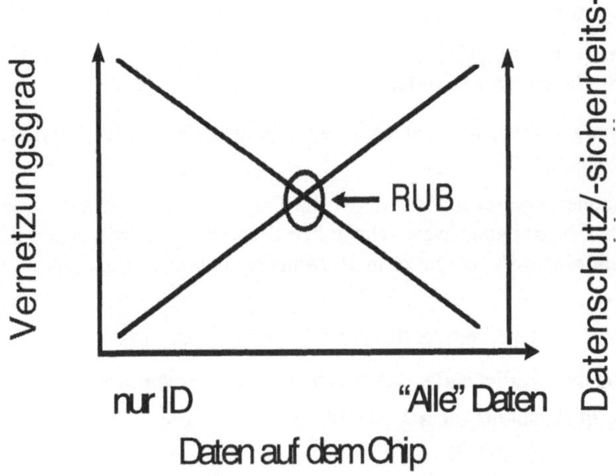

Abb. 5: Auswirkungen der Datenmenge auf dem Chip

Darin enthalten sind zwar die Studiengänge und Fächer, nicht aber die erzielten Abschlüsse oder Noten. Der Ausleihstatus der Bibliothek gibt nur wieder, ob Belastungen (z.B. durch Ausleihen) bestehen, aber nicht welcher Art diese sind.

Die Netz-Zugangsdaten dienen dem Zugriff auf die zentrale Mailbox. Die (zur Zeit) zwei Wertmarken-Zähler sind in einem separaten DF untergebracht. Der dickste Brocken sind das Schlüsselpaar für die digitale Signatur und die benötigten Teile des Zertifikats. Die Einschränkung auf einen 768-bit Schlüssel mit CRT-Kodierung wurde von der Karte vorgegeben.

Damit jede potentielle „offline"-Akzeptanzstelle auch nur die Daten zur Verfügung hat, die sie benötigt, wurde das von der AG UniversCard entwickelte mehrstufige Zugriffsschutzkonzept auf die Gegebenheiten in Bochum adaptiert und gemäß den Mechanismen der eingesetzten Karte implementiert. Alles zusammen, in Kombination mit der Tag-Length-Value-Kennung, hat die Karte bis auf 5 Byte gefüllt.

2.5 Sicherheitsmechanismen

Jeder Akzeptanzstelle werden eine oder mehrere Zugriffsmöglichkeiten gewährt. Diese verteilen sich nach folgenden Schutzklassen.

⇒ **freier Lesezugriff**
Zugriff auf Gültigkeit und VRR-Berechtigung

⇒ **Hochschulinterner Lesezugriff**
Name, Matrikelnummer, Adresse
Fächer und Netzaccount mit PIN

⇒ **Dezentraler Schreibzugriff**
  Bibliothek & Rechenzentrum

⇒ **Zentraler Schreibzugriff**
  nur das Studierendensekretariat

Innerhalb der aufgeführten Klassen wird eine „Feinsteuerung" der Zugriffsrechte durch Secret-Codes gewährleistet.

In den Diskussionen kommt immer wieder die Frage nach der Sicherheit der Karte auf. Natürlich muß der Chipkartenausweis sehr hohen Sicherheitsanforderungen genügen. Daß der alte Ausweis mit einem Farbkopierer in 10 Minuten perfekt zu fälschen ist, wird aber nur zu gerne ignoriert.

Folgende technische Sicherheitsmechanismen kommen zum Einsatz:

⇒ **Trennung** der **Studierendendaten** von den **Geldbörsendaten**,

⇒ **gegenseitige Authentisierung** von Terminal und Karte,

⇒ Absicherung durch **3DES**,

⇒ **verschlüsselte Übertragung**,

⇒ **Session Keys**,

⇒ **abgeleitete Schlüssel**

2.6 Selbstbedienungsterminals

Die Selbstbedienungsterminals, auch UFOs genannt, bestehen aus 4 Arbeitsplätzen und sind in Abb. 6 dargestellt. Von diesen UFOs sind zunächst insgesamt 5 in der Universität aufgestellt. Die geplante Anzahl ist erheblich höher, aber selbst diese Zahl müßte ausreichen, um den Rückmeldeprozeß aller Studierenden um den Faktor zwei zu verkürzen.

Der Einsatz von Standardhardware gewährt eine hohe Flexibilität in Bezug auf die zur Verfügung stehenden Dienstleistungen. Die Alarmsicherung gestattet es, eine offene Arbeitsatmosphäre zu schaffen und die Arbeitsplätze nicht „einmauern" zu müssen. Dies trägt auch zu einem hohen Wiedererkennungswert bei, so daß ein Terminal auch in unerwarteter Umgebung als UFO erkannt wird.

Bei der Konzeption der Arbeitsplätze war Multifunktionalität stets Trumpf. Nur durch die Integration unterschiedlichster Anwendungen und deren Verknüpfung mit dem Identifikationsmedium Chipkarte kann ein echter Mehrwert erzeugt werden.

Die Terminals stellen folgende Funktionalitäten zur Verfügung:

⇒ **Internet-Zugang E-Mail**
  automatische Konfiguration auf die persönliche Mailbox

⇒ **Digitale Signatur von Dokumenten**
  a) E-Mail
  b) Anstoß (rechts-)verbindlicher Vorgänge

⇒　**Studentenverwaltung**

　　a)　Rückmeldung,

　　b)　Adreßänderung

　　c)　Bescheinigungsdruck

⇒　**Bibliotheksfunktionen**

⇒　**Bargeldlose Zahlung von Gebühren**

Alle Funktionen sind nur mit einer gültigen Chipkarte nutzbar. Die PIN wird im Bedarfsfall abgefragt.

Abb. 6: Selbstbedienungsterminal (UFO)

2.7　Erstellung der Ausweise

Das Chipkartenprojekt ist vollständig in den Immatrikulationsprozeß eingebettet: Die Studierenden entscheiden über die Teilnahme am Pilotprojekt und daraufhin wird ein entsprechender Ausweis online erstellt. Die Studierenden können die Chipkarten sofort mitnehmen. Der Ablauf der Immatrikulation wird in Abb. 7 schematisch dargestellt.

Die Erfassung des Bildes mit einem Scanner hat sich gegenüber dem Einsatz einer Digitalkamera als erheblich praktischer erwiesen. Die zunächst eingesetzte Software der Firmen IntraProc und InterCard hat sich als wenig geeignet für eine solch enge Einbindung in das System erwiesen. Aus diesem Grund kommt jetzt eine Software zum Einsatz, die komplett von der Ruhr-Universität Bochum erstellt wurde.

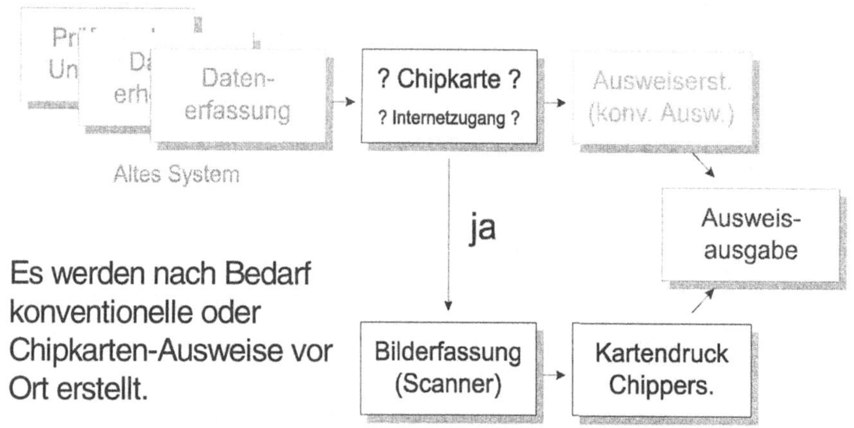

Abb. 7: Ablauf einer Immatrikulation

3 Resultate

Im nachfolgenden Kapitel werden die Zwischenresultate des Pilotprojekts zusammengefaßt und ein Ausblick auf weitere Entwicklungen gegeben.

3.1 Erfahrungen

⇒ **Akzeptanz**
Die Akzeptanz dieses neuen Mediums bei den Studierenden ist sehr hoch. Alleine an die Erstsemester – diese haben noch nicht in einer Schlange zur Rückmeldung gestanden – sind mehr als 6000 Ausweise ausgegeben worden. Das entspricht einer Quote von etwa 90%.

⇒ **Versachlichung der Datenschutzdiskussion**
Positiv zu bemerken ist auch die Versachlichung der Datenschutzdiskussion. Die Chipkarte ist nicht mehr nur „Teufelswerk" und „Hilfsmittel zur Einführung von Studiengebühren", sondern ein Medium mit Stärken und Schwächen, über das sachlich diskutiert wird.

⇒ **Öffnung für das Medium Chipkarte auch in anderen Bereichen der Universität**
Von vielen Seiten wurden und werden wir auf Einsatzmöglichkeiten dieser oder anderer Chipkarten in neuen, bisher nicht berücksichtigen Gebieten angesprochen.

3.2 Kosten

Auch im öffentlichen Dienst gewinnt das Thema Kostenrechnung zunehmend an Bedeutung. Ein Ergebnis des Pilotprojektes war eine Kostenabschätzung für eine vollständige Umstellung (36000 Ausweise, 20 SB-Terminals) auf Chipkartenausweise.

Kosten „Alter Ausweis"		Kosten „Chipkartenausweis"	
Personal		Personal	
	276.000 DM		23.000 DM
Terminals		Terminals	
	7.000 DM		65.000 DM
Verbrauchsmaterial		Verbrauchsmaterial	
	17.000 DM		11.000 DM
Ausweise		Ausweise	
	3.000 DM		93.000 DM
Gesamt	**303.000 DM**	**Gesamt**	**192.000 DM**

Allein schon die Einsparung von Personalkosten bei der Umstellung der Rückmeldung auf Selbstbedienung rechtfertigt den Kostenaufwand für die Chipkartenausweise und die anzuschaffende Hardware.

3.3 Verbleibende Aufgaben

Die Startphase des Pilotprojektes kann als überwunden angesehen werden. Es sind jedoch noch weitere Probleme zu lösen. Die Implementierung der digitalen Signatur in der Karte ist nur der erste Schritt zu einer Umgebung, die das Werkzeug „Chipkarte" umfassend nutzen kann. Auf dem Gebiet der Schlüsselverwaltung und der Verzeichnisdienste bleiben noch eine Reihe von Aufgabe zu bewältigen. Auch ist die Frage, ob und wie den Anforderungen des Gesetzes zur Digitalen Signatur Rechnung getragen werden kann, noch zu klären.

Auf Dauer wird sich eine Chipkarte nur dann durchsetzen, wenn sie im Universitätsbereich umfassend als Identifikationsmedium eingesetzt werden kann. Wir haben uns selbst einen Richtwert von einem neuen Dienst pro Semester vorgegeben.

3.4 Ausblick

Im Rahmen eines TEN Telekom Projekts sollen die Probleme im Umfeld der digitalen Signatur gelöst und ein europäischer Standard für Universitäts-Chipkarten geschaffen werden. Ebenso ist eine Vermarktung der Ergebnisse des Pilotprojektes und des angesammelten Know-Hows im Gespräch.

Last but not least: Bislang wurde stets von einem Studierendenausweis auf Chipkartenbasis gesprochen. Die Ausgabe von entsprechenden Ausweisen an Bedienstete ist das erklärte Ziel der Universität. Allein die bestehenden Regelungen des Personalvertretungsrechts des Landes NRW erweisen sich hier als etwas hinderlich. Die Umsetzung dieses Punktes wird in Kürze in Angriff genommen.

Chipkarteneinsatz im Hochschulbereich
Eine kritische Bestandsaufnahme

Frank Ziemke

InterCard Projektbüro
Wittenberger Str. 50, D-30179 Hannover
Frank.Ziemke@t-online.de

Zusammenfassung

Unter Chipkarteneinsatz im Hochschulbereich wird die multifunktionale Nutzung von Chipkarten sowohl für Studierende als auch Bedienstete verstanden. Mögliche Anwendungen sind die Ausweisfunktion, Identifikation für SB-Anwendungen, Zutrittskontrolle, bargeldlose Bezahlung, digitales Signieren von Dokumenten, Benutzung des ÖPNV etc. [1].

Zweck der Multifunktionalität ist u. a. die z. T. bereits vorhandene Kartenflut einzudämmen bzw. gar nicht entstehen zu lassen und die Synergieeffekte zwischen den einzelnen Anwendungen zu nutzen. Dieser Beitrag vermittelt einen Überblick über den aktuellen Entwicklungsstand an deutschen Hochschulen und versucht den Status quo anhand eines Vier-Phasen-Modells in den Gesamtentwicklungsprozeß einzuordnen.

1 Einleitung

Im Jahre 1993 begann man sich in Deutschland erstmals ernsthaft mit dem Thema Chipkarteneinsatz im Hochschulbereich zu beschäftigen. Es fanden erste hochschulübergreifende Diskussionsrunden – beispielsweise in Nordrhein-Westfalen – statt. Auch die HIS-GmbH nahm sich der Thematik an. Die Industrie bot erste noch rudimentäre Systemlösungen auf Basis einer halboffenen elektronischen Geldbörse an.

Im Rahmen des seit 1994 laufenden bayerischen Forschungsprojektes 'Optimierung von Universitätsprozessen' befaßte sich eine Arbeitsgruppe mit der Konzeptionierung einer multifunktionalen Universitäts-Chipkarte (MUCK). Im März 1996 schließlich konstituierte sich die Harmonisierungsinitiative UniversCard [2], die im Rahmen ihrer Arbeit erste hochschulübergreifende Lösungsansätze formulierte, die in die aktuellen Realisierungen der Hochschulen auch teilweise Eingang gefunden haben.

2 Entwicklungslinien

Zum allgemeinen Verständnis des Entwicklungsprozesses des Chipkarteneinsatzes an deutschen Hochschulen erscheint es mir sinnvoll, zunächst kurz die vier idealtypischen Phasen von Technikgeneseprozessen im allgemeinen darzustellen [3]. Anschließend wird der Versuch unternommen, die aktuellen Entwicklungen den vier Phasen zuzuordnen.

2.1 Theoretische Grundlagen

In einer ersten Explorationsphase werden zahlreiche, meist isolierte Projekte in dem noch unbesetzten Feld durchgeführt. In der sich anschließenden Phase kristallisiert sich allmählich ein Mainstream heraus. Manche der realisierten Varianten werden als untauglich verworfen, einige können sich behaupten. In der dritten Phase, der Diffusionsphase, erfolgt eine Verbreitung derjenigen Systeme, die sich als Mainstream herauskristallisiert haben. Das Verhalten der Mitakteure kann der Systemanbieter nicht vorhersehen, möglicherweise gibt es Reaktionen, die den Systemanbieter veranlassen, sein Produkt zu variieren, damit es den Erfordernissen des Marktes besser gerecht wird (Phase vier). Die folgende Abbildung faßt die vier Phasen noch einmal kurz tabellarisch zusammen.

Phase	Beschreibung
Phase 1 Explorationsphase	Isolierte Projekte in einem noch unbesetzten Feld
Phase 2 Sondierungsphase	Herauskristallisieren eines Mainstream
Phase 3 Diffusionsphase	Verbreitung der Mainstream-Systeme
Phase 4 Konsolidierungsphase	Produktanpassungen an die Marktanforderungen

Abb.1: Vier-Phasen-Modell

2.2 Aktuelle Entwicklungen

Die Entscheidung, ob ein Chipkartensystem einführt wird, hängt wesentlich von den folgenden Voraussetzungen ab:

- Vorhandene technische, infrastrukturelle, organisatorische, rechtliche, finanzielle und personelle Rahmenbedingungen

- Innovationspotential und auch Risikobereitschaft der Hochschule

- Zwang zur Rationalisierung aufgrund finanzieller Restriktionen der Hochschule

- Existenz eines "Leidensdrucks", der ggf. sogar eine kurzfristige Einführung eines Chipkartensystems notwendig macht.

Sind an der Hochschule optimale Rahmenbedingungen vorhanden und auch das notwendige Innovationspotential sowie eine Motivation zur Einführung eines Chipkartensystems gegeben – das können Marketingaspekte oder ein gewisser "Leidensdruck" sein (z. B. lange Schlangen bei der Semesterrückmeldung, die man in Zukunft vermeiden möchte oder der Zwang zur Abrechnung von Dienstleistungen), dann sind die Voraussetzungen für die Einführung eines Chipkartensystems gegeben. Im Normalfall ist auch Zeit genug für eine wohlüberlegte Systementscheidung vorhanden (Beispiele: Ruhr-Universität Bochum, Universität Trier oder Universität Würzburg).

Besteht akuter Handlungsbedarf ein Kartensystem einzuführen, neigen manche Hochschulen dazu, sich für vermeintlich preisgünstige, monofunktionale und damit nicht zukunftsweisende Adhoc-Lösungen zu entscheiden. Solch ein akuter Handlungsbedarf kann beispielsweise sich häufender Diebstahl von wertvollen Geräten aus zu schützenden Räumen oder die Kündigung von Wachpersonal sein, die die kurzfristige Installation eines Zutrittskontrollsystems notwendig machen. Diese Situation könnte für den Einstieg in ein multifunktionales Chipkartensystem genutzt werden, wählt man jedoch ein Low-cost-System, ist diese Möglichkeit oft für längere Zeit verbaut. Hier kann der Hochschule eine kompetente, neutrale Beratung helfen, auch kurzfristig eine sinnvolle Systementscheidung zu treffen.

Hat sich die Hochschule für die Einführung eines Chipkartensystems entschieden, hängt die Systemauswahl – wie die Erfahrung gezeigt hat – u. a. von der Gewichtung der folgenden Faktoren ab:

- Anwendung der Digitalen Signatur (Motivator: Inkrafttreten des Signaturgesetzes)

- Bargeldlose Bezahlfunktion (Geldkarte, Finanzclearing)

- Steigende Bedeutung des Internet als Kommunikations- und Informationsmedium für Hochschulen (Stichwort: "virtuelle Universität")

- Anforderungen des Datenschutzes

- Empfehlungen der AG UniversCard

- Finanzierungsaspekte.

Die folgende Abbildung vermittelt einen Überblick über die bisher in Deutschland getroffenen Systementscheidungen:

Intelligente Speicherkarte	Prozessorkarte	Kryptoprozessorkarte
2 Fachhochschulen (1 x kontaktbehaftet, 1 x kontaktlos)	7 Hochschulen; davon 4 Fachhochschulen hochschuleigene Karte und 3 Universitäten Geldkarte (2 x kontoungebunden, 1 x kontogebunden)	6 Hochschulen; davon 5 Universitäten und 1 Fachhochschule. 1 Universität TwinCard (Krypto/kontaktlos)

Abb.2: Systemenscheidungen nach Chipkartentyp

Die Entscheidung für eine intelligente Speicherkarte haben die jeweiligen Fachhochschulen bereits im Jahr 1996 bzw. vorher getroffen. Man kann davon ausgehen, daß heute kaum noch eine Hochschule die intelligente Speicherkarte für den multifunktionalen Einsatz wählen würde. Im Gegensatz dazu reicht die intelligente Speicherkarte beispielsweise für die Anforderungen der Studentenwerke – im wesentlichen bargeldlose Bezahlvorgänge und ggf. Zutrittskontrolle – auch heute noch aus, und wird dementsprechend auch weiterhin als Chipkartentyp gewählt.

Bei der Wahl der Prozessorkarte fällt auf, daß gerade die Fachhochschulen die hochschuleigene Prozessorkarte gewählt haben. Diese Einheitlichkeit ist – bezogen auf drei Fachhochschulen – zum einen bedingt durch eine im entstehen begriffene landesweite Lösung, zum anderen spielte der Finanzierungsaspekt eine gewichtige Rolle. Man mißt der Anwendung der digitalen Signatur z. Z. noch nicht die Bedeutung zu, die den höheren Preis der Kryptoprozessorkarte gerechtfertigt hätte.

Zwei Universitäten haben sich für den Einsatz einer kontoungebundenen Geldkarte als multifunktionalen Studierendenausweis entschieden, wobei die Universität Trier die Karte bereits im Herbst 1997 eingeführt hat. Ein wesentlicher Grund für die Entscheidung der Universität Trier war, daß der bargeldlosen Bezahlfunktion ein hoher Stellenwert beigemessen wurde und die Sparkasse das Finanzclearing durchführt und somit die Universität in diesem Punkt entlastet.

Die Universität Würzburg plant im Rahmen des MUCK-Projektes die kontogebundene Geldkarte – d. h. die ec-Karte – als Identifikationsmedium zu benutzen. Als Ausweis kann die Karte damit per se nicht benutzt werden. Ausschlaggebend für die Entscheidung war, bereits existierende Karten ohne wesentlichen finanziellen Mehraufwand mitbenutzen zu können.

Die Hochschulen, die sich für eine Kryptoprozessorkarte entschieden haben, betrachten den Einsatz der digitalen Signatur als essentiellen Bestandteil einer multifunktionalen Hochschulkarte. Hier sind Investitionen auch im Hinblick auf die "virtuelle Universität" getätigt worden.

Die kontaktlose Technologie wird vermutlich in den nächsten beiden Jahren nur in Kombination mit einem kontaktbehafteten Chip im Hochschulbereich eine Rolle spielen (TwinCard), wobei der kontaktlose Teil nur für spezielle Anwendungen wie Zutrittskontrolle und Zeiterfassung genutzt wird. Später wird die CombiCard die Rolle der TwinCard übernehmen.

2.3 Einordnung der Entwicklungen in das Phasenmodell

In welcher Phase des Technologiegeneseprozesses befinden wir uns also? Mehrere Indikatoren sprechen dafür, daß wir uns im Übergang von der Explorationsphase zur Sondierungsphase befinden.

Für alle im Grunde denkbaren technischen Varianten von multifunktionalen Chipkartensystemen gibt es inzwischen Implementierungen oder sie stehen unmittelbar bevor. Die intelligente Speicherkarte scheidet dabei – wie gesagt – bereits im Vorfeld als Lösungsvariante für die Zukunft aus, da sie den Anforderungen an eine multifunktionale Chipkarte in mehrfacher Hinsicht nicht genügt.

Wahrscheinlich kann man davon ausgehen, daß der Anteil der Hochschulen, der die Kryptoprozessorkarte wählt, in Zukunft tendenziell eher steigen wird. Die Ankündigung der Sparkassenverbände mit der nächsten Geldkartengeneration ggf. optional die Kryptoprozessortechnologie anzubieten, weist in die gleiche Richtung. Wie schnell sich die Kryptokarte durchsetzen wird, wird u.a. davon abhängen, ob sich das Signaturgesetz zügig und praxisnah umsetzen läßt.

Da die Wahl einer hochschuleigenen Prozessorkarte gegenüber der Kryptoprozessorkarte von den jeweiligen Hochschulen nicht zuletzt aus Kostengründen getroffen wurde, kann man davon ausgehen, daß bei einem höheren Verbreitungsgrad der Kryptokarte dieses Argument zunehmend an Bedeutung verliert, z. Z. spielt dieser Kartentyp aber noch eine wesentliche Rolle.

Wie weit sich die anonyme Geldkarte als primäre Zahlungskarte mit einem vorkonfektionierten Bereich für Zusatzapplikationen am Hochschulmarkt durchsetzen wird, wird davon abhängen, inwieweit die Banken/Sparkassen flexibel auf die Anforderungen der Hochschulen in Zukunft reagieren (Gültigkeitsdauer, Speicherkapazität) und natürlich, ob die Geldkarte insgesamt genügend Akzeptanzstellen gewinnt.

Bei allen Unwegbarkeiten kann man für die überschaubare Zukunft die beiden Mainstreams "hochschuleigene Kryptoprozessorkarte" (Modell Ruhr-Universität Bochum) und "Geldkarte" (Modell Universität Trier) prognostizieren, wobei die Sparkassen als Geldkartenanbieter sich anscheinend gegenüber der Deutschen Bank durchsetzen.

Wir befinden uns also in der Übergangsphase von der Explorationsphase zur Sondierungsphase. Es werden Tendenzen sichtbar, aber zu viele Features sind in der Praxis noch nicht erprobt, als daß sich – global betrachtet – Entwicklungen bereits stabilisiert haben könnten. Im Detail sind hingegen bereits konkrete Festlegungen getroffen worden. In allen bisherigen Chipkartenprojekten wurde die ThermoChromic-Folie – ein in den Plastikkörper eingebetteter, im Thermotransfer-Verfahren wiederbeschreibbarer silbriger 1 cm breiter Streifen – als Medium für die optische Validierung gewählt. Semesterweise aufklebbare Hologramme sind also für diesen Zweck bereits obsolet. Ebenso gehört ein ThermoChromic-Drucker mit integriertem Chipkartenleser, der die Folie löschen und wiederbeschreiben und gleichzeitig den Chip updaten kann, zur Grundausstattung der Hochschulen.

Zum Vergleich möchte ich die wesentlichen Merkmale der beiden Mainstream-Modelle gegenüberstellen.

2.4 Beschreibung der Mainstream-Modelle

Merkmale des Bochumer Modells:

- Hochschuleigene Karte; freie Wahl der Chipkartentechnologie, hier: Krypto-Prozessorkarte

- Die Hochschule hat völlig freien kreativen Gestaltungsspielraum für das Gesamtsystem, nur begrenzt durch die technischen Möglichkeiten und die gesetzlichen und rechtlichen Rahmenbedingungen

- Die Hochschule ist für die Gesamtfunktionalität des Systems verantwortlich.

- Abgeschlossene Geldbörse (Wertmarkenzähler) für einzelne bargeldlose Abrechnungen (Kopieren, Gebühren Bibliothek).

- Semestergebühren können durch zusätzlichen Einsatz bereits vorhandener bargeldloser Zahlungssysteme (ec-Karte, Kreditkarte) beglichen werden.

- Das Finanzclearing einer halboffenen elektronischen Geldbörse – zur Zeit noch nicht aktiviert – muß von der Hochschule organisiert werden, die gesamte Infrastruktur ist von der Hochschule zur Verfügung zu stellen.

- Die Rahmenbedingungen für den Einsatz der digitalen Signatur müssen von der Hochschule geschaffen werden, d. h. unter anderem muß eine Sicherheitsinfrastruktur, die letztendlich den Anforderungen des Signaturgesetzes genügt, sukzessive aufgebaut werden.

- Flexibilität bei veränderten Rahmenbedingungen.

- Anlehnung an das UniversCard-Konzept.

- Finanzierung durch Ministerium und Universität.

Merkmale des Trierer Modells:

- Kontoungebundene Geldkarte (Prozessorkarte) der Sparkasse Trier mit universeller Zahlungsfunktion (bundesweit), relativ geringer Speicherplatz, vorkonfektionierter Speicherplatz für Zusatzanwendungen.

- Technologisch ist der Rahmen durch den ZKA vorgegeben, die Hochschule kann auf veränderte Rahmenbedingungen nicht direkt und autonom reagieren.

- Layout-Gestaltung der Chipkarte: Vorderseite Sparkasse Trier, Rückseite Universität Trier.

- Die Hochschule trägt nur eine Teilverantwortung für die Funktionsfähigkeit des Systems (Zusatzanwendungen).

- Die Infrastruktur für die bargeldlose Bezahlung stellt die Sparkasse Trier in Form von Geräten (Aufwerter) und als Dienstleistung (Finanzclearing) zur Verfügung. Die Sparkasse übernimmt die Zahlungsgarantie.

- ZKA-Zertifizierung der Chipkartenperipherie (mit Zahlungsfunktion) notwendig.

- Beschränkte Gültigkeitsdauer der Geldkartenfunktion (3 Jahre + Ausgabejahr).

- Geldkartenfunktion übertragbar, Studierendenausweisfunktion per se nicht.

- Die digitale Signatur kann nicht realisiert werden, da die Geldkarte keinen Krypto-Coprozessor besitzt .

- Finanzierung durch die Universität mit finanzieller Unterstützung durch die Sparkasse Trier.

Für die Zukunft bleibt die Frage offen, ob es _einen_ Mainstream geben wird oder die beiden dargestellten Ansätze koexistent nebeneinander auf Dauer existieren können

3 Ausblick

Alle Anzeichen sprechen dafür, daß sich die Chipkarte im Hochschulbereich etablieren wird. Es wird sich allerdings keineswegs notwendigerweise die technologisch beste Lösung durchsetzen. Die Finanzlage an den Hochschulen hingegen, die Bereitschaft der Sparkassen und deren übergeordneter Verbände, die Hochschulprojekte zu subventionieren, werden mit ausschlaggebend dafür sein, welches Lösungsmodell gewählt wird. In diesem Kontext ist die Phantasie der Hochschulen gefragt – insbesondere im Rahmen der Globalhaushalte – neue Wege zur Finanzierung von Chipkartenprojekten zu finden, um die Systemauswahl nicht allein von der Finanzierung abhängig zu machen. Ferner wird entscheidend sein, inwieweit sich aktuelle Harmonisierungsbemühungen (in einigen Bundesländern werden landesweite Lösungen angestrebt) realisieren lassen, um durch Rahmenverträge und die Nutzung von Mengenstaffeln günstigere Konditionen zu erhalten. In diesem Zusammenhang wird die Etablierung der Geldkarte davon abhängen, inwieweit die Sparkassen auf die Anforderungen der Hochschulen in Zukunft flexibel reagieren können

Man darf auf die Entwicklung in den nächsten beiden Jahren gespannt sein.

Literatur

[1]   Ziemke, F.: Chipkartenanwendungen für Studierende, In: Tagungsband 5. GMD-SmartCard Workshop, Darmstadt 1995

[2]   Lender, F.; Schloßer, K.-H.; Schmidt, W.; Ziemke, F.: UniversCard Arbeitsunterlage, 3. Arbeitstreffen in München, HIS GmbH, 1995

[3]   Forschungsgruppe Telekommunikation – Universität Bremen: Die Verbreitungschancen von Wertkarten im kartengestützten Zahlungsverkehr. In: Vortrag ausgewählter Projektergebnisse auf dem Experten-Workshop am 24. Nov. 1993

MAJA
Konzept eines Multiple Application JavaCard Environment

Sachar Paulus

Institut für Theoretische Informatik
TU Darmstadt
Alexanderstraße 10
64283 Darmstadt

Zusammenfassung

Wir stellen die Ziele des aktuellen Chipkartenprojekts an der TU Darmstadt vor. Dabei geht es um den Einsatz einer multifunktionalen Chipkarte mit der Möglichkeit, dynamisch Anwendungen durch den Benutzer zu laden und zu entladen.

1 Motivation

Die Nutzung der Chipkartentechnologie steckt noch in den Anfängen. Bedingt durch den technischen Aufbau werden spezielle Anforderungen an die Funktionalität von in Chipkarten eingesetzten elektronischen Bausteinen gestellt. Dies hat insbesondere einen sehr eingeschränkten Speicherplatz zur Folge. Das ist der Hauptgrund, weshalb es heute überwiegend Chipkartenlösungen mit einer einzigen Anwendung pro Karte gibt.

Zur Zeit arbeiten mehrere Hersteller an einer Spezifikation für ein offenes Chipkartenbetriebssystem, welches mehrere Anwendungen unterstützen soll [4]. Erste Implementierungen sind schon auf dem Markt. Die Verwaltung mehrerer Applikationen und die entsprechende notwendige Sicherheitsinfrastruktur führt zu einem noch wesentlich größeren Speicherverbrauch auf der Karte.

Das Speicherproblem wird deshalb im Zusammenhang mit Multiapplikationschipkarten für einige Zeit eines der schwierigsten Probleme bleiben, auch wenn man die Entwicklung der Hardwaretechnologie in die Überlegungen miteinbezieht. Daher ist es berechtigt, nach Lösungen zu suchen, welche in der Lage sind, Applikationen auf der Karte dynamisch auszutauschen.

Dieser Ansatz hat aber noch einen weiteren, ethischen Vorteil: der Benutzer kann selbst entscheiden, welche Applikationen er auf seine Karte lädt; damit wird die persönliche Freiheit des Individuums gewahrt.

In diesem Aufsatz besprechen wir zuerst bestehende Multiapplikations-Lösungen. Danach werden wir argumentieren, warum die Sprache JAVA für einen unseren Zweck ideal ist und die von uns vorgeschlagene dynamische Lösung genauer spezifizieren. Schließlich schildern wir die geplante Realisierung im Rahmen des UniCard Projekts der TU Darmstadt.

2 Heutige Realisierungen

2.1 Fehlende Standards

Die Entwicklung der SmartCard-Technologie hat, ähnlich wie vor Jahren die Entwicklung von Software für PCs, größtenteils abseits der Universitäten stattgefunden. Das Wissen über Internas des Entwicklungsprozesses bedeutet bares Geld, so ist der Informationsfluß zu den öffentlichen Forschungseinrichtungen relativ dünn. Selbst sogenannte "öffentliche Standards" wie die ISO Norm 7816 sind kostenpflichtig!

Durch eine fehlende öffentliche de-facto-Standardisierung gibt es eine Handvoll von unterschiedlichen, proprietären Lösungen, die untereinander inkompatibel sind, d.h. die Applikation von einer Chipkarte kann nicht auf eine Chipkarte von einem anderen Hersteller geladen werden. Dies ist zwar im Sinne der derzeit existierenden Anbieter, verhindert aber eine Erweiterung des Marktes. Es ist einem Anbieter von Sicherheitssoftware nicht ohne weiteres möglich, eine Chipkartenapplikation zu vermarkten ohne sich definitiv für eine spezielle Chipkarte zu entscheiden. Es ist also im Sinne einer Dezentralisierung und Flexibilisierung des Marktes, einen gemeinsamen Standard zu finden.

2.2 Multiapplikationsbetriebssysteme

Der Übergang von Single-Application SmartCards hin zu Multi-Application SmartCards scheint der richtige Zeitpunkt für einen solchen Sprung zu sein. Motivation ist, die Programmierung der Applikationen an andere Anbieter "outzusourcen". Aus diesem Grund gibt es mehrere Initiativen, einen gemeinsamen Standard zu finden. Bekanntester Vertreter ist hierbei `Multos`. Dabei handelt es sich um eine Spezifikation für ein Multi-Application Betriebssystem des Maosco-Konsortiums. Das Maosco-Konsortium besteht aus Firmen wie Mondex, Mastercard, Gemplus et. al. Ziel ist es, einen sog. *offenen* Industriestandard festzulegen, d.h. durch Veröffentlichung der Spezifikation (und Bereitschaft, Vorschläge zu überdenken) wird ein de-facto Standard festgelegt.

Auch die `SOScard`, ein offenes Multiapplikationschipkartenprojekt der EU, erlaubt jedem, die Spezifikation des Betriebssystems einzusehen und an der Entwicklung teilzunehmen. Im Gegensatz zur `Multos` Initiative, welche von einem reinen Industriekonsortium geleitet wird, wird das `SOScard` Projekt auch teilweise an einer Hochschule entwickelt.

3 JAVA

Seit diesem Jahr ist nun eine vollkommen andere Lösung in aller Munde: die JavaCard. Mit diesem Konzept soll es nun möglich sein, mit der relativ einfachen objektorientierten Hochsprache JAVA kleine Programme zu entwickeln, welche man auf die Karte in einem kompilierten Byte-Code laden und interpretieren kann. Besonders reizvoll ist diese Vorstellung aus zwei Gründen: erstens, weil das Konzept der Sprache in gewisser Weise Sicherheit "eingebaut" hat und zweitens, weil die Entwicklungskosten für Applikationen sehr niedrig sein werden.

3.1 Das Sicherheitskonzept von JAVA

Im Unterschied zu den meisten anderen modernen Programmiersprachen ist JAVA eine interpretierte Sprache. Damit sind Sicherheitsüberprüfungen zu Laufzeit möglich. Andererseits bietet Java dem Programmierer die Vorteile einer objektorientierten Sprache, was es aufgrund der erzwungenen hierarchischen Gültigkeitsbereiche von Daten leicht macht, diese zu überprüfen. Das zugrundeliegende Konzept ist das sog. *Sandkastenmodell*: Jedes Kind (also jede Klasse, jeder Prozess) darf nur mit seinen Werkzeugen (also mit dem ihm zugeordneten Ressourcen) in seinem eigenen Sandkasten (in seinem Speicherbereich) spielen. Die Einhaltung dieses Prinzips wird wie folgt erreicht:

Erst wird aus dem C++-ähnlichen Sourcecode durch Kompilation ein sog. *Byte-Code* erstellt, welcher dann von einem speziellen Interpreter, der *Java Virtual Machine* (JVM) interpretiert wird. Während der Kompilation werden die üblichen statisch möglichen Überprüfungen vorgenommen. Wird nun der erzeugte Byte-Code von der JVM geladen, wird zuerst ein Byte-Code-Verifier aufgerufen, der den erzeugten Code dynamisch überprüft, z.B. interne Zeiger auf Gültigkeit überprüft.

Da die Sprache JAVA sowohl Netzwerkelemente enthält als auch Threads, also Prozessgenerierung unterstützt, ist sie für zukunftsorientierte Anwendungen auf einer Multiapplikationschipkarte prinzipiell bestens geeignet. Eine einzelne (Chipkarten-) Applikation ist dann ein *applet*, also ein kleines Code-Fragment, das nicht selbständig ablaufen kann, sondern eine ausführende (Java-) Applikation braucht. Die Vorstellung ist dabei, daß "einfach" eine Java Virtual Machine und die ausführende Java-Applikation auf die Chipkarte gebracht werden müssen, idealerweise in Hardware.

Nicht zu verwechseln mit JAVA sind angelehnte Skriptsprachen wie JSkript von Sun oder Javaskript von Microsoft. Beide haben mit dem Sicherheitskonzept von Java wenig zu tun.

3.2 Die JavaCard: das Medium der Zukunft

Zum Zeitpunkt des Entstehens dieses Artikels gibt es eine einzige JavaCard, nämlich die Cyberflex von Schlumberger. Dabei wird die Java Virtual Machine nicht in Hardware implementiert, sondern auf einer "klassischen" Karte emuliert. Aus Platzgründen wurde dabei auf den Byte Code Verifier verzichtet. Diese Lösung hat zwei Nachteile: einerseits ist die Karte durch diese Emulation, wie auch interpretierter Byte Code in einer JVM auf einem Rechner, sehr langsam. Andererseits wird der Bytecode nicht mehr dynamisch überprüft. Der zweite Nachteil kann z. B. durch eine Zertifizierungsinfrastruktur ("signed" applets) aufgefangen werden.

Sun microsystems hat auf der Systems 97 einen "echten" JavaChip präsentiert, welcher Byte-Code in Hardware interpretieren kann. Es gibt zwar noch starke thermische Probleme wegen der CISC-Architekur des Prozessors, jedoch ist dies sicherlich die Lösung der Zukunft. Die Entwicklungskosten für Applikationen sind auf ein Minimum beschränkt, und auch die funktionale Sicherheit der Karte ist, bei entsprechender Realisierung der JVM in Hardware, größtenteils gewährleistet. D.h. auch die Kosten für Evaluation einer Applikation ist minimiert, sofern die Karte mit JVM schon zertifiziert ist.

4 Das Ziel

Ziel ist es, eine Chipkarteninfrastruktur zu entwickeln, welche das Laden und Entladen von Applikationen auf einer Chipkarte unterstützt.

4.1 Akzeptanzprobleme?

Die öffentliche Meinung hat aus verständlichen Gründen ein sehr zwiespältiges Verhältnis zu Multiapplikationschipkarten. Um so wesentlicher ist es, gewährleisten zu können, daß verschiedene Applikationen gegeneinander abgesichert sind. Dies kann durch den richtigen Einsatz von Kryptographie erreicht werden. Ein bewußter Zugriff auf den globalen Event-Handler, den Supervisor auf der Karte, darf nur nach erfolgreicher Authentifikation erfolgen. Durch Einsatz moderner Kryptographie-Techniken wird für einzelne Applikationen zudem ein Grad an Anonymität erreicht, der im wirklichen Leben nicht erreichbar ist. Diese technischen Möglichkeiten stellen nicht nur einen Fortschritt in der Umsetzung der Freiheit des Einzelnen dar, sondern auch eine Grundvoraussetzung in die Akzeptanz einer Multiapplikationschipkarte.

Nicht die Chipkarte, sondern die einzelne Applikation muß dabei als Anwendung im Vordergrund stehen. Die Chipkarte ist nur das Medium, das die verschiedenen Applikationen trägt. Die einzelne Applikation verdient Würdigung oder Kritik, je nachdem wie sie mit den Daten des Benutzers umgeht. Die Chipkarte muß dafür eine sichere Infrastruktur zur Verfügung stellen.

4.2 Die Applikationen

Die möglichen Chipkartenapplikationen gliedern sich unserer Meinung nach in vier Gruppen:

Als erstes zentrale Applikationen, die sensible Daten verwalten oder nichtanonyme Finanzdienstleistungen zur Verfügung stellen. Hierzu gehört z.B. die Kreditkartenfunktion. Auch ein Gesundheitsprofil würde in diese Gruppe gehören. Wesentliches Merkmal von Applikationen dieser Gruppe ist die Voraussetzung, daß andere Anwendungen aus (datensicherheits-) rechtlichen Gründen auf keinen Fall die Daten einer solchen Anwendung auslesen können dürfen. Diese Daten müssen verschlüsselt auf der Karte gespeichert sein und erst eine entsprechende Authentifikationsprozedur darf die Entschlüsselung und Freigabe der Daten erlauben.

In der nächsten Gruppe finden sich Applikationen, die zwar eine gewisse Authentifikationsfunktion haben, deren Offenlegung jedoch nicht mit einem prinzipiellen Nachteil für den

Kartenbesitzer verbunden ist. In dieser Gruppe wären z.B. Personalausweis, Führerschein, Mitgliedskarten, usw. anzusiedeln. Ein Zugriff muß authentifiziert werden, die Daten müssen nicht verschlüsselt werden.

Die dritte Gruppe besteht aus Applikationen, die anonyme Dienstleistungen zur Verfügung stellen. Hier fänden sich die Geldkartenfunktionalität, ein Kinoabo oder eine MensaCard. Die Applikation darf keinen Rückchluß auf den Benutzer der Karte erlauben. Typischerweise wird hierbei ein "Guthaben" auf der Karte gespeichert, welches nach und nach verbraucht wird.

In keiner dieser drei Fälle darf der Anwender in der Lage sein, die Daten der Anwendungen zu manipulieren.

Die letzte Gruppe schließlich beinhaltet Anwendungen, die der Benutzer selbst auf die Karte bringen kann und die z.B. die Funktion einer Visitenkarte haben. Hierbei ist es gerade erwünscht, daß man die Daten der Applikation auslesen kann.

Zusammenfassend muß der Event-Handler der Karte (also das allgemeine Zugriffsprogamm) insbesondere im Hinblick auf die o.g. Notwendigkeit, die Spezifikation offenzulegen, in der Lage sein, folgende Aktionen durchzuführen:

- Zuordnung einer Applikation nach öffentlich bekanntem Schema,

- Authentifizierung einer Applikation und

- Verschlüsselung einer Applikation.

4.3 Push-in, Pull-off

Unser Ziel ist es, aus Speicherplatzgründen eine Applikation von der Karte herunterzuladen (Pull-off), und statt dessen eine andere draufzuladen (Push-in). Dies wird durch eine spezielle Applikation, den Load-Handler durchgeführt. Das Herunterladen geschieht durch eine 1-1 Kopie des der Applikation zugeordneten Speichers auf den Rechner. Gehört die Applikation zu einer der Gruppen 1-3, so darf der Benutzer diese Daten nicht ändern, wohl aber einsehen. D.h. beim Herunterladen einer Applikation dieses Typs wird die Kopie durch die Karte signiert. Beim erneuten Push-in erkennt die Karte eine eventuelle Veränderung und handelt entsprechend.

Die Signatur hat einem MAC oder einer Hash-funktion gegenüber den Vorteil, daß sie von einer dritten Instanz, etwa dem Hersteller der Applikation, etwa bei Verlust einer Karte mit dem öffentlichen Schlüssel ebenfalls überprüft werden kann. Würde man einen MAC benutzen, könnte der Hersteller prinzipiell die Daten manipulieren, was nicht im Sinne des Benutzers ist.

Dem Benutzer steht ein Loader Tool zur Verfügung, welches sich gegenüber dem Load Handler authentifiziert und die Verwaltung der "off" Applikationen übernimmt. Dies kann plattformunanbhängig in JAVA geschrieben sein. Läuft das Tool auf dem eigenen PC, so ist die oben geschilderte Signatur ausreichend. Steht ein Frontend online zur Verfügung, so daß alle "off" Applikationen von einem öffentlichen Terminal aus geladen werden könnten, muß der Load Handler eine Verschlüsselung mit dem Public Key des Loader Tools vornehmen. Somit kann eine höchstmögliche Portabilität erreicht werden.

4.4   Die Karte als Netzwerkkomponente

Durch das neue Internet-Protokoll IPv6 wird es möglich sein, jeder Chipkarte eine IP-Nmmer zuzuordnen, so daß gesicherte Kommunikation mit dem Heimrechner, der die "off"-Daten verwaltet, möglich ist. Somit ist eine einfache, portable Lösung nach obigem Schema zum Verwalten der Kartenapplikationen möglich.

So wie ein Bürger heutzutage selbst entscheiden kann, ob er einen Ausweis, eine Kreditkarte usw. bei sich führt, so kann er hier selbst entscheiden, welche Applikation er auf die Karte lädt. Durch diese Variabilität ist der Schutz der Persönlichkeit gewahrt. Die Karte wird zu einem hochintelligenten Medium, welches die persönliche Freiheit des Individuums unterstützt, ohne die Dienstleister zu benachteiligen.

5   Das UniCard-Projekt an der TU Darmstadt

Da unser Projekt erst angelaufen ist, können wir bisher leider nicht mit einer Implementation aufwarten. Wir werden nun die aktuelle Diskussionsgrundlage für das UniCard-Projekt an der TU Darmstadt schildern.

Vor ein paar Jahren gab es eine Initiative der Verwaltung, als eine der fortschrittlichsten Hochschulen Deutschlands eine Studentenkarte einzuführen. Diese konnte aufgrund politischer Differenzen nicht durchgesetzt werden. Inzwischen hat sich das Studentenwerk für eine eigene MensaCard entschieden und das Rückmeldeverfahren wurde dadurch, daß keine Krankenversicherungsbestätigung mehr vorgelegt werden muß, drastisch vereinfacht. Somit sind zwei potentielle Anwendungen vorerst "aus dem Rennen". Die Konzentration liegt bei diesem neuen Projekt auf der Kombination der folgenden Anwendungen:

1. eine GeldKarten-Funktionalität,

2. die sichere Verwaltung der erworbenen Studentenleistungen innerhalb eines Fachbereichs,

3. Digitale Signatur,

4. die Zugangsberechtigung zu den elektronischen Ressourcen und

5. evtl. eine Zugangslösung für einige Gebäude der Universität.

Diese Anwendungen gehören nach dem obigen Schema folgenden Gruppen an:

1. Gruppe 1: Verwaltung der Studentenleistungen,

2. Gruppe 2: Zugangsberechtigungen, digitale Signatur

3. Gruppe 3: GeldKarte.

Zur Zeit der Erstellung dieses Artikels wird im Rahmen eines Praktikums an der Implementation der Zugangsberechtigung für die Rechnernetze und an der digitalen Signatur gearbeitet. Der Load Handler ist in der Designphase.

Literatur

[1] *Public Key Cryptography Standard # 11: Cryptographic Token Interface Standard PKCS-11.*
http://www.rsa.com/standards.

[2] *ISO 7816 — Identification cards - Integrated circuit(s) cards with contacts. Parts 1-8.*
International Organization of Standardization. Geneve.

[3] *The Java Card Specification.*
http://java.sun.com/products/javacard.

[4] *Multos - the smartcard gets smarter.* The Maosco Consortium.
http://www.multos.com.

[5] *SOSCARD — Secure Operating System Smart Card.*
http://www.digicash.com/projects/soscard.

[6] *Handbuch der Chipkarten.* W. Rankl, W. Effing. 2. erw. Auflage. München, Wien: Hanser 1996.

[7] *Security Reference Model for the Java Developer's Kit 1.0.2.* Marlena Erdos, Bret Hartman, Marianne Mueller.
http://java.sun.com/security/SRM.html

Einsatz von Chipkarten im Trust Center
Vor- und Nachteile des Einsatzes
von Chipkarten
zur Ausstellung von Zertifikaten

Jobst Biester[1] · Alfred Scheerhorn[1] · Matthias Gärtner[2]

[1]CCI GmbH, Lohberg 10, D-49716 Meppen
{biester,adscheer}@cci.de

[2]Concord Eracom GmbH, Talstr. 11, D-72218 Wildberg
mgaertner@concord-eracom.de

1 Einleitung

Chipkarten werden zur Zeit als ideales Medium zur Speicherung und Anwendung privater Schlüssel angesehen, insbesondere zur digitalen Signierung von Daten. Der entscheidende Vorteil der Chipkarte gegenüber anderen Schlüsselträgern ist dabei die Kombination aus mobiler Einsatzfähigkeit und sicherer Schlüsselverwahrung. Bei einer weiten Verbreitung von Chipkarten liegt es nahe, dieselben auch als Schlüsselträger für den Einsatz der Schlüssel zu Zertifizierungszwecken im Trust Center zu nutzen.

Bei der Zertifizierung im Trust Center ist jedoch für Schlüsselträgermedien ein anderes Anforderungsprofil vorgegeben als bei deren Nutzung durch Teilnehmer. In dem Beitrag wird die Verwendung von Chipkarten einerseits und von Sicherheitsmodulen andererseits zum Zweck der Zertifizierung im Trust Center diskutiert. Als Sicherheitsmodul wird dabei ein kompaktes Hardware-Modul verstanden, das z.B. in Form einer PC-Einsteckkarte ausgeführt sein kann.

Ziel dieses Beitrags ist die Bewertung der unterschiedlichen Lösungen zum Zweck der Zertifizierung im Trust Center.

In der Diskussion werden neben der Speicherung und Anwendung eines privaten Schlüssels weitere sicherheitsrelevante Funktionen berücksichtigt, die im Trust Center benötigt werden. Dies sind

- der Schlüsselwechsel des Trust Centers,
- die Zugriffskontrolle auf den Zertifizierungsschlüssel,
- die revisionssichere Protokollierung aller Aktionen, die den Zertifizierungsschlüssel verwenden
- und die Schlüsselgenerierung für Teilnehmer.

Daneben werden auch

- die Vertrauenswürdigkeit des Schlüsselträgers und der Einsatzumgebung,
- die Anpassung an neue Algorithmen und Schlüssellängen sowie
- die Nutzung mehrerer Zertifizierungsschlüssel betrachtet.

Der Einsatz von Chipkarten und Sicherheitsmodulen wird bezüglich der verschiedenen Anforderungen und Aufgaben diskutiert. Abschließend erfolgt eine zusammenfassende Darstellung und Bewertung.

2 Anforderungen und Aufgaben

In den folgenden Abschnitten wird der Einsatz von Chipkarten einerseits und der Einsatz von Sicherheitsmodulen (SM) andererseits für verschiedene im Trust Center erforderliche Aufgabenstellungen und Anforderungen diskutiert. Vor- und Nachteile werden tabellenartig gegenübergestellt.

2.1 Technischer Entwicklungsstand

Die Charakteristika der beiden Systeme unterscheiden sich signifikant, wobei mehrere Bereiche unterschieden werden können.

Chipkarte	Sicherheitsmodul
(+) Physikalische Schutzmechanismen sind weit entwickelt und können hohe Ansprüche erfüllen.	(+) Die physikalischen Schutzmechanismen sind mit denen einer Chipkarte vergleichbar.
(+) Logischer Zugangsschutz ist gut untersucht, weitgehend standardisiert und hinreichend flexibel.	(+) Der logische Zugangsschutz ist nicht einheitlich. Es gibt Gestaltungsmöglichkeiten.
(-) Speicherkapazität ist sehr beschränkt.	(+) Die Speicherkapazität ist um mehrere Größenordnungen höher als bei einer Chipkarte.
(-) Prozessorkapazität ist sehr beschränkt.	(+) Die Prozessorkapazität ist um mehrere Größenordnungen höher als bei einer Chipkarte.
(-) Funktionalität ist spezialisiert und nicht leicht um komplexere Funktionen erweiterbar.	(+) Die interne Funktionalität läßt sich während des Betriebs (gesichert) verändern und erweitern. Die Mechanismen sind verfügbar, so daß die Erweiterungen nicht unbedingt vom Hersteller, sondern ggf. von dritter Seite erstellt werden können.
(+) Eine Erweiterung ist jedoch grundsätzlich im Rahmen der Kapazitätsbeschränkungen möglich.	(+) Durch diese Flexibilität kann schnell auf geänderte Bedingungen/Spezifikationen eingegangen werden.

(+) Enthält dedizierte kryptographische Koprozessoren.	(+) Enthält leistungsfähige dedizierte kryptographische Koprozessoren.
(-) Die Kommunikationsleistung ist beschränkt.	(+) Durch die direkte Einbindung in den Computer ist eine hohe Kommunikationsleistung gegeben.
(-) Zusätzliche Kommunikationswege sind schwierig zu realisieren.	(+) Es existieren vom steuernden Computersystem unabhängige und manipulationssichere Kommunikationswege, die zum Aufbau von Backup-Konzepten, Anzeigekomponenten etc. dienen können.

2.2 Zertifikats- und Sperrlisten-Signierung

Die Zertifikats- und Sperrlistensignierung ist der zentrale Einsatzzweck des privaten Zertifizierungsschlüssels.

Chipkarte	Sicherheitsmodul
(-) Verhinderung der Entwendung der Chipkarte erforderlich. (Eine temporäre Entwendung kann z. B. für eine „Differential Fault Analysis" genutzt werden)	(+) Beim Versuch der Entwendung werden alle geheimen Schlüssel durch das SM aktiv gelöscht.
(-) Hashfunktion wird im allgemeinen nicht vollständig auf der Chipkarte durchgeführt.	(+) Im SM ist die vollständige Behandlung von Zertifikaten möglich: Vorprüfung (Format- und Plausibilitätsprüfung, z. B. Namen), Einfügung von festen Bestandteilen (z.B. automatische Seriennummer, CA-Name), Signierung und Korrektheitsprüfung.
	(+) Entsprechend der höheren Rechenleistung laufen die Signierungen schneller ab.

2.3 Schlüsselgenerierung für CA-Schlüsselpaare

Die Schlüsselgenerierung für die CA muß hohen Anforderungen genügen, welche u.a. im Maßnahmenkatalog [MK] aufgezeigt werden.

Grundsätzlich kann das Schlüsselpaar in beiden Umgebungen auch von außen eingebracht werden, wobei jedoch der Manipulations- und Zugriffsschutz berücksichtigt werden müssen. Die interne Erzeugung insbesondere von Signaturschlüsseln (wichtigste Komponente im Trust Center) ist jedoch vorzuziehen.

Chipkarte	Sicherheitsmodul
(-) Die interne Erzeugung von Schlüsseln ist bislang nur für wenige Karten verfügbar.	(+) Die interne Erzeugung von Schlüsseln ist möglich.
(-) Nicht immer ist ein echter physikalischer Zufallszahlengenerator auf der Karte vorhanden.	(+) Enthält häufig einen physikalischen Zufallszahlengenerator.
(+) Die beweissichere Vernichtung solcher Schlüssel ist durch Vernichten der ganzen Chipkarte sehr einfach.	(+) Die Qualität der Schlüssel ist ggf. höher. Aufgrund der höheren Rechenleistung können die Schlüssel besser getestet werden.

2.4 Schlüsselwechsel des Trust Centers

Für den Schlüsselwechsel einer Zertifizierungsstelle (Certification Authority, CA) gibt es technisch verschiedene Lösungen.

Der neue Zertifizierungsschlüssel kann durch eine übergeordnete Zertifizierungsstelle (i. allg. die PCA der Infrastruktur) generiert und der CA auf einem Schlüsselträger geliefert werden. Hieraus resultieren Interoperabilitätsanforderungen: Die PCA muß den Schlüssel in genau der Form liefern können, in der ihn die CA nutzen und in den gesicherten Schlüsselspeicher einbringen kann.

Alternativ kann die CA sich ihren neuen Schlüssel selbst generieren. Bis die Zertifizierung des Schlüssels durch die PCA erfolgt ist, muß der neue Schlüssel sicher zwischengespeichert werden.

Bei der Speicherung des Schlüssels auf einer Chipkarte muß der Mißbrauch der Chipkarte mit dem neuen Schlüssel für den Übergangszeitraum bis zu deren Einsatz geeignet verhindert werden. Falls die eingesetzte Chipkarte nicht die Möglichkeit der Generierung und Zwischenspeicherung eines zweiten Schlüssels unterstützt, bedingt jeder Schlüsselwechsel technisch einen Chipkartenwechsel.

Bei Verwendung eines Sicherheitsmoduls kann die Schlüsselgenerierung im SM erfolgen. Der neu generierte Schlüssel kann bis zu seinem Einsatz unproblematisch derart gespeichert werden, daß er genauso wie der aktuelle Zertifizierungsschlüssel geschützt ist. Die Speicherung und Nutzung mehrerer Schlüssel ist möglich.

Es muß nach einem Wechsel sichergestellt werden, daß der alte Schlüssel nur bestimmungsgemäß genutzt werden kann. Im allgemeinen wird eine Vernichtung des geheimen Schlüssels erfolgen.

2.5 Vertrauenswürdigkeit der Schlüsselträger

Zur objektiven Bestätigung der Vertrauenswürdigkeit eingesetzter Komponenten sieht das Signaturgesetz eine Zertifizierung der Komponenten nach ITSEC vor. Eine andere Art der Referenz für Vertrauenswürdigkeit ist z. B. ein weitverbreiteter Einsatz von Komponenten über einen längeren Zeitraum, ohne daß wesentliche Probleme oder Sicherheitslücken bekannt werden.

Chipkarte	Sicherheitsmodul
(+) ITSEC-E4 hoch Evaluierung und Zertifizierung gemäß Signaturverordnung ist aufgrund noch handhabbarer HW/SW-Komplexität relativ leicht möglich.	(-) ITSEC-E4 hoch Zertifizierung aufwendiger, da mehr Sicherheitsfunktionen zu berücksichtigen sind als bei einer Chipkarte. (+) Hohe Vertrauenswürdigkeit durch weitverbreiteten Einsatz in sicherheitskritischen Bereichen (z. B. dem Bankbereich).

2.6 Vertrauenswürdigkeit der Einsatzumgebung

Aufgrund ihrer begrenzten physikalischen Komplexität weist die Chipkarte nur eine begrenzte Funktionalität auf. Entsprechend müssen viele Funktionen in die Einsatzumgebung der Chipkarte ausgelagert werden, die bei Einsatz eines Sicherheitsmoduls von diesem mit übernommen werden können.

Chipkarte	Sicherheitsmodul
(-) Da die Chipkarte nur die direkte Anwendung des Zertifizierungsschlüssels übernimmt, muß die Chipkarte innerhalb einer vertrauenswürdigen Einsatzumgebung betrieben werden.	(+) Zertifikatsverarbeitung kann weitgehend im SM erfolgen. (+) Schlüsselverwaltungsfunktionen können vom SM übernommen werden. (+) Es sind weniger Sicherheitsanforderungen an die Einsatzumgebung notwendig.

2.7 Zugriffskontrolle auf den Schlüsselträger

Bei der Dienstleistung der Zertifizierung von Signaturschlüsseln liegt es nahe, digitale Signaturen auch zur Authentisierung gegenüber der Hardware zu verwenden. Dieses Vorgehen hat den weiteren Vorteil, daß die zur Authentisierung verwendete Signaturschlüssel gleichzeitig zur revisionssicheren Protokollierung der Anwendung des Zertifizierungsschlüssels genutzt werden können. Der Einsatz von Signaturschlüsseln für verschiedene Zwecke ist im Rahmen eines Sicherheitskonzepts zu berücksichtigen.

Chipkarte	Sicherheitsmodul
Die Authentisierung gegenüber der Chipkarte erfolgt i. allg. über PIN. (-) Authentisierung gegenüber der Chipkarte mittels digitaler Signaturen nur aufwendig realisierbar, da Schlüsselverwaltungsfunktionen auf der Chipkarte implementiert werden müssen. (-) PINs müssen meist noch durch das steuernde Computersystem transferiert werden.	Es gibt Sicherheitsmodule, die mit einer seriellen Schnittstelle ausgestattet sind: (+) Eine Zugriffskontrolle zum SM über sichere Chipkarten ist möglich (z. B. Authentisierung mittels asymmetrischer digitaler Signatur, Challenge-Response). Die Verwaltung von Zugriffsrechten im SM ist möglich. Die Authentisierung unterliegt der vollständigen Kontrolle durch das SM.

(+) Es gibt jedoch gesicherte Chipkartenleser, welche eine sichere PIN-Eingabe ermöglichen.	(+) Es ist möglich z. B. das 4-Augen-Prinzip umzusetzen.

2.8 Revisionssichere Protokollierung

Bei der Protokollierung der Anwendung des Zertifizierungsschlüssels muß sichergestellt werden, daß anhand des Protokolls nachträglich festgestellt werden kann, durch wen der Schlüssel genutzt wurde. Beim Einsatz einer Chipkarte als Schlüsselträger erfolgt die Protokollierung durch die Einsatzumgebung der Chipkarte. Als Nutzerauthentisierung für die Protokollierung kann daher entweder die Authentisierung des Nutzers gegenüber der Einsatzumgebung oder dessen Authentisierung gegenüber der Chipkarte genutzt werden.

Chipkarte	Sicherheitsmodul
(-) Die Protokollierung bedarf einer sicheren Einsatzumgebung. Die Protokollierung kann entweder an die Authentisierung gegenüber der Chipkarte gekoppelt sein oder an die Authentisierung gegenüber der Einsatzumgebung. Im letzteren Fall ist die Kopplung der Authentisierungen sicherzustellen.	(+) Protokollierung ist innerhalb des Sicherheitsmoduls realisierbar. (+) Bei Zugriffskontrolle zum SM über Chipkarten mittels digitaler Signaturen ist eine digital signierte Protokollierung möglich. (+) Protokolldaten können verschlüsselt und ggf. vom SM signiert abgelegt werden. (+) Protokolldaten können gesichert ausgelagert werden.

2.9 Schlüsselgenerierung für Teilnehmer

Die Schlüsselgenerierung für Teilnehmer betrifft das Medium der Speicherung des Zertifizierungsschlüssels selbstverständlich nur, falls dieses zu diesem Zweck genutzt wird.

Generell sind Chipkarten vorteilhaft, bei denen die Schlüsselgenerierung direkt auf der Karte erfolgt, um die Problematik des beweissicheren Löschens von geheimen Teilnehmer-Signaturschlüssel zu umgehen.

Unabhängig davon ist die Generierung sicherer Schlüssel sowohl in Chipkarten als auch in Sicherheitsmodulen möglich. Werden Schlüssel zentral generiert, ist das Problem des Ladens der Schlüssel in die Schlüsselträger und des Löschens der Schlüssel sowohl für die Chipkarte als auch für das SM als zentrale Schlüsselgenerierungskomponenten gegeben.

Die Forderung des Maßnahmenkatalogs [MK] zur ständigen Überwachung des Zufallsgenerators für die Schlüsselgenerierung durch statistische Tests kann jedoch einfacher beim Einsatz eines Sicherheitsmoduls zur Schlüsselgenerierung realisiert werden, als beim Einsatz einer Chipkarte.

Chipkarte	Sicherheitsmodul
(-) Die Schlüsselerzeugung auf der Chipkarte dauert lange, wenn überhaupt möglich.	(+) Die Schlüsselerzeugung im SM geht aufgrund der höheren Rechenleistung schneller vonstatten.
(-) Die Schlüsselerzeugung startet erst auf Anfrage.	(+) Schlüssel können im SM parallel zu den Operationen des steuernden Computers „auf Vorrat" erzeugt werden und stehen dann je nach Bedarf schneller zur Verfügung.
(+) Die Schlüsselgenerierung für den Teilnehmer mit dessen designierter Chipkarte vermeidet den ansonsten notwendigen Transport der Schlüsseldaten.	(+) Es besteht die Möglichkeit die Güte des Zufallsgenerators potentiell ständig zu kontrollieren.

2.10 Anpassung an neue Signaturverfahren

Neue Signaturverfahren implizieren Änderungen der bestehenden Funktionalität der Signaturkomponente.

Wie o.g. bieten beide System grundsätzlich die Möglichkeit eines Nachladens von Funktionalität, jedoch auf einem anderen Niveau.

Bei Chipkarten kann die Funktionalität nach der Produktion, jedoch vor Auslieferung an den Kunden verändert werden. Nach der Auslieferung ist eine Änderung nicht mehr vorgesehen. Es wird angenommen, daß das SM die Möglichkeit eines gesicherten Software-Downloads bietet. Hierbei akzeptiert das SM Softwareänderungen nur, falls der zu ladende Kode verschlüsselt und korrekt signiert ist. Zur Sicherung dieses Nachladens sollten dedizierte Schlüssel zur Verfügung stehen.

Es kann bei beiden Signaturkomponenten davon ausgegangen werden, daß der Einsatz neuer Algorithmen keine Auswirkung auf die Sicherheit der Einsatzumgebung hat.

Chipkarte	Sicherheitsmodul
(-) Für den Einsatz neuer Signaturverfahren sind im allgemeinen neue Chipkarten erforderlich.	(+) Flexible Erweiterbarkeit durch Möglichkeit eines gesicherten Software-Downloads.
(-) Neue Algorithmen werden ggf. nicht durch die HW der Chipkarte unterstützt.	(-) Neue Algorithmen werden ggf. durch die HW des SM nicht optimal unterstützt. Dies kann jedoch meist durch die höhere Leistungsfähigkeit ausgeglichen werden.
(-) Durch die Spezialisierung sind ggf. auch bei nur einem Signaturverfahren Beschränkungen gegeben, etwa eine festgelegte Schlüssellänge, von der nicht abgewichen werden kann.	(+) Das SM enthält meist einen leistungsfähigen generischen Hauptprozessor.

2.11 Anpassung an erhöhte Schlüssellängen

Aufgrund neuer Forschungsergebnisse in der Kryptanalyse und aufgrund der Erhöhung der Rechnerleistung ist es von Zeit zu Zeit erforderlich, die Schlüssellängen veränderten aktuellen Bedingungen anzupassen.

Chipkarte	Sicherheitsmodul
(-) Es können nur Schlüssellängen genutzt werden, die die jeweilige Chipkarte unterstützt. Diese sind i. a. begrenzt wegen der begrenzten Komplexität der HW und SW der Chipkarte. (-) Bei Überschreitung der Begrenzung ist der Einsatz einer leistungsfähigeren Chipkarte erforderlich, sofern überhaupt verfügbar.	(+) Unterstützung hinreichend großer Schlüssellängen (z. B. RSA bis 2048 Bit). (+) Keine Anpassungen erforderlich. (+) Ggf. nur Software-Änderung erforderlich, sofern die HW doch nicht ausreicht. (+) Der gleichzeitige Einsatz mehrerer Schlüssel auch mit unterschiedlichen Schlüssellängen (Schlüsselwechsel) ist problemlos möglich.

2.12 Nutzung mehrerer Zertifizierungsschlüssel

Zum Betrieb mehrerer Zertifizierungsstellen ist es vorteilhaft, wenn mehrere Zertifizierungsschlüssel innerhalb derselben Einsatzumgebung genutzt werden können.

Chipkarte	Sicherheitsmodul
(-) Die Speicherung und logische Trennung mehrerer Zertifizierungsschlüssel auf einer Chipkarte kann problematisch sein. (-) Eine Chipkarte unterstützt meist nur ein Signaturverfahren.	(+) Sichere Speicherung und Verwaltung mehrerer Zertifizierungsschlüssel unproblematisch. (+) Anwendung unterschiedlicher Signaturverfahren unproblematisch.

3 Zusammenfassung

Zur Speicherung und Anwendung privater Schlüssel für Anwender ist die Chipkarte ein geeignetes Medium, insbesondere wegen ihrer Kombination aus sicherer Schlüsselverwahrung und mobiler Einsatzfähigkeit. Für den Einsatz von Schlüsseln zur Zertifizierung im Trust Center ergibt sich jedoch ein verändertes Anforderungsprofil. Neben der reinen Anwendung des privaten Schlüssels stehen dabei Schlüsselverwaltungsaufgaben im Vordergrund. Die der Chipkarte durch ihre Größe inhärente, begrenzte Komplexität ihrer Hard- und Software wirkt sich dabei nachteilig aus. Viele Funktionalitäten müssen in die Einsatzumgebung der Chipkarte ausgelagert werden. Beim Einsatz eines Sicherheitsmoduls, z. B. ausgeführt als PC-Einschubkarte, kann ein Großteil aller sicherheitsrelevanten Funktionen vollständig durch das Sicherheitsmodul übernommen und kontrolliert werden. Die folgende Abbildung verdeutlicht die Aufgabenverteilung der verschiedenen Lösungen.

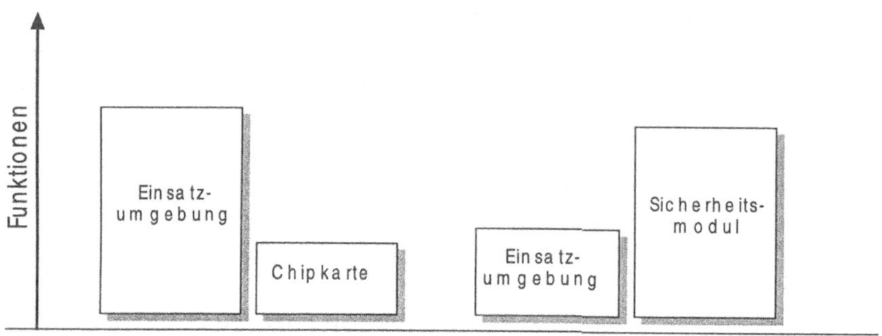

Der Nachteil des Einsatzes eines Sicherheitsmoduls ist die Aufwendigkeit und Schwierigkeit der Prüfung des Moduls (z. B. Evaluierung nach ITSEC), aufgrund der hohen Hard- und Softwarekomplexität eines Sicherheitsmoduls. Der weit verbreitete Einsatz dieser Module in sicherheitskritischen Bereichen (z. B. bei Banken) kann jedoch als Referenz für deren Vertrauenswürdigkeit dienen.

Chipkarten sind demgegenüber aufgrund ihrer klaren Spezifikation aller Baugruppen und der geringeren Komplexität der Software-Funktionen einfacher evaluierbar. Da beim Einsatz von Chipkarten jedoch viele sicherheitsrelevante Funktionen in die Einsatzumgebung ausgelagert werden, muß auch die Einsatzumgebung entsprechend vertrauenswürdig und geprüft sein, um hohen Sicherheitsanforderungen zu genügen. Da des weiteren das Sicherheitsmodul bezüglich vieler anderer Anforderungen wie dem Schlüsselwechsel, der Erweiterbarkeit und der Protokollierung Vorteile gegenüber der Chipkarte aufweist, erscheint insgesamt zum Zweck der Zertifizierung im Trust Center der Einsatz eines Sicherheitsmoduls vorteilhaft gegenüber dem Einsatz einer Chipkarte.

Java™ für Chipkarten

Thomas Stocker

Giesecke & Devrient
thomas.stocker@gdm.de

Zusammenfassung

Java ist zur Zeit ein sehr aktuelles Thema. Diese relativ neue Programmiersprache wird im ersten Kapitel kurz umrissen. In den nächsten beiden Kapiteln werden Gründe für eine Java Karte ermittelt und die Arbeit des Java Card Forums vorgestellt, das sich die Standardisierung der Java Karte zur Aufgabe gemacht hat. Das letzte Kapitel befaßt sich dann mit dem aktuelle Java Card Standard.

1 Überblick über Java™

Es gibt wohl wenige andere Themen in der IT-Welt, dessen in den letzten Jahren soviel Beachtung geschenkt wurde wie Java. Ein Blick in Fachzeitschriften genügt, um zu erkennen, wie brandaktuell diese Thematik ist.

1.1 Historie von Java™

Die Geschichte von Java™ begann ziemlich unspektakulär. Anfang der 90er entwickelte Sun Microsystems eine Programmiersprache, damals noch unter dem Namen "Oak", die für den Einsatz im Consumer-Bereich (Set-Top-Boxen, Kaffeemaschine) gedacht war. Die Idee dabei bestand darin, plattformunabhängigen Code zu generieren, der auf unterschiedlichen Mikroprozessoren ausgeführt werden kann - "Write Once, Run Anywhere". Der Java-Bytecode war geboren.

Allerdings erst 1995 gelang der Durchbruch, als Sun die Bedeutung des Konzepts für heterogene Netze erkannte. Aus Oak wurde Java. Nicht zuletzt die Einsatzmöglichkeiten, die sich durch die wachsende weltweite Vernetzung ergaben, waren mit ein Grund für den Erfolg.

1.2 Merkmale von Java™

Die Programmiersprache Java soll anhand ein paar ausgewählter Schlagwörter näher charakterisiert werden.

Java ist eine **objektorientierte** Sprache.

Java ist **einfach** zu erlernen. Zum einen liegt es daran, daß sich die Syntax stark an C++ anlehnt. Beherrscht man C++, so findet man sich innerhalb kürzester Zeit auch in Java zurecht. Ein Hemmnis beim Erlernen einer neuen Programmiersprache stellen oft die Änderungen in der Syntax dar, die am Anfang häufige Fehlerquellen darstellen.

Zum anderen fehlen darin einige der Elemente, die in C++ häufig zu Problemen geführt haben, wie z. B. Überladen von Operatoren, Zeiger, Freigeben von Speicher (wird vom System durch Garbage Collector erledigt).

Nicht zuletzt das Fehlen dieser potentiellen Fehlerursachen und vor allem die vielfältigen Laufzeitüberprüfungen tragen erheblich zur hohen **Sicherheit** und **Robustheit** von Java-Programmen bei. Diese Eigenschaften waren mit ein Design-Ziel von Java und der virtuellen Maschine.

Java wird **interpretiert**. Im Gegensatz zu C++ wird Java nicht in Maschinencode übersetzt, sondern in einen Zwischencode, den sog. Java Bytecode. Dieser wird von einer virtuellen Maschine ausgeführt.

Aufgrund jenes Zwischencodes und einer einheitlichen Bibliothek von low-level Routinen (zusammengefaßt unter der sog. Java API) ist Java Bytecode **plattformunabhängig** und leicht **portierbar**. Grundvoraussetzung dafür ist, daß auf der entsprechenden Hardware die Java Virtual Machine und die Java API verfügbar sind.

Java unterstützt **Multithreading.** Darunter versteht man Applikationen, die mehrere Dinge scheinbar gleichzeitig erledigen können

Die Java Virtual Machine unterstützt **dynamisches** Linken. Klassen werden erst dann geladen, wenn sie wirklich benötigt werden. So können beispielsweise auch Klassenbibliotheken zur Laufzeit erweitert werden, ohne daß die Benutzer davon betroffen sind.

2 Warum eigentlich Java für Chipkarten?

Die Anwendungen die mittlerweile auf Chipkarten laufen werden zusehends komplexer. Immer seltener reicht die auf der Karte befindliche Funktionalität aus, um eine Anwendung zu realisieren. Es muß zusätzlicher Programmcode in die Karte gebracht werden. Darüber hinaus befinden sich aber auch genügend Funktionen auf der Karte, die nie benutzt werden. Dieses Problem kann man nur umgehen, wenn die Karte nicht nur als Datenspeicher fungiert, sondern sich frei programmieren läßt.

Letzteres können die Chipkartenhersteller natürlich schon lange. Aber auch die Kunden wollen nicht mehr länger nur auf bereits existierende Funktionen zurückgreifen bzw. die zusätzliche Funktionalität von den Kartenherstellern entwickeln lassen. Sie wollen ihre Anwendungen selbst erstellen.

Dem zunehmenden Wunsch der Kunden, selbst ausführbaren Programmcode in die Chipkarten zu bringen, kann nur entsprochen werden, wenn der Kunde nicht in der Lage ist, die Sicherheitsstruktur des Betriebssystems zu unterlaufen. Dies läßt sich entweder durch entsprechende Hardware realisieren oder softwareseitig, wie dies durch die Java Virtual Machine der Fall ist, indem der Code interpretiert ausgeführt wird.

Es muß dabei der Zugriff auf Code- und Datenbereiche verhindert werden, für die der Benutzer keine Rechte besitzt. Die Java Virtual Machine eignet sich durch ihre konzeptionelle Sicherheitsstruktur dafür sehr gut.

3 Java Card Forum

Obwohl Java eigentlich für den Embedded-Bereich konzipiert war, sind die Anforderungen an die Hardware doch so groß, daß die Java Virtual Machine in heutigen Chipkarten nicht uneingeschränkt realisiert werden kann. Man stößt recht schnell an Grenzen, sei es beim Speicher für die Objekte (Heap) und Operanden (Stack), sei es beim Programmspeicher für die Java Virtual Machine oder bei den Prozessoren, die zur Zeit vorwiegend 8-Bit orientiert sind.

Es müssen daher Einschränkungen am eigentlichen Java Befehls- und Funktionsumfang vorgenommen werden. Um diese einheitlich zu halten und die Belange der meisten Betriebssystemhersteller zu berücksichtigen, wurde das Java Card Forum ins Leben gerufen.

Das Java Card Forum wurde im Februar 1997 gegründet. G&D war von Anfang an mit dabei. Inzwischen zählen 8 Firmen zu den ständigen Mitgliedern des Forums. Voraussetzung dafür ist der Besitz einer Java Card Lizenzierung durch JavaSoft.

3.1 Aufgaben des Java Card Forum

- Bei der Definition der Java Card API mitzuwirken. Die Interoperabilität zwischen unterschiedlichen Chipkartenbetriebssystemherstellern soll dadurch gewährleistet werden („Write Once, Run Anywhere").

- Die Java Card VM und ihre Eigenschaften zu definieren.

- Im Chipkartenmarkt und bei den Verfassern von Spezifikationen für die Java Card API zu werben und sie zu verbreiten.

- Technische Dokumente zur Unterstützung der Java Card API zu erstellen (application notes u. ä.).

- Technische Informationen zwischen den Mitgliedern bezüglich der Java Card API und ihrer Implementation auszutauschen

- Chiphersteller zu unterstützen, damit die JVM bzw. Teile davon in Hardware realisiert werden können.

- Kontakt pflegen zu Softwareentwicklern (bezüglich der Java Card API).

3.2 Anforderungen an eine Java Card

Bevor man sich mit der eigentliche Definition der Java Karte befassen kann, müssen zunächst einmal Anforderungen klar sein, die eine solche Karte erfüllen soll:

- Eine Java Card muß natürlich Java kompatibel sein. Das heißt, es muß für die Entwicklung von Java Card Applets jede handelsübliche Entwicklungsumgebung (Editor, Compiler, Debugger) verwendbar sein. Zum anderen muß natürlich die Struktur von Java erhalten bleiben, damit die Merkmale von Java auch für die Karte gültig sind.

- Eine Java Card sollte natürlich auch auf einer Karte realisiert werden können. Dies gilt sowohl für den Speicherbedarf, der dafür benötigt wird als auch für die Ausführungszeit.

- Eine Java Card muß in bestehenden Systemen lauffähig sein, in denen Chipkarten verwendet werden.

- Daten müssen auch über eine Stromunterbrechung hinweg erhalten bleiben. Dafür sollte allerdings nicht auf ein unterliegendes Kartenbetriebssystem zurückgegriffen werden müssen, sondern dies muß die Java Card automatisch bereit stellen.

3.3 Ziele und Ergebnisse des Java Card Forum

- Der Java Card 2.0 Standard (Version 1.0) konnte von JavaSoft zum 15. Oktober 1997 veröffentlicht werden.

- Für 1998 soll für die Bereiche Telekommunikation (GSM), Zahlungsverkehr und Informationstechnologie jeweils eine domänenspezifische API definiert werden.

4 Java Card 2.0 Standard

Der Java Card 2.0 Standard definiert die Grundfunktionalität einer Java Karte. Diese Grundfunktionalität ist gegeben durch die Java Card Runtime Environment, die sich zum einen aus der Java Card Virtual Machine (Java Card VM) und der Java Card API zusammensetzt.

Eine Applikation einer Java Karte wird als Applet bezeichnet und benötigt zur Ausführung eine Java Card 2.0 kompatible Runtime Environment.

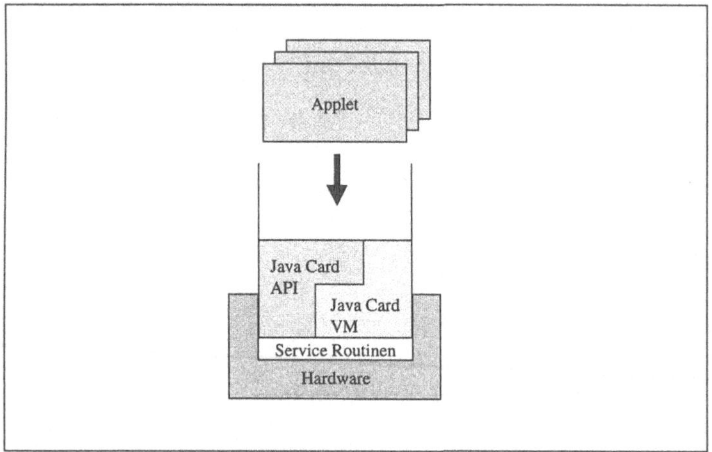

Abb. 1: Aufbau einer Java Card

4.1 Java-Sprachumfang

Für Chipkartenanwendungen ist es nicht erforderlich, den gesamten Java-Sprachumfang zur Verfügung zu haben. Deshalb wurde nur eine Untermenge daraus für Java Card 2.0 hergenommen. Dadurch vereinfacht sich die virtuelle Maschine (VM) und reduziert sich ihr Speicherbedarf.

4.1.1 Nicht unterstützte Elemente

Nicht unterstützt werden in Java Card 2.0 das dynamische Nachladen von Klassen, Multithreading, mehrdimensionale Felder, das Clonen von Objekten. Im Clonen von Objekten verbirgt sich sogar eine Sicherheitslücke. Man stelle sich vor, man könnte das Objekt einer aufgeladenen Geldbörse clonen ... . Darüber hinaus wird ein Garbage Collector nicht vorausgesetzt, wenngleich er auch nicht explizit ausgeschlossen wird. Aufgrund der Komplexität wird es allerdings, wenn überhaupt, nur ein partieller Garbage Collector sein können.

Zu den nicht unterstützen Typen gehören neben *char* (Unicode-Format!) auch *float*, *double*, und *long*.

4.1.2 Unterstützte Elemente

Variablen vom Typ *int* werden in der Java Card als 32-bit Werte interpretiert, damit sich die Arithmetik einer Java Karte analog zu anderen Java-Systemen verhält. Die Aufwärtskompatibilität (von kleineren zu größeren Systemen) muß gewahrt bleiben. Das bedeutet aber, daß dafür ein 32-bit breiter Stack vorhanden sein muß. Da allerdings die meistens Chipkartencontroller nach wie vor 8-bit orientiert sind, ist es nicht zwingend vorgeschrieben, daß eine Java Card 2.0 kompatible Karte *int* als 32-bit Werte unterstützen muß. Dies würde zu unnötig hohen Hardware-Anforderungen führen und damit die Entwicklung einer Java Karte mit geringem Speicherbedarf verhindern. Verwendet die Plattform einen 16-bit breiten Stack, so können auch nur 16-bit Integer Werte verarbeitet werden. Wenn die Karte 32-bit Werte nicht unterstützt so muß der Applet-Entwickler beim Laden darauf hingewiesen werden.

Desweiteren unterstützen Java Karten Packages, dynamisches Instantiieren von Objekten, Polymorphie, Interfaces und Exceptions.

Durch die Verwendung nativer Methoden für Klassen, die in die ROM-Maske bzw. bei der Produktion fest in die Karte eingebracht werden, erzielt man einen deutlichen Gewinn bei der Ausführungszeit. Daraus ergibt sich die klare Forderung, native Methoden bis zur Beendigung der Produktion auf jeden Fall zu unterstützen. Während des Nachladens im Feld jedoch, stellt nativer Code ein klares Problem dar. Zum einen opfert man durch Verwenden von nativem Code die Plattformunabhängigkeit. Zum andern könnte ein Angreifer damit eine Hintertür einbauen, mit der er die Sicherheit der Karte unterlaufen könnte. Ob in der eigentlichen Betriebsphase native Methoden unterstützt werden, wurde deshalb dem Ermessen der einzelnen Kartenhersteller überlassen.

4.1.3 Bytecode

Aufgrund des reduzierten Sprachumfangs, wird beim Compilieren lediglich eine Untermenge des definierten Java-Bytecodes benötigt. Die in Java Card verwendeten Bytecodes sind also exact dieselben wie im Sprachstandard; es werden auch keine neuen Bytecodes hinzugefügt. Dies ist notwendig, damit die Class-Datei auch in anderen Java Umgebungen als der Java Card ausgeführt und getestet werden kann (Aufwärtskompatibilität!).

4.2 Java Card Virtual Machine

Die komplette VM kann auch trotz der Einschränkungen noch nicht auf eine Chipkarte gebracht werden. Bedenkt man, daß eine optimierte Realisierung immer noch zwischen 100k und 200k Codebedarf besitzt, auf heutigen Chipkartencontrollern jedoch lediglich ca. 32k-48k Programmspeicher vorhanden sind, so wird dies relativ schnell klar.

Die eigentliche Java VM wird darum aufgeteilt in einen Teil, der außerhalb der Karte vor der Code-Interpretierung abgearbeitet wird (Software für einen PC). Dabei wird der Java Bytecode in einen Java Card Bytecode konvertiert (siehe hierzu auch das Kapitel über Laden und Linken, 4.2.2). Der andere Teil befindet sich wirklich auf der Java Card. Nur beides zusammen ist die eigentliche virtuelle Maschine.

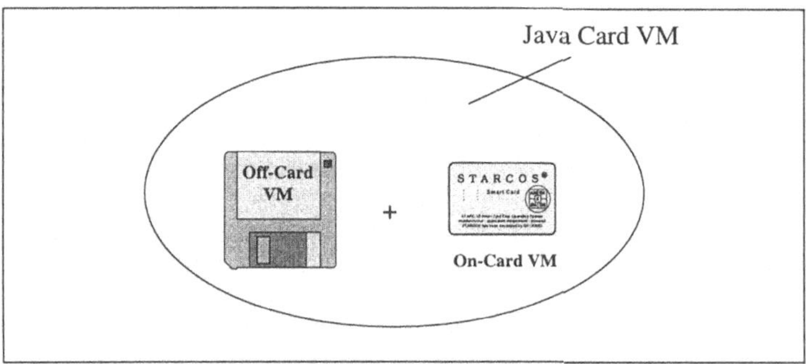

Abb. 2: Java Card Virtual Machine

4.2.1 Applet

Eine Java-Anwendung für eine Java Karte wird als Applet bezeichnet. Dies ist die kleinste Einheit, die auf einer Karte selektiert und ausgeführt werden kann bzw. die Sicherheit bereitstellen kann. Es können sich beliebig viele Applets auf einer Karte befinden (vorausgesetzt der Speicher reicht hierfür aus).

Jedes Applet besitzt eine AID (= application identifier, definiert in ISO/IEC 7816-5, [ISO95]) Mit Hilfe dieser AID kann das gewünschte Applet identifiziert und selektiert werden.

4.2.2 Laden und Linken

Statische Arbeiten können bereits durch den Off-Card Teil erledigt werden. Dazu zählen unter anderem die Kontrolle, ob nur gültige Bytecodes verwendet wurden, die (teilweise) Auflösung des Konstantenpools, die Datenflußkontrolle u. ä. Bei erfolgreicher Ausführung wird aus der Class Datei eine für die Java Card ausführbare Class Datei (= Card Class Datei) erzeugt.

In der Java Card dient diese Card Class Datei als Eingabeformat. Im Gegensatz zu Standard Java, erhält das Laden in der Java Card eine besondere Bedeutung. Die Sicherheit von Java kann nur die gesamte virtuelle Maschine garantieren. Die virtuelle Maschine wurde allerdings in zwei Teile zertrennt. Somit muß beim Laden mittels Kryptographie sichergestellt werden, daß die geladenen Daten auch von einer authentischen Off-Card VM erzeugt wurden. Es läßt sich ja nicht ausschließen, daß ein Angreifer die Card Class Datei zu manipulieren bzw. vielleicht sogar in diesem Format zu programmieren versucht. Für das Format dieser Card Class Datei gibt es zum augenblicklichen Zeitpunkt noch keinen einheitlichen Standard.

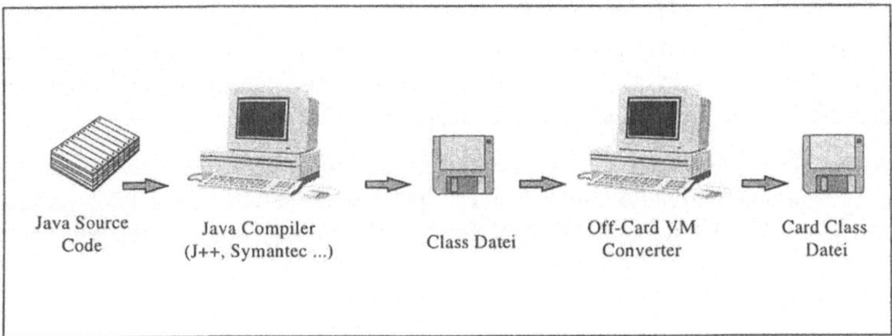

Abb. 3: Erzeugen einer Card Class Datei

Das Laden erfolgt in Form von Packages. Eine Card Class Datei muß mindestens ein Package, kann aber auch mehrere Packages enthalten. In einem Package darf sich maximal eine von der Klasse *Applet* abgeleitete Klasse befinden. Diese von *Applet* abgeleitete Klasse beinhaltet die eigentliche Anwendung. Das Laden einzelner Klassen ist genauso wenig möglich wie das dynamische Nachladen von Klassen während der Ausführung des Java Bytecode. Alle benötigten Klassen müssen durch den Ladevorgang auf die Karte gebracht werden, damit das anschließende Linken erfolgreich sein kann.

4.2.3 Firewall

Die einzelnen Applets einer Karte sind voneinander hermetisch abgeriegelt. Alle Objekte sind einem Applet zugeordnet, und zwar dem, der sie erzeugt hat. Normalerweise ist ein Zugriff auf ein Objekt aus einem anderen Applet nicht möglich, da hierzu die Referenz auf dieses Objekt fehlt. Aber selbst wenn eine Referenz gegeben wäre, verhindert eine Firewall Zugriffe auf Objekte eines fremden Applets.

Nun gibt es allerdings einen Bedarf, daß Objekte von verschiedenen Applets gemeinsam nutzen zu können (z. B. Master PIN, personenbezogene Daten u. ä.). Hierfür wurde ein Mechanismus eingerichtet, der es ermöglicht, bestimmte Objekte eines Applets mit allen oder mit bestimmten Applets zu teilen. Nur danach ist ein Zugriff eines fremden Applets möglich.

4.2.4 Error und Exception Handling

Beim Auftreten von Fehlern während der Ausführung wird die Programmabarbeitung sofort beendet, da dies weitreichende Auswirkungen auf die Korrektheit nachfolgender Programmteile hätte. Dadurch könnte die Sicherheit der Karte kompromittiert werden.

Die Ausnahmebehandlung erfolgt analog wie in Standard Java (mit *try{}* und *catch{}*). Man unterscheidet zwischen zwei Arten: "checked" und "unchecked exceptions".

"Checked exceptions" sind solche, die im Java Code explizit ausgelöst und an einer anderen Stelle für die Ausnahmebehandlung "gefangen" werden. Zum Auslösen einer Ausnahme muß man eine Instanz einer von *Exception* abgeleiteten Klasse erzeugen. Da ein Garbage Collector auf einer Java Card nicht vorgeschrieben ist, belegen die so erzeugten Objekte festen

Speicherplatz. Dieser Speicherplatz ist verloren, selbst wenn die Objekte überhaupt nicht mehr benutzt werden (keine Referenz darauf mehr vorhanden ist). Darum gibt es in Java Card 2.0 eine Möglichkeit, alle benötigten Exception-Objekte zu Anfang zu erzeugen und sie mittels einer statischen Methode wiederzuverwenden.

Unter "unchecked exceptions" versteht man unerwartete Ausnahmen, die erst zur Laufzeit auftreten (z.B. Division durch 0). Diese werden von der virtuellen Maschine selber ausgelöst und müssen nicht unbedingt im Java Code gefangen werden.

Es gibt noch eine spezielle Art der "unchecked exceptions", die *ISOException*. Eine davon abgeleitete Exception beinhaltet immer einen spezifischen Status SW1SW2 (definiert in ISO/IEC 7816-4 [ISO95]), der dann als Teil der Antwort an das Terminal gesandt wird.

4.2.5 Lebenszeiten

Die Lebenszeit der Java Card VM erstreckt sich über den gesamten Lebenszyklus der Karte, d. h. sie startet mit dem Ende der Produktion und endet erst mit dem Recycling der Karte. Anders als die VM auf einem PC, deren Lebenszeit sich nur über die Einschaltdauer ausdehnt, überdauert die Karten-VM Stromunterbrechungen. Jede Stromunterbrechung wird von der Java Card als besondere Exception behandelt. Die davor angelegten Objekte existieren bei der nächsten Kontaktierung weiter als hätte diese Stromunterbrechung gar nicht stattgefunden.

Da es in Java Card 2.0 keinen Mechanismus zum Deaktivieren bzw. Löschen von Applets gibt, ist die Lebenszeit der Applets gleich der Lebenszeit der Karte.

Die Lebenszeit von Objekten ist klar definiert. Sie startet beim Erzeugen mit *new()*. Solange eine Referenz auf dieses Objekt besteht, existiert das Objekt. Verweist keine Referenz mehr darauf, so ist die Lebenszeit beendet. Das Objekt ist für das System nicht mehr sichtbar, d. h. quasi "nicht mehr vorhanden". Der Speicherplatz für dieses Objekt bleibt allerdings belegt. Sollte sich in der Zukunft ein Garbage Collector auf der Karte befinden, so könnte dieser dann den belegten Speicher freigeben.

4.3 Java Card API (Application Programming Interface)

Die Java Card API stellt eine Bibliothek dar von häufig benötigten Grundfunktionen einer Chipkarte. Sie läßt sich in zwei Bereiche aufteilen - "Core" und "Standard Extension". Während die Core-Bereiche der API auf jeder 2.0 kompatiblen Karte vorhanden sein müssen, gilt für die Extension, daß, falls sich Teile davon auf der Karte befinden, sie genau den standardisierten APIs zu entsprechen haben.

Außer Kryptographie und dem Filesystem gehören alle in diesem Kapitel aufgelisteten Funktionalitäten zur Core-API.

4.3.1 Kommunikation

Für die Kommunikation zwischen Java Card 2.0 und Terminal sind die Transport Protokolle T=0 und T=1 (definiert in ISO/IEC 7816-3, [ISO95]) vorgegeben, da diese Protokolle heutzutage üblicherweise in Terminals Verwendung finden. Es wurde zwar vom Java Card Forum der Bedarf nach leistungsfähigeren Protokollen erkannt - die Karte entwickelt sich langsam zu einem eigenständigen Knoten eines heterogenen Netzwerks - jedoch ist es dafür noch zu früh. Die Java Card muß sich erst einmal in bestehenden Systemen etablieren.

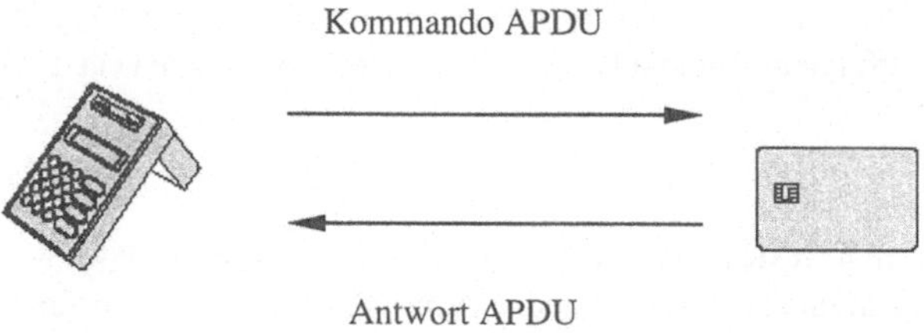

Abb. 4: Kommunikation zwischen Terminal und Karte

Um eine Integration von Java Karten in bereits bestehende Systeme zu gewährleisten, wird auf das bestehende Challenge-Response-Verfahren (definiert in ISO/IEC 7816-4 [ISO95]) zurückgegriffen. Das Terminal sendet ein Kommando (Kommando APDU) zur Karte, diese führt die damit verbundene Aktion aus und liefert das Ergebnis (Antwort APDU) zum Terminal zurück. Dabei agiert das Terminal als "Master", die Karte als "Slave".

Für diesen Mechanismus wurde eine API definiert, die vollkommen protokollunabhängig gestaltet ist. Ein Applet-Entwickler muß nun mit Hilfe dieser API die Kommunikation nach außen aufbauen. Alle Kommandos werden direkt zum Applet durchgereicht, das dann die Kommandos interpretieren und bearbeiten muß. Benötigte Daten sind gegebenenfalls anzufordern.

4.3.2 Sharing von Objekten

Wie bereits im Kapitel über die Firewall (4.2.3) erwähnt wurde, gibt es die Möglichkeit Objekte mit anderen Applets explizit zu teilen. Hierfür stellt die API eine statische Methode zur Verfügung. Nur das Applet, das ein Objekt erzeugt hat, ist auch in der Lage es mit anderen zu teilen, sprich diese Methode auszuführen. Der Vorgang ist irreversibel. Ein einmal freigegebenes Objekt kann nicht mehr zurückgenommen werden.

4.3.3 Integrität von Transaktionen

Durch die Lebenszeit der Java Card VM über Stromunterbrechungen hinweg, kommt der Integrität von Transaktionen besondere Bedeutung zu. Es gilt zu gewährleisten, daß trotz einer unerwarteten Stromunterbrechung die Karte mit ihren Daten in einen konsistenten Zustand beibehält.

Man unterscheidet hierbei zwei Arten. Die erste ist für den Applet Programmierer unsichtbar. Falls sich während einer Änderung eines einzelnen Objekt oder Feldes eine Stromunterbrechung ereignet, so muß sich diese Änderung wie eine atomare Aktion darstellen. Entweder sie geschah, oder eben nicht. Ein Zwischenzustand (teilweises Änderung) darf nicht auftreten.

Nun gibt es aber auch den Fall, daß die Konsistenz erst mit Änderung mehrerer unterschiedlicher Felder von unterschiedlichen Objekten gewahrt ist. Hierfür hat der Programmierer zu sorgen, daß alle Änderungen in einem Transaktionsblock gekapselt sind. Damit ist auch für diesen Fall ein atomares Verhalten gewährleistet. Sollte eine Unterbrechung der Versorgungsspannung geschehen, bevor die Transaktion für gültig erklärt worden ist, werden alle

Änderungen innerhalb dieses Transaktionsblockes verworfen und der alte Zustand wiederhergestellt. Nach erfolgtem Abschluß dieser Transaktion werden alle neuen Werte auf einmal übernommen.

4.3.4 PIN

Die API unterstützt Aufgaben, die im Zusammenhang mit PINs benötigt werden. Die PIN-Klassen beinhalten Methoden für das Ändern, Authentisieren und Setzen von PINs.

4.3.5 Utility

Die unter Utility zusammengefaßten Methoden beschränken sich auf Typkonvertierungen und Funktionen auf Felder (kopieren, vergleichen von arrays).

4.3.6 Kryptographie

Die Kryptographie zählt zu den Erweiterungen, da sie immer mit exportrechtlichen Problemen und Restriktionen behaftet ist.

Aufgrund eben jener scheidet eine sehr flexible Schnittstelle aus, die eine effektive Programmierbarkeit eines Algorithmus erlauben würde (z.B. Schnittstelle zum Kryptocontroller). Es dürfen nach Auslieferung keine neuen Algorithmen hinzugefügt werden. Natürlich bedeutend das Fehlen noch lange nicht, daß dies nicht doch geschehen könnte. Eine Java Card läßt sich schließlich frei programmierbaren. Allerdings wäre ein in Java geschriebener Algorithmus nicht sonderlich effektiv, schon gar nicht auf einer Chipkarte.

Die Kryptographie-API stellt Methoden sowohl für symmetrische als auch asymmetrische Kryptoalgorithmen zur Verfügung, sowie zur Hash-Bildung und zur Generierung von Pseudo-Zufallszahlen. Als Algorithmen werden für den 2.0 Standard DES, DES3, RSA und SHA1 verwendet.

4.3.7 Dateisystem

Für die Emulation eines Dateisystems (siehe ISO/IEC 7816-4 [ISO95]) wird ein Framework bereitgestellt, der ebenfalls zur Erweiterung zählt. Im Applet muß dieses Dateisystem erst aufgebaut, sprich implementiert werden, mitsamt der ganzen Struktur und der Zugriffskontrolle. Die API beinhaltet eine Schnittstelle, um eine sehr einfache Zugriffskontrolle für externe Lese- und Schreibzugriffe aufzubauen. Das Applet allerdings darf jederzeit darauf zugreifen, d. h. das Dateisystem ist direkt dem Applet zu- bzw. untergeordnet.

Für die Einordnung des Dateisystems als Erweiterung gibt es zwei Gründe:

Erstens gibt es Anwendungen und Systeme, die bereits heute ohne ein Dateisystem auskommen (z.B. Mondex). Für solche Systeme würde unnötig Speicher belegt werden, falls das Dateisystem Voraussetzung für eine 2.0 kompatible Karte wäre. Und Speicher ist schließlich nach wie vor das wertvollste Gut einer Karte.

Zweitens bietet Java auch die Möglichkeit, Daten anderst abzuspeichern als im Dateisystem - in Objekten. Objekte besitzen die Möglichkeit eine Zugriffskontrolle auf Daten und Methoden aufzubauen (entsprechend ihrer Definition *private, protected, public*). Es bietet sich förmlich an, Daten, die vor unberechtigten Zugriffen geschützt werden sollen, in Objekten zu repräsentieren. Das Verhalten eines Dateisystems kann dadurch vollständig emuliert werden.

Es ist also prinzipiell gar nicht nötig, ein Dateisystem auf einer Java Card zu haben. Damit allerdings nicht jedes Applet den gleichen Code beinhalten muß bzw. um auch komplexere Systeme leicht emulieren werden können und wegen der Kompatibilität zu alten Systemen, wurde diese API geschaffen. Darüber hinaus kann die API in Maschinencode vorliegen und damit die Ausführungszeit wesentlich verbessern.

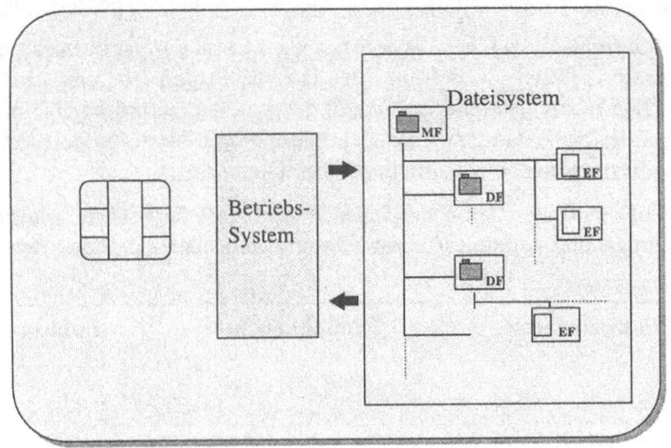

Abb. 6: Herkömmliche Chipkarte

Abb. 7: Java Card

Zur Verdeutlichung sei allerdings noch mal gesagt: Es handelt sich bei dieser API nicht um eine Schnittstelle zu einem Dateisystem eines gängigen Chipkarten-Betriebssystems. Es stellt lediglich eine objektorientierte Emulation eines solchen dar.

4.4 Entwicklung von Java Card Applets

Die Entwicklung von Java Card Applets gestaltet sich dreistufig.

In einer ersten Stufe wird das Applet mittels handelsüblichen Entwicklungswerkzeugen (Editor, Compiler, Debugger) entwickelt. Es kann wirklich jeder Java-Compiler zum Erzeugen der Class Datei verwendet werden, vorausgesetzt, man besitzt die Java Card API. Während dieser Phase wird die Software bereits auf ihre Korrektheit hin geprüft. (Funktionstest).

In einer zweiten Stufe wird die Class Datei in das Card Class Format übersetzt und mit einem Java Card Simulator getestet, da sich im Simulator die Fehlersuche einfacher gestaltet wie auf einer wirklichen Karte. Diese Stufe ist dafür gedacht, zu überprüfen, daß man sich an die Einschränkungen des Sprachumfangs gehalten hat und daß die Software auch auf der stark eingeschränkten Kartenumgebung lauffähig ist (Simulationstest).

In der letzten Stufe wird die Card Class Datei in eine echte Java Karte geladen und auf der wirklichen Zielumgebung getestet. Hier ist allerdings nur noch ein Black Box Test möglich (Endtest).

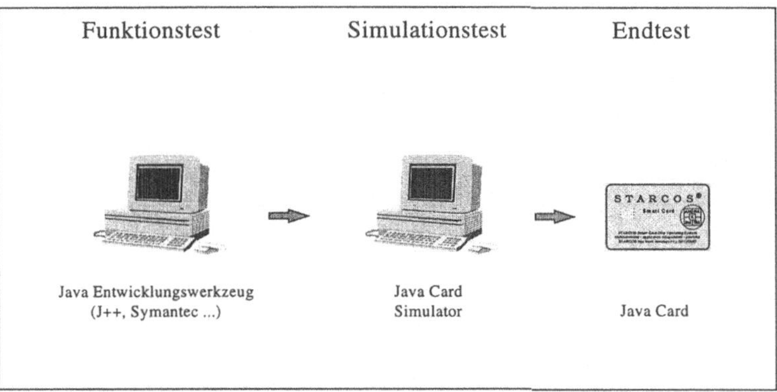

Abb. 8: Testprozeß für Java Card Applets

Literatur

[dal97]   Dalheimer, Matthias Kalle: Java Virtual Machine. Herausgeber: O'Reilly, 1997.

[lin97]   Lindholm, Tim & Yellin, Frank: Java - Die Spezifikaiton der virtuellen Maschine. Herausgeber: Addison-Wesley, 1997.

[JCL97]   Java Card 2.0 Language Subset und Virtual Machine Specification, Sun Microsystems, Revision 1.0 Final, 1997.

[JCP97]   Java Card 2.0 Programming Concepts, Sun Microsystems, Revision 1.0 Final, 1997.

[JCA97]   Java Card 2.0 API, Sun Microsystems, Revision 1.0 Final, 1997.

[ISO95]   ISO/IEC 7816/1-5, 1995.

NetCentre Company Smart Card Issuance for IBM's NetCentre Network Computers

E. Hechler[1] · J. McKeon[2]

[1]IBM Deutschland Entwicklung GmbH
Smart Card Development
Postfach 1380, D-71003 Böblingen
ehechler@de.ibm.com

[2]IBM Corporation
Smart Card Solutions
Research Triangle Park, NC 27709, USA
mckeon@us.ibm.com

Key Words

NetCentre Company Card, NetCentre Network Computers (NCs), IBM Multi-function Card (MFC), Trusted Third Party (TTP), Card Issuing System (CIS), Card Management System (CMS), Application Management System (AMS), Key Management System (KMS), Public Key Infrastructure (PKI), Trusted Card Issuing Schemes, Card Issuer, Application Provider, Certificate Authority (CA), Digital Certificate Issuing, Certification Revocation List (CRL), Local Registration Authority (LRA), Personalization Application (PA), Chip Initialization, Card Personalization Station (CPS), IBM Net.Data, InterNet, IBM Smart Card (SC) Toolkit.

1 IBM NetCentre Network Computers

Network Computers (NCs) have generated a lot of press, both positive and negative, over the last year. The NC's promise of low cost of ownership, ease of administration, and high security has captured the interest of many I/T professionals. Since its debute at fall '96 Comdex, the NetCentre NC platform has received rave reviews at numerous international trade shows and OEM/customer engagements. It has been recognized by I/T experts and analysts as a cutting-edge platform for building a variety of NC devices. One reason for this response can be found in the NetCentre's use of smart cards (SCs).

Why use a smart card with the NetCentre NC? The use of a smart card with a NC represents an ideal marriage of complimentary technologies. Designed to be a multi-user device, the NetCentre contains no persistently stored software or user data (it doesn't even have a hard drive) and provides a general purpose NC platform. The smart card, on the other hand, is a very personal device. It contains personal data, on the plastic and stored in the chip, that is unique to its owner. By using personal data from a smart card the NetCentre can be transformed from a multi-user, general-purpose platform to a single-user, customized device.

Easily carried in a wallet or purse, the smart card is also very portable. The NetCentre user can take advantage of this portability by taking his or her computing environment with them as they move from device to device. The smart card grants them access to their own desktop, applications, and data - turning any available NetCentre into a familiar, personal workstation. Smart cards and NCs each offer unique security functions in their own right. The NetCentre guarantees the integrity of its platform through a variety of security measures, including secure network boot and clean desktop startup. The smart card also enforces a number of security features, including user authentication and secure data storage on the card. Working together, they can provide a comprehensive security system for the NetCentre user.

So it is in these three areas - customization, portability, and security - that a smart card adds value to the NetCentre platform. In allowing the user to customize the NetCentre into their personal device, take their computing environment with them, and providing them with advanced security features, the smart card makes the NetCentre a better network computing platform.

Just as the smart card enables unique features on the NetCentre, so too the NetCentre was designed to be an enabling platform for smart card applications. Along with this enablement comes an important opportunity for smart card technology and solutions. Consider the following:

- The NetCentre platform will be built and branded by a number of leading computer equipment manufactures around the world. Significant volumes are predicted.

- The NetCentre hardware integrates a smart card reader into every reference board delivered with the NetCentre OEM Development Kit.

- The NetCentre software integrates IBM MFC (Multi-function Card) smart card support into every software load delivered with the NetCentre OEM Development Kit.

- A login application is shipped as part of the NetCentre software. It requires the use of an MFC smart card for access to the NetCentre NC.

- The NetCentre smart card support is designed to use a minimal amount of real estate on the MFC smart card. This is to allow room for other smart card applications that might want to share the same card.

- The NetCentre Company smart card is issued world-wide to any NetCentre NC user through IBM's card/certificate issuing service.

The result is a large install base of MFC-enabled client devices and many NetCentre NC users in posession of an MFC smart card.

The primary reason for proposing a turnkey smart card solution for the NetCentre is to simplify, for the OEM and customer, the task of deploying NetCentre NCs into customer environments. Although the use of a NetCentre smart card appears simple, the underlying infrastructure is rather complex. All the elements of a Public Key Infrastructure (PKI) and Card Issuing System (CIS) must be in place for the solution to work efficiently. The NetCentre Company Card solution will hide that complexity while providing the benefits the underlying infrastucture affords.

A secondary reason for a turnkey solution is to solidify the smart card opportunity presented in the NetCentre platform. OEMs can and will make modifications to the reference implementation as they bring the NetCentre platform to market in their own branded devices. The OEM can use the integrated MFC support in the NetCentre, without making major changes. A smart card solution that ads value to the overall NetCentre platform, and depends on the integrated MFC support, will help encourage this.

2 Trusted Card Issuing Scheme

The card issuing scheme describes the logistics and the cooperation between the various institutions, which are involved in the card and also the certificate issuing process. Cooperation is the details on the split and scope of work and the flow of data and information between these companies and institutions. Thus, this trusted card/certificate issuing scheme deals with the different tasks (e.g. data collection, chip initialization, certificate issuing, chip and plastic personalization, etc.) and who performs these tasks (e.g. the card issuer, a Trusted Third Party (TTP), a services organization like IBM Global Services, etc.).

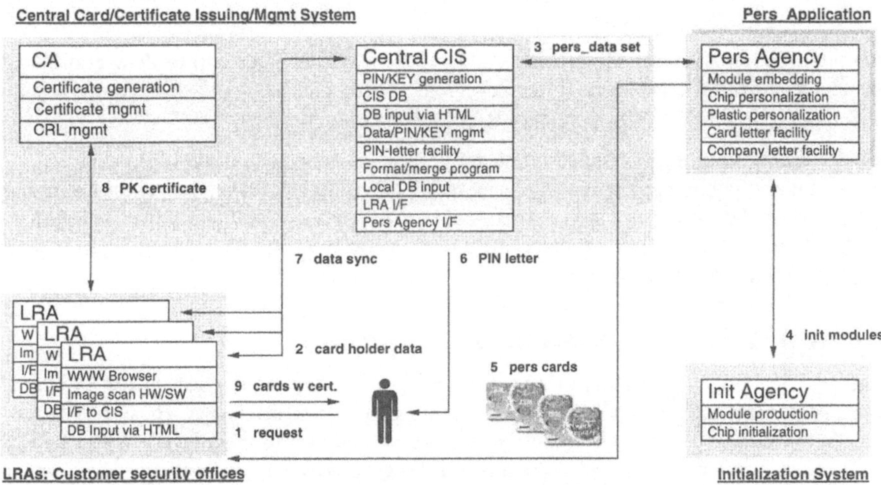

Fig. 1: Centralized Trusted Card/Certificate Issuing Scheme

Logistic deals with the following questions: how many initialized modules have to be available by when? Where does the plastic and the artwork comes from? What is the timing between data capturing and final card issuing? What is the timing for re-issue of lost and stolen cards? In general, card issuing logistics answers the question: who has to exchange what kind of data/information with whom in what time frame and in what fashion? A centralized card/certificate issuing scheme was choosen for this NetCentre Card Issuing System (CIS) project. Thus, there will be a central services organization per region (e.g. North America, EMEA ...), like IBM Global Services. The following list describes the various components and institutions or companies, which are involved in the proposed centralized card/certificate issuing scheme and figure 1 illustrates this proposed scheme for the NetCentre Company card:

Central CIS: Given the requirements and the necessary integration of the IBM World Registry for certificate issuing and management, a centralized card issuing scheme was chosen. Thus, the customers of the OEMs will not operate their own CIS system; all necessary card and card-holder data is stored in a central CIS DataBase (DB). Also, all CIS functions, such as generation of PINs, building of initialization and personalization scripts, personalization orders incl. personalization data sets, and PIN-letters are performed centrally in each region (e.g. North America). The PIN-letters are sent to each card-holder. A PIN-letter contains the PIN to access the card and notifies a card-holder, that his card can be picked up at his responsible or authorized Local Registration Authority (LRA).

LRAs: Capturing of card-holder data (e.g. name, picture, ...) is done locally at the client level. These customer security offices are the Local Registration Authorities (LRAs) in this card/certificate issuing scheme. Each customer may entertain several LRAs. The future card-holder will appear at one of the customer's LRA office for data capturing and to take delivery of the card as well.

Initialization Agency: An initialization agency will perform the mass initialization of the chips during module production.

Personalization Agency: Module embedding into the plastic cards will be done at different Personalization Agencies in the different regions. These agencies will also send the personalized cards back to the corresponding LRA of a particular customer.

Certification Authority: The public/secret key pair is generated at the LRA level and the corresponding public key certificate is generated at the CA level. The key pair can either be generated on the card (using IBM's MFC 4.2) or via Netscape Communicator 4.0. The management of certificates incl. management of the Certification Revocation List (CRL) is done through IBM World Registry SW.

3 High Level CIS System Description

The advantage of the NetCentre Company CIS system is the relationship to IBM's overall Card Management System (CMS) and Card Issuing System (CIS) design. Thus, the CIS system follows the separation of issuing functionality at the <u>management layer</u>, where the CIS DB resides and the personalization orders will be generated, etc., the <u>network layer</u>, which does the routing of personalization orders from the management layer down to the Card Personalization Station (CPS), and the <u>device layer</u>, which does the actual chip and plastic personalization. IBM's SC Toolkit will be used by the card issuing SW to allow for card transparency. It follows a brief description of the major CIS components and platforms:

CIS Database: This central CIS DB stores all card, application, and card-holder related data for all employees of all companies, regardless which OEM they will receive the NC Development Kit from. Thus, all personal data, will be captured (e.g. card-holder name and image) or generated (e.g. PINs and cryptographic keys) and then stored in the DB. The plastic artwork or image is brought onto the card prior to the plastic personalization process. As a consequence, this data will not be stored in the CIS DB. For pilot projects with smaller card volumes, this CIS DB will hold the card-holder relevant data temporarily in the card issuing/personalization DB. However, other non-personal data elements, such as the initialization and personalization scipts will always be stored in the DB.

IBM Net.Data: IBM Net.Data will be used to connect the client system at the LRA level with the central CIS DB, which will be based on DB2/6000. This eliminates platform dependent client application development and allows for enhancement to even utilize Java applets for data capturing and display in the future through Net.Data's JDBC I/F to any DB2 data. Also, access to other data sources, such as Oracle and Sybase of existing customer DBs is guaranteed through Net.Data capability.

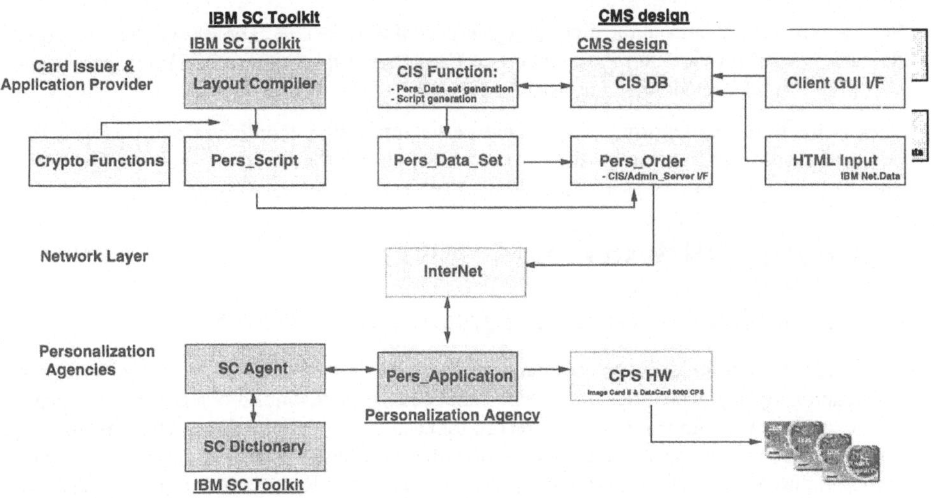

Fig. 2: High-level Smart Card Issuing Concept

Client GUI I/F: Data input and retrieval to and from the central CIS DB will eventually be done through HTML pages. Translation into UPDATE or SELECT queries against the CIS DB will be done through Net.Data applications. The JDBC I/F allows for Java applets to run at the client level as well (in the future). This client GUI is synchronized with the data input facilities of IBM's World Registry SW.

CIS Function (personalization order & personalization data set generation): This program is one of the core components of the entire CIS system. It builds the personalization order, which is composed of the personalization script and the personalization data sets (for chip and plastic personalization), and complemented by some administrative data (e.g. order number, date, status, comment, etc.). The system is flexible enough, to store plastic personalization data other than the Company image and logo or OEM logo (e.g. card-holder picture, barcode data, ...) in the CIS DB.

Layout Compiler: This component of the IBM SC Toolkit will be used for generation of the initialization and personalization scripts.

CIS/CPS I/F: This is the I/F routine, which allows for communication between the CIS system and the Card Personalization Station (CPS). Secure communication features of the Internet (SSL V.2 and SSL V.3) will be used to route data between the central services organization and the personalization agencies.

Crypto Functions: These are basic cryptographic functions, which are primarily aimed for PIN (CHV_1, CHV_2), loginKey, and other DES key generation. The asymmetric cryptographic keys and the public key certificate are generated and stored onto the card through IBM's World Registry SW using PKCS #11.

Image Scanning HW/SW: Low-cost equipment shall be used to scan the card-holder picture during the data capturing process at the LRA. This image will be contained within the personalization data set generation.

SC Toolkit: The agent and the dictionary component of IBM's SC Toolkit will be utilized at the device level for ease and most important Card Operating System (COS) transparent access to cards and also command translation purposes.

Personalization-Application: A personalization application coordinates and performs the chip and plastic personalization. IBM's personalization building block will be used to interpret the personalization data sets along with the personalization scripts.

4 Detailed CIS System Description

4.1 Personalization Order & Personalization Data Set

The personalization order is the communication vehicle between the CIS system and the personalization agency, which operates the personalization application. It contains the personalization data set with several records, 1 record for each card to be processed. The personalization order and the personalization data set are build according to the TLV structure. The following figure 3 shows the structure of the personalization order and also outlines, where the various data or components come from. The order will be kept in the CIS DB after it was sent to the personalization agency and processed there. Since the personalization order contains mostly ids as pointers to other data elements, there are only limited storage concerns. It has the advantage, that the order processing history can still be queried, even after the cards are already sent to the Companies. Because the structure of the personalization order and the personalization data set is according to the TLV standard, it will be easy to modify or enhance all components to include additional data elements or information. Also, it gives the advantage that logos, scripts, etc. can be included in the order, but they don't have to be included if the data is already present at the personalization agency (PA). Figure 4 shows the structure of the personalization order and the personalization data set all the way down to the individual data elements, such as the User_Profile_URL, CHV_2, etc. Since the elements in the personalization order (e.g. the C_Script_TLV) and the personalization data set (e.g. the Chip_Data_TLV) have to be processed by IBM's SC Toolkit and the Personalization Building Block (PBB), the structure of those Data_Entry_TLVs is designed in such a way, that they contain an Id_TLV and a Data_TLV. Only those data elements, which have to be referenced by either the C_Script or the P_Script do contain such an Id_TLV and a Data_TLV. In order to be conform to the SC Toolkit, there are additional layers embedded between the Chip_Data_TLV, Plastic_Data_TLV, Card_Letter_TLV, CID_TLV and the various Data_Entry_TLVs, which are not shown in figure 4.

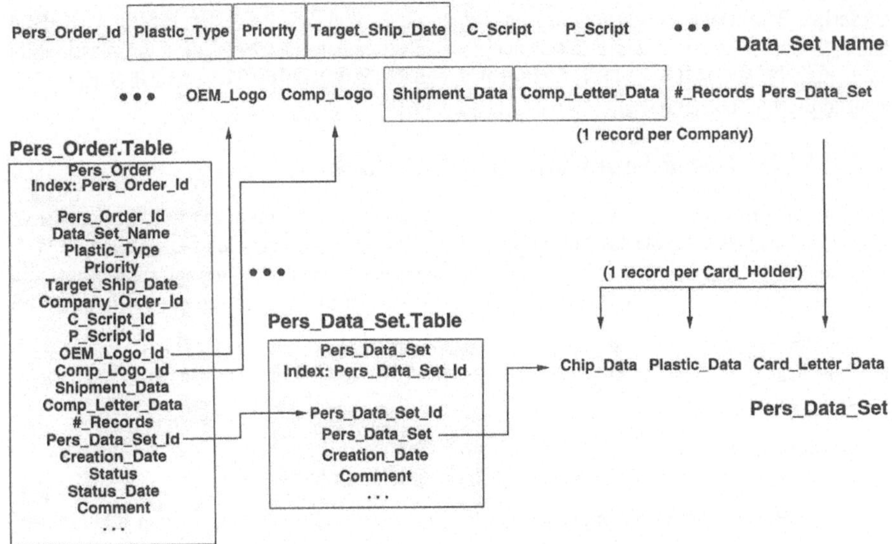

Fig. 3: Structure of Personalization_Order

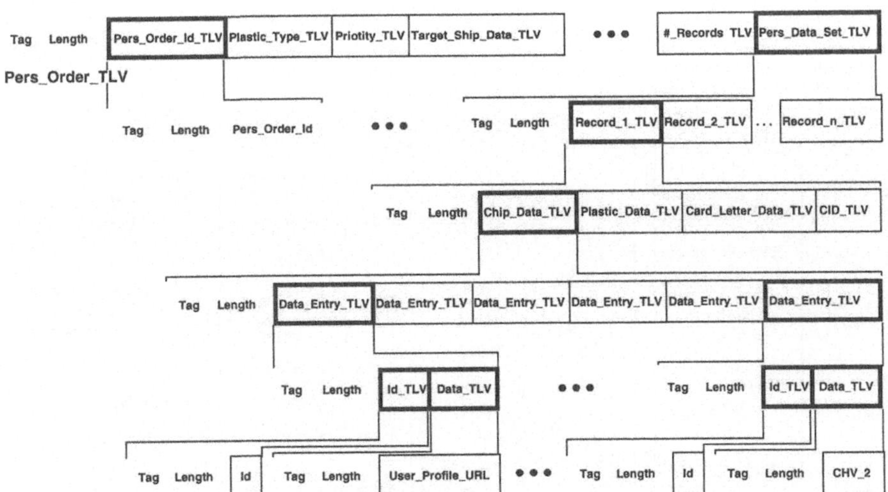

Fig. 4: TLV Structure of Personalization Order & Personalization Data Set

The Id_TLV contains an Id or an index, which will be referenced during interpretation of the C_Script. The Data_TLV contains the actual data, such as the User_Profile_URL, the CHV_2, etc. In order to apply this concept to all other data elements (also the plastic data elements), the OEM_Logo_TLV, Company_Logo_TLV, C_Script_TLV, and P_Script_TLV is structured in the same way.

4.2 HTML-based Input via IBM Net.Data

The interface to insert data into the CIS DB from any client station at the CIS service site or LRA level will be a simple JavaScript and a RFC1867 (form based file upload) enabled Web Browser (e.g. Netscape Navigator 2.02 or higher) using standard internet technology and protocols. The infrastructure of the environment required to build up the system is based on the following:

- An Intranet or Internet network to connect the LRA or CIS services clients to the card is-suing server,

- an Internet server running IBM Net.Data to support CGI (Common Gateway Interface) to handle and restrict requests to the HTML documents and CGI scripts,

- DB2/6000 (or a DBMS supporting ODBC) to run the CIS database, which manages the data itself, and

- IBM Net.Data applications and scripts controlling and processing the data flow.

The IBM Net.Data product provides ease of use and effective usage and access to CGI scripts and is used as an I/F to DB2/6000. IBM Net.Data will process macros and a set of different functions before customizing the HTML page. Thus, the HTML documents and forms are not static as in usual HTML but dynamic.

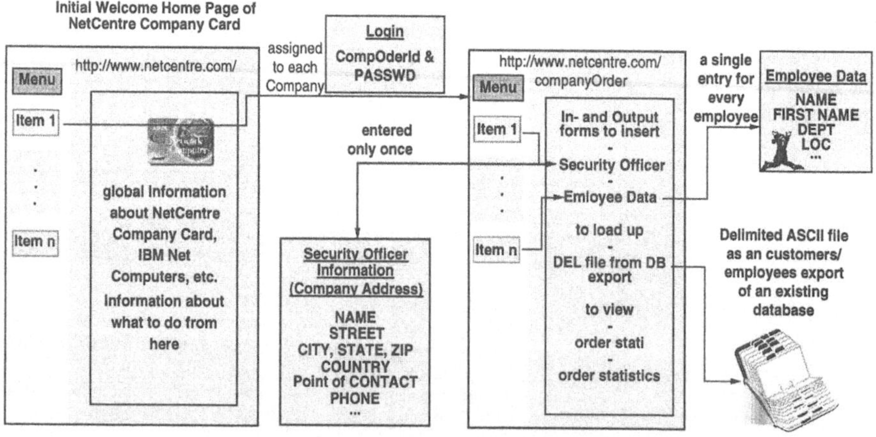

Fig. 5: HTML-based Data Input

To be positioned for future functional enhancements to the CIS system, such as Java-based data capturing with sophisticated applet-based consistency checks at the LRA level, the system uses the IBM Net.Data product as a CGI process and macro preprocessor to call other CGI scripts or other functions, access the CIS DB and to customize the HTML pages. For scalability reasons, the system uses only standard protocols, interfaces and products that are available on a wide range of systems.

The process flow of entering information into the CIS DB is similar for the CIS service site administrator and the LRA security officer. Figure 5 shows the main path of execution.

In all cases, the NetCentre Card Issuing home page will be accessed first, showing some general information about the CIS system and links to other corresponding IBM pages. The main home page is created in frames containing a header always present over the whole working session, the main working page and an index. This index is used for selecting further issuing tasks. The CIS system includes 2 main topics: 'administration' and 'process order'. By selecting 'administration', the CIS services personnel can administer OEM- and Company-Data and can create Company-orders, which are processed from a purchase order. By selecting 'process order', one specific company order is processed. It will be linked to a specific LRA security officer or even a CIS services person by the login userid, which has to be entered on the HTML-page.

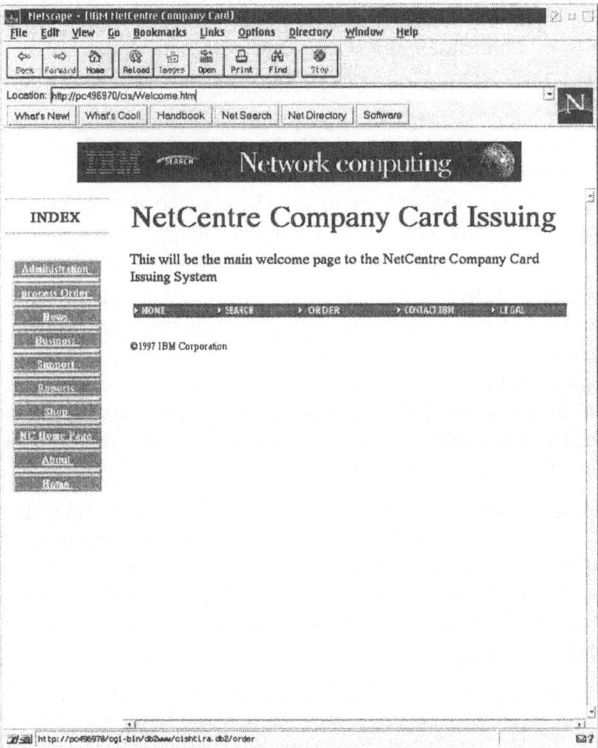

Fig. 6: NetCentre Main Welcome Page

Both application scenarios are separately protected by userid and password. Thus, a company order cannot be processed with a working CIS administration userid and password and vice versa. As the web-browser remembers the login for each scenario, it is sufficient to enter the userid and password combination once per session.

There is only 1 example of the HTML pages given in this paper: Figure 6 shows the **NetCentre Main Welcome Page** of the CIS system. The main welcome page contains information about the application and possibilities of the NC CIS and also links to other IBM information, which is related to the NC. A CIS services site administrator can select and login to the administration scope, and an LRA security officer can select and login to process a specific company order.

It follows a short description on the processing and data input flow:

The **NetCentre Admin Welcome Page** displays information on functions that can be processed, such as creation and change of OEMs, Companies or Company-orders and an index, which leads to these functions. The index changes again when going back to the main Welcome page.

The CIS service site administrator will **Create a new OEM** entry, which is based on information from this OEM. Information such as the OEM address and the representative can be added or modified. An OEM logo can be uploaded into the CIS DB after the OEM has been created. If a Company is not stored yet in the CIS DB, the administrator can also create a new Company entry from a purchase order. Company address or representative information can be administered and a Company logo can be uploaded into the CIS DB after the Company has been created.

For an existing Company, a **New Company Order** can be created. The information from this Company purchase order will be used. This purchase order includes the number of cards, the number of LRAs, and the priority of the order. The number of cards can be within a set of predefined ranges. The priority determines the terms and conditions, such as the pricing for this order.

The **NetCentre LRA Welcome Page** of the CIS system displays information about the Company (logo and Company name), the Company-order, and the number of cards left to this particular order. The LRA security officer can select functions from an index, such as adding a new card-holder with corresponding card information, changing existing card-holder or card information, or upload a Company logo (JPEG or GIF). All functions that have influences on the Company-order (e.g. new cards) can only be accessed after the LRA security officer has entered his address information.

With **Add new Card Holder**, card-holder data can be entered and administered. As the card-holder belongs to a specific Company, there might already be card-holder information inserted into the CIS DB. There may also be new card-holders, which the LRA security officer intends to issue cards for. The created card-holder will be accessible from every other Company-order initiated by that Company and might therefore be changed by any other LRA security officer of this Company.

All information entered in this screen may be uploaded as a delimited ASCII file. The LRA security officer can export information from his own employee DB, add card information (e.g. the User_Profile_URL) to it and process this potentially large data set in one step.

4.3 Personalization Application

The personalization application will be executed on an ImageCard II station and on the Da-taCard 9000 CPS depending on the card volume. The application will follow the conceptual design, which is outlined in figure 7: the input to the application are the personalization orders, containing the personalization data sets. These personalization orders will be send securily over the net (e.g. InterNet, leased lines, etc.).

The CIS I/F module has to capture the orders and has to forward them to the personalization application start/logistics module. This routine will analyze the orders and will prioritize them according to the priority field in each order. All orders, which have been received on a particular day, will be brought into the right sequence for further processing by the personalization application.

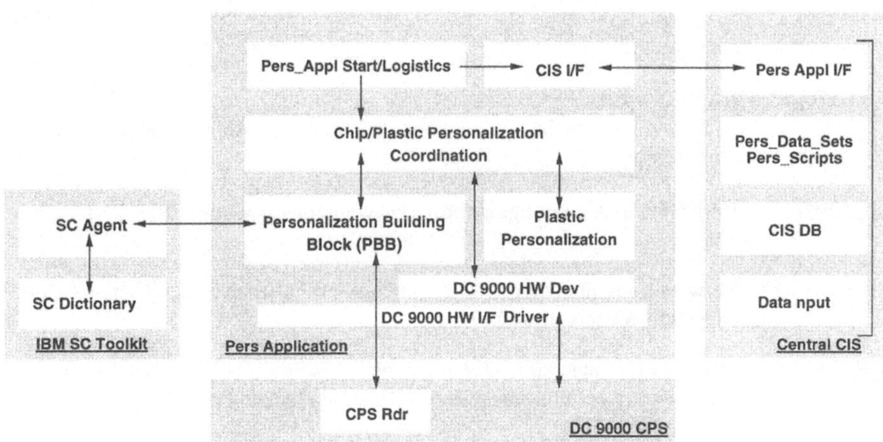

Fig. 7: High-level Personalization Application Concept

The next larger routine of the personalization application is the chip/plastic personalization corodination module. This routine will manage and coordinate the chip and plastic personalization using the chip personalization script and plastic personalization script. This coordination module will process the personalization data set, which contains the chip-data, plastic-data, and the card-letter data. The chip-data is referenced by the chip personalization script, the plastic-data by the plastic personalization script, and the card-letter-data is used as input into a local Word Processing program for letter processing purposes. The chip personalization script is generated by IBM's Layout Compiler; the plastic personalization script is a simple table, which contains the instructions for processing all plastic-data. This can be illustrated by using a simple example: the plastic-data contains the employee number and the script contains the instruction, that this employee number is to be transformed into a barcode, that this barcode has to have a certain specified size, and that it has to be placed onto the plastic at a specific location.

The processing of the chip personalization script (referencing the chip-data) is done through the Personalization Building Block (PBB) from IBM. This PBB is using the Agent and the

Dictionary of the IBM SC Toolkit for card access purposes. The processing of the plastic personalization script (referencing the plastic-data) is done through the plastic personalization module of the application. The coordination module ensures that chip-data and the corresponding plastic-data are processed correctly using the same smart card.

References

[Hechler97]   Hechler, Eberhard & McKeon, John: NetCentre Company Card. In: http://mckeon.raleigh.ibm.com.CompanyCard.CompanyCard.html, IBM Corporation, 1997.

[IBM97a]   IBM Net.Data. In: http://www.software.ibm.com/data/net.data, IBM Corporation, 1997.

[IBM97b]   IBM Multi-function Card (MFC). In: http://www.chipcard.ibm.com, IBM Corporation, 1997.

[IBM97c]   IBM Registry and World Registry. In: http://www.internet.ibm.com/commercepoint/registry/index.html, IBM Corporation, 1997.

[IBM97d]   IBM Network Computing. In: http://www.internet.ibm.com, IBM Corporation, 1997.

[Net97]   Netscape Communicator 4.0. In: http://www.netscape.com, Netscape Communications Corporation, 1997.

[PKCS97]   PKCS #11 Standard (Cryptographic Token Interface Standard). In: http://www.rsa.com/rsalabs/pubs/PKCS, RSA Laboratories, 1997.

Prüfung von Chipkartensicherheit

Stefan Rother

TÜV Informationstechnik GmbH
Essen
S.ROTHER@TUVIT.CUBIS.DE

Zusammenfassung

Zur Prüfung der Sicherheit von Chipkarten stehen dem Evaluator zwei praxisbewährte Methoden zur Verfügung. Die europaweit standardisierte ITSEC-Methodik ist aufgrund ihres generischen Charakters für nahezu jede Security-Evaluation geeignet. Zusammen mit einer anschließenden Zertifizierung lassen sich damit z.B. durch das Signaturgesetz gestellte staatliche Anforderungen erfüllen. Bei ITSEC handelt es sich um eine reine Dokumentenprüfung. Für die Zulassung zum elektronischen Zahlungsverkehr in Deutschland zwingend erforderlich ist eine Untersuchung gemäß der 13 vom Zentralen Kreditausschuß (ZKA) festgelegten Kriterien. Ein System gilt als erfolgreich geprüft, wenn es diese Anforderungen nachweislich erfüllen kann. Da beide Verfahren keine expliziten Anforderungen an die Vorgehensweise einer Sourcecodeprüfung stellen, konnte sich hierfür bei dem unabhängigen Prüfinstitut TÜV Informationstechnik GmbH (TÜViT), Essen, ein praxisorientiertes Prüfschema bewähren.

1 Einleitung

Mit dem Einzug von Chipkarten in viele Bereiche des alltäglichen Lebens hat sich ein neues Umfeld von sicherheitstechnischen Anwendungen (z.B. Zugangsschutz durch Chipkarten, elektronischer Ausweis, elektronische Geldbörse, Signaturkarte etc.) von beachtlichem Ausmaß aufgetan. Es stellt sich immer häufiger - nicht nur in Fachkreisen, sondern auch in breiten Schichten der Bevölkerung - die Frage, ob und inwieweit man diesem Medium vertrauen kann und darf.

Die Überprüfung der Informationssicherheit (information security) von Mikroprozessor-Chipkartensystemen und deren Umfeld (Chipkartenleser, Terminals etc.) bedeutet deshalb kurz-, mittel- und langfristig eine neue Herausforderung an Struktur und Einsatz von Security-Prüfkonzepten.

Geprüfte Chipkartensysteme haben aus betriebswirtschaftlicher Sicht bei der Vermarktung wegen ihrer höheren Akzeptanz Wettbewerbsvorteile, denn unbestrittenermaßen wächst die Sicherheit im Bereich der Produkthaftung. Sowohl unvorhergesehene Bugs im Programmcode, als auch fahrlässig oder absichtlich installierte Trapdoors können ein komplettes Chipkartensystem im Wirkbetrieb kompromittieren und - aufgrund der Offline-Struktur des Systems - den damit verbundenen Schaden in unkalkulierbare Höhen treiben. Zusätzlich werden Kosteneinsparungen durch sinkende Reklamationsraten - sowohl von Kunden, als auch Betreibern - und den damit verbundenen verkürzten Fehlerbehebungszyklen erreicht.

Der Begriff **Prüfung** ist definiert gemäß der Norm DIN EN 45001 als *technischer Vorgang, der aus dem Bestimmen eines oder mehrerer Kennwerte eines bestimmten Erzeugnisses, Verfahrens oder einer Dienstleistung besteht und gemäß einer vorgeschriebenen Verfahrensweise durchzuführen ist.*

Die Prüfung ist dabei grundsätzlich den folgenden Prinzipien verpflichtet:

- Unvoreingenommenheit,

- Objektivität,

- Wiederholbarkeit und

- Reproduzierbarkeit.

Die Prüfung darf nicht - auch nicht von gut gemeinten - Vorurteilen belastet sein (Unvoreingenommenheit) und sollte im Ergebnis nicht von subjektiven Überzeugungen, Kenntnissen oder Einschätzungen abhängen (Objektivität). Eine erneute Prüfung am selben Ort bzw. eine Prüfung an anderer Stelle sollte zum selben Resultat gelangen (Wiederholbarkeit bzw. Reproduzierbarkeit).

Prinzipiell läßt sich festhalten, daß es keinen fest definierten Standard für die Prüfung von Sourcecode hinsichtlich des Aspektes **Security** gibt. Um dennoch ein objektives, o.g. Prüfprinzipien verpflichtetes Prüfprozedere realisieren zu können, hat sich bei der TÜViT neben dem Einsatz diverser Software-Analysetools das Verfahren „Sourcecodeprüfung mittels Walkthrough-Methode" für statische Sourcecodeprüfungen erfolgreich etabliert. Als Prüfgrundlage können hierfür die generische ITSEC[1]-Methodik bzw. die 13 ZKA-Kriterien in Form einer Checkliste dienen.

2 Chipkartenprüfung nach ITSEC

2.1 Evaluierung und Zertifizierung

Die Evaluierung nach ITSEC-Kriterien ist für die Prüfung komplexer Systeme auf Dokumentenebene die in Europa derzeit umfassenste Methodik, um zu Aussagen über die Security eines Produkts oder Systems im Wirkbetrieb zu kommen. ITSEC definiert Gefährdungen der Informationssicherheit nach drei grundsätzlichen Anforderungen:

- **Verfügbarkeit:** Daten und Dienstleistungen müssen für autorisierte Benutzer jederzeit verfügbar sein.

- **Vertraulichkeit:** Informationen dürfen nur den dazu legitimierten Personen zugänglich sein.

- **Integrität:** Informationen müssen vor unbeabsichtigter bzw. unberechtigter Änderung geschützt werden.

Dem potentiellen Verlust obiger Anforderungen soll mit einer vertrauenswürdigen Implementierung von Sicherheitsfunktionen entgegengewirkt werden. Die ITSEC-Kriterien setzen den Schwerpunkt ihrer Anforderungen auf die Dokumentenprüfung. Der Hersteller eines

---

[1] Information Technology Security Evaluation Criteria, europäisch harmonisierte Kriterien für die Bewertung der Security von Produkten und Systemen der Informationssicherheit.

Prüfobjektes fertigt Dokumente, in denen er den sicherheitstechnischen Aufbau des zu prüfenden Objekts beschreibt. Dieses Konzept wird von einem unabhängigen Prüfinstitut hinsichtlich konzeptioneller Schwachstellen auf Anwendungsebene untersucht, d.h. es wird eine Dokumentenprüfung durchgeführt.

ITSEC behandelt dabei die Aspekte **Wirksamkeit** und **Korrektheit**. Hinsichtlich der Wirksamkeit begründet der Hersteller mittels der von ihm zu dem Punkten **Konstruktion** (*Eignung/Zusammenwirken der Funktionalität, Stärke der Mechanismen* sowie *Konstruktionsschwachstellen)* und **Betrieb** (*Benutzerfreundlichkeit* und *operationelle Schwachstellen)* getätigten Aussagen, daß sein Produkt bzw. System mit den realisierten sicherheitsspezifischen Funktionen und Mechanismen auch wirklich die vorgegebenen Sicherheitsziele erreichen kann. Zusätzlich wird bei der Bewertung der Wirksamkeit die Fähigkeit der Sicherheitsmechanismen des Prüfobjektes, Widerstand gegen einen direkten Angriff zu leisten, bewertet (Aspekt **Stärke des Mechanismus**). Für die Stärke der Mechanismen des Prüfobjektes sind drei Stufen definiert - niedrig, mittel, hoch -, die ein Maß für das Vertrauen sind, inwieweit die Sicherheitsmechanismen des Prüfobjektes in der Lage sind, direkten Angriffen zu widerstehen.

Bei der Bewertung der Korrektheit wird untersucht, ob die sicherheitsspezifischen Funktionen und Mechanismen bis hinunter auf den Quellcode korrekt implementiert sind. Hinsichtlich der **Korrektheit** existieren die Punkte **Konstruktion/Entwicklungsprozeß** (*Sicherheitsvorgaben, Architekturentwurf, Feinentwurf* sowie *Implementierung)*, **Konstruktion/Entwicklungsumgebung** (*Konfigurationskontrolle, Programmiersprachen/ Compiler und Sicherheit beim Entwickler)*, **Betrieb/ Betriebsdokumentation** (*Benutzerdokumentation und Systemverwalterdokumentation)* und **Betrieb/Betriebsumgebung** (*Auslieferung/Konfiguration* und *Anlauf/Betrieb)*.

Hierzu wurden sieben Evaluationsstufen (E0 bis E6) definiert, welche verschiedene Schärfen des Vertrauens in die sicherheitstechnische Korrektheit darstellen. E0 bedeutet unzureichendes Vertrauen. E1 steht für einen Einstiegspunkt, unterhalb dessen kein sinnvolles Vertrauen aufrechtzuerhalten ist und E6 steht für die höchste Stufe des Vertrauens. Mit den sechs erfolgreichen Evaluationsstufen E1 bis E6 wird ein weiter Bereich von möglichem Vertrauen erfaßt. Nicht alle diese Stufen werden notwendigerweise für alle Arten von Chipkartensystemen, welche eine unabhängige Evaluation der technischen Sicherheitsmaßnahmen erfordern, benötigt oder sind für sie geeignet. Die Evaluationsstufen werden im Zusammenhang mit den Korrektheitskriterien definiert. Die Anforderungen an die Wirksamkeit (einschließlich der Stärke der Mechanismen) ändern sich zwischen den Stufen nicht. Bei der Bewertung der Wirksamkeit wird auf der Bewertung der Korrektheit aufgebaut und es werden die vom Hersteller zur Verfügung gestellten Dokumente verwendet.

Damit eine Chipkarten-Evaluation wirkungsvoll und mit minimalen Kosten durchgeführt werden kann, muß der Prüfer eng mit dem Entwickler zusammenarbeiten. Ideal ist eine Zusammenarbeit von Anbeginn der Entwicklung an, um ein gutes Verständnis für die Sicherheitsvorgaben zu entwickeln und um auf die Auswirkungen bestimmter Entscheidungen für die Evaluation hinweisen zu können. Der Prüfer muß aber unabhängig bleiben und darf nicht vorschlagen, wie ein Prüfobjekt entworfen oder implementiert werden soll. Dies ist vergleichbar mit der Rolle eines externen Finanzrevisors, der eine gute Arbeitsbeziehung mit der Finanzabteilung aufbauen muß und der in vielen Fällen von deren internen Aufzeichnungen und Kontrollen nach ihrer Prüfung Gebrauch macht. Trotzdem muß auch er unabhängig und kritisch bleiben.

Die Forderungen nach Sicherheitstests und Analysen in den Kriterien verdienen besondere Erwähnung; in allen Fällen liegt die Verantwortung für Test und Analyse beim Hersteller. Für alle Evaluationsstufen, mit Ausnahme von E1, wird der Prüfer vorrangig die Test- und Analyseergebnisse überprüfen, die der Hersteller zur Verfügung gestellt hat. Der Prüfer muß eigene Test- und Analysearbeiten nur durchführen, um die gelieferten Ergebnisse zu überprüfen, die gelieferten Nachweise zu ergänzen und um Schwachstellen zu untersuchen. Bei der Evaluierungsstufe E1 ist es freigestellt, ob Testergebnisse zur Verfügung gestellt werden oder nicht. Ist dies nicht der Fall, so muß der Prüfer zusätzlich Funktionalitätstests gegen die Sicherheitsvorgaben durchführen.

Aufgrund ihrer Beschaffenheit, auf Basis von Dokumenten zu prüfen, eignen sich die ITSEC-Prüfkriterien in der Praxis jedoch oftmals nur bedingt zur Prüfung von (Chipkarten-) Sourcecode. Zum einen stellen die ITSEC-Kriterien ausschließlich generische Anforderungen an die Vorgehensweise und Ziele einer Sourcecodeprüfung, zum anderen ist das Verhältnis von aufzubringendem Prüfaufwand für die logische Korrektheit und Wirksamkeit auf Anwenderebene (Bedrohungen, Sicherheitsziele etc.) zur eigentlichen Codeinspektion von Mikroprozessorchipkarten unausgewogen. Hinzu kommt, daß sich heutige Chipkartensystem praktisch immer an zuverlässigen Normen (z.B. ISO7816, prEN1546 etc.) orientieren, so daß eine tiefgehende Prüfung der Struktur der Applikationsschicht nicht erforderlich ist. Viel bedeutender ist die Analyse der Realisierung des Systems, also des Sourcecodes, wodurch z.B. technische Bugs (Programmierfehler, Trapdoors, Endlosschleifen etc.), aber auch prinzipielle Strukturfehler gefunden werden können. Im folgenden werden einige Beispiele genannt, die den generischen Charakter der ITSEC hinsichtlich des Aspektes Sourcecodeprüfung wiederspiegeln:

- **ITSEC E3.11**: „...muß die Übereinstimmung zwischen Quellcode ... und den Basiskomponenten des Feinentwurfs beschreiben".

- **ITSEC E3.18**: „Alle Programmiersprachen, die für die Implementierung benutzt werden, müssen klar definiert sein..".

- **ITSEC E3.19**: „Die Definition der jeweiligen Programmiersprache muß die Bedeutung aller Anweisungen, die im Quellcode benutzt werden, eindeutig festlegen...".

- **ITSEC E4.11**: „Die Zuordnungsbeschreibung muß die Übereinstimmung zwischen Quellcode ... und den Basiskomponenten des Feinentwurfs beschreiben".

- **ITSEC E5.11**: „Der Quellcode und die Hardware-Konstruktionszeichnungen müssen vollständig in kleine, verständliche und getrennte Teile aufgegliedert sein".

- **ITSEC E5.14**: „Der Hersteller muß die folgende Dokumentation bereitstellen: ...Quellcode aller benutzten Laufzeitbibliotheken"

- **ITSEC E6.1**: „Der Hersteller muß die folgende Dokumentation bereitstellen: ... (die) Werkzeuge, welche dazu verwendet werden können, Inkonsistenzen zwischen Quellcode und ausführbarem Code zu entdecken, sofern es sicherheitsspezifische oder sicherheitsrelevante Quellcodekomponenten gibt (z.B. Disassembler und/oder Debugger).

- **ITSEC E6.11**: „Die Zuordnungsbeschreibung muß die Übereinstimmung zwischen den Sicherheitsmechanismen, wie im Quellcode ... dargestellt, und der formalen Spezifikation der sicherheitsspezifischen Funktionen in den Sicherheitsvorgaben erklären".

2.2 Konformität zum Signaturgesetz

Soll mittels einer Chipkarte eine Rechtsverbindlichkeit für digitale Signaturen hergestellt werden, dann regelt das Signaturgesetz (SigG), welches als Artikel 3 des Informations- und Kommunikationsdienste Gesetz (IuKDG) seit dem 01.08.97 in Kraft ist, die Rahmenbedingungen, unter denen eine digitale Signatur als vertrauenswürdig betrachtet werden kann. Für den Aspekt Chipkartenprüfung bedeutet dies, daß solche technischen Komponenten einer hinreichenden Prüfung mit einer Bestätigung der Erfüllung der Anforderungen zu unterziehen, die für

(1) die Erzeugung, Speicherung und Anwendung von Signaturschlüsseln

(2) die Erzeugung und Prüfung von digitaler Signaturen

(3) die Darstellung zu signierender Daten

(4) die Überprüfung signierter Daten

(5) die Nachprüfbarkeit und Abrufbarkeit von Signaturschlüssel-Zertifikaten

verwendet werden.

Als Prüfgrundlage **muß** die ITSEC verwendet werden. Demzufolge sind für die Komponenten eines professionellen Signatursystems auf Chipkartenbasis, sofern eine Konformität zum Signaturgesetz angestrebt wird, Evaluationen gemäß ITSEC, Stufe E4, erforderlich. Die Bestätigung der Erfüllung der Anforderungen nach SigG/SigV hat durch anerkannte Stellen zu erfolgen, welche i.w. mit den zugelassenen Zertifizierungsstellen (z.Z. Bundesamt für Sicherheit in der Informationstechnik (BSI), TÜViT GmbH und IT Security Services GmbH identisch sein werden.

3 ZKA-Sicherheitsuntersuchung

Die Prüfung der Sicherheitsstruktur von Chipkartenspezifikationen sowie der Implementierung des Sourcecodes (Chipkartenbetriebssystem und Applikationen in beliebigen Programmiersprachen wie z.B. Assembler, C, Java etc.) erfolgt in Anlehnung an die 13 deutschen ZKA-Kriterien (siehe unten). Sie basieren auf den Sicherheitsanforderungen für das Electronic Cash - System in Deutschland. In der Praxis hat sich bei TÜViT als Prüftiefe bzw. -schärfe ein De-Facto-Standard bewährt, welcher sich auf Basis o.g. europäisch harmonisierter Prüfmethodik ITSEC, **Level E3 hoch**, etablieren konnte.

Zur Validierung in Anlehnung an die ZKA-Kriterien werden die Chipkarten-Sicherheitsuntersuchungen wie folgt durchgeführt:

* **Prüfung der bereitgestellten Dokumente**
  Die umgesetzte Sicherheits-Architektur wird hinsichtlich der Aspekte Funktionalität und Wirksamkeit analysiert. Da heutige Chipkartensysteme i.d.R. auf der Norm ISO7816 respektive prEN1546 beruhen, können diese als funktionaler Standard für die Untersuchung der herstellerspezifischen Chipkartenspezifikation herangezogen werden. Wie bereits erwähnt, ist eine explizite ITSEC Untersuchung z.B. hinsichtlich der Aspekte Bedrohungen, abwehrenden Sicherheitsmechanismen und Sicherheitszielen hier (außer wenn eine Evaluierung **mit anschließender Zertifizierung** beabsichtigt wird, z.B. zur Erfüllung des Signaturgesetzes mit **ITSEC E4 HOCH**) nicht erforderlich.

- **Prüfung der Software**
  Zur Beurteilung der Korrektheit und Wirksamkeit der implementierten Routinen des Betriebsystems und der dazugehörigen Applikationen sowie zur Validierung auf Konformität zu bestehenden Normen, Standards und Spezifikationen, insbesondere der ZKA-Spezifikation, wird im Rahmen der Sicherheitsbegutachtung der Quellcode untersucht, wie im Kapitel „Sourcecodeprüfung mittels Walkthrough-Methode" beschrieben.

Im einzelnen wird anhand der vorgelegten Spezifikationsdokumente untersucht, ob das zu prüfende Chipkartensystem zu den im folgenden aufgeführten **13 ZKA-Kriterien** konform ist:

(1) **Komponenten-Authentikation:**
Die komponentenweise Authentikation muß durch Aktionen wie 'interne Authentisierung', 'externe Authentisierung' bzw. 'gegenseitige Authentisierung' erfolgen. Diese müssen im Quellcode korrekt und wirksam mittels eines 'State-of-the-Art' - Algorithmus implementiert sein. Dies könnte z.B. ein Tripple-DES- bzw. RSA-Verfahren mit entsprechender Schlüssellänge sein. Der Standard DES mit 56 Bit Schlüssellänge ist nicht mehr 'State-of-the-Art'!

(2) **Nachrichtenintegrität:**
Alle sicherheitsrelevanten Informationen in den Nachrichten sind vor und nach der Übertragung in den Komponenten des Chipkartensystems mit geeigneten Verfahren gegen unberechtigte Veränderung zu schützen, wobei sowohl Veränderungen an sicherheitsrelevanten Informationen während der Übertragung zwischen den beteiligten Komponenten, als auch das unautorisierte Einspielen von Nachrichten erkannt werden müssen. Die Nachrichtenintegrität kann von der Chipkarte durch Verwendung von Prüfsummen über die übertragenen Daten, ausgetauschten Zufallszahlen und durch Message Authentication Codes gewährleistet werden. Entsprechende Reaktionen auf Integritätsverletzungen bzw. auf wiedereingespielte Nachrichten müssen erfolgen. Reaktionen auf Integritätsverletzungen bzw. wiedereingespielte Nachrichten während der Übertragung sind Fehlermeldungen und der Abbruch des aktuellen Kommandos.

(3) **Benutzer-Authentikation:**
Wenn das chipkartengestützte Zahlungssystem die Authentikation des Benutzers durch Verwendung seiner PIN erfordert, muß sichergestellt sein, daß bestimmte Funktionen nur dann ausgeführt werden können, wenn die richtige PIN bekannt ist. Die interne Sicherheit muß auf der korrekten Implementierung von Access Conditions basieren.

(4) **Geheimhaltung von PINs und kryptographischen Schlüsseln:**
Die PIN darf außerhalb von gesicherten Bereichen nie im Klartext übertragen werden. Wird sie in Systemkomponenten bearbeitet oder gespeichert, so muß sie gegen unautorisiertes Auslesen und Verändern geschützt sein. Kryptographische Schlüssel dürfen auf elektronischen Übertragungswegen nie im Klartext übertragen werden. Werden sie in Systemkomponenten benutzt oder gespeichert, so müssen sie gegen unautorisiertes Auslesen und Verändern geschützt sein. Keine Systemkomponente darf eine Möglichkeit zur Bestimmung einer PIN aufgrund einer erschöpfenden Suche bieten.

(5) **Protokollierung:**
Alle sicherheitsrelevanten Unstimmigkeiten müssen in den beteiligten Systemkomponenten protokolliert und die Protokolle gegen unbefugte Manipulationen geschützt werden. Dabei müssen die auf der Chipkarte gespeicherten Protokolldaten authentisch und unverfälschbar bleiben und sich an eine auswertende Instanz übertragen lassen.

(6) Schlüsselmanagement:
Zur Verteilung, Verwaltung und eventuell zum turnusmäßigen Wechsel und Ersetzen von Schlüsseln sind Regelungen zu treffen. Schlüssel, für die mindestens der Verdacht auf Kompromittierung besteht, dürfen im gesamten chipkartengestützten Zahlungssystem nicht mehr verwendet werden. Für symmetrische Verschlüsselungsverfahren sind zwischen allen Komponenten des chipkartengestützten Zahlungssystems Verfahren zur Dynamisierung von MAC-Bildung und Verschlüsselung, insbesondere PIN-Verschlüsselung, zu verwenden. Für die Verschlüsselung der PIN und zur MAC-Bildung sind unterschiedliche Schlüssel zu verwenden.

(7) Forderungen an die Hardware:
Alle Verschlüsselungen, Entschlüsselungen, Umschlüsselungen, die MAC-Bildung und kryptographische Prüfungsprozeduren werden in gegen unberechtigte Zugriffe besonders geschützten Komponenten (Sicherheitsmodulen) durchgeführt. Die zugehörigen Schlüssel sind ebenfalls in solchen Sicherheitsmodulen abgelegt. Der Schutz von Daten und Programmen gegen Veränderung bzw. Auslesen in Sicherheitsmodulen muß so hoch sein, daß während der Lebensdauer des Moduls Angriffe mit vertretbarem Aufwand nicht möglich sind, wobei der für einen erfolgreichen Angriff nötige Aufwand gegen den hieraus zu ziehenden Nutzen abzuwägen ist. In Sicherheitsmodulen müssen sicherheitsrelevante Daten und Abläufe (z. B. Schlüssel, Programme) gegen unberechtigte Veränderung und geheime Daten (z. B. Schlüssel, PINs) gegen unberechtigtes Auslesen geschützt sein. Dies muß durch folgende Maßnahmen gewährleistet werden: Bauart des Sicherheitsmoduls, eventuell im Zusammenwirken mit Sicherheitsmechanismen der Software des Sicherheitsmoduls, Laden von Programmen in Sicherheitsmodule nur bei der Herstellung oder kryptographische Absicherung des Ladevorgangs, kryptographische Absicherung des Ladens von sicherheitsrelevanten Daten, insbesondere von kryptographischen Schlüsseln. Auch vor dem Auslesen mittels Angriffen, die Zerstörung des Moduls in Kauf nehmen, müssen geheime Daten in Sicherheitsmodulen geschützt sein. Die Chipkarten des jeweiligen Zahlungssystems müssen den Anforderungen dieses Kriteriums genügen. Insbesondere muß durch Chip-Hardware und -Software sichergestellt werden, daß die Chipkarte nur über die spezifizierte Schnittstelle in der festgelegten Weise kommunizieren kann.

(8) Organisatorische Maßnahmen bei der Herstellung und Personalisierung:
Dieser Punkt betrifft die organisatorische Sicherheit von Chipkartensystemen und gehört daher nicht in den Rahmen der Sourcecodeprüfung. Der Hersteller garantiert, ausschließlich den zur Prüfung vorgelegten Code im Wirkbetrieb unter Anwendung der beschriebenen Parametrisierung einzusetzen.

(9) Ablaufsicherung:
Durch die Sourcecodeprüfung wird sichergestellt, daß die Abläufe und das Handling der Daten innerhalb des Programms (Zustandsautomat) nicht manipuliert werden können.

(10) Verarbeitung anderer Anwendungen:
Es muß sichergestellt werden, daß die implementierten Anwendungen nicht miteinander kollidieren. Jeder andere Systemablauf wird hier als abgesichert vorausgesetzt.

(11) Verschlüsselungsverfahren:
Es dürfen nur Verschlüsselungsverfahren verwendet werden, die einer Kryptoanalyse mit ausgewähltem Klartext widerstehen, also dem aktuellen Stand der Technik entsprechen. Die Sicherheit darf nicht auf der Geheimhaltung der Verfahren beruhen, sie muß

durch die Geheimhaltung der Schlüssel gewährleistet sein. Es muß sich um ein Verfahren handeln, welches dem aktuellen Stand der Technik entspricht (siehe [1]). Erwartet wird zudem das Vorlegen der Burries Cases vom Hersteller zum Nachweis der korrekten und wirksamen Implementierung des DES-Algorithmus, soweit dieser im Einsatz ist. Das ZKA favorisiert ausschließlich den symmetrischen DES-Algorithmus; jedoch auch andere Verfahren haben ihre Berechtigung.

(12) Eindeutige Repräsentation:
Jede Sicherheitskomponente eines Chipkartensystems muß im Gesamtsystem eindeutig identifiziert sein, z.B. durch die individuelle Seriennummer von Chipkarten.

(13) Personelle Forderung:
siehe Punkt [8].

Hierauf basierend findet eine Untersuchung der folgenden Programmkomponenten auf Spezifikations- und Realisierungsebene statt:

- sämtliche Betriebssystem- und Applikationskommandos.

- Ver- und Entschlüsselung.

- Schlüsselableitungen, Signaturerzeugungen und Signaturprüfungen.

- Zufallsgeneratoren.

- Kommunikationsmechanismen.

- Speicher- und Ressourcenverwaltung.

- Schlüsselverwaltung und -sicherung bzw. Schlüsselsuche.

- Zugriffskontrollmechanismen (Access Conditions).

- Attribute und Completion Codes.

- Authentisierungmechanismen und Checksummenbildungen.

- Dokumente über die eingesetzte Hardware.

4 Sourcecodeprüfung mittels Walkthrough-Methode

Wie prüft man 10.000 Zeilen Assembler-Sourcecode hinsichtlich des Aspektes Security? Wenn man es nicht macht, ist die Fehlerwahrscheinlichkeit relativ hoch. Prüft man stichprobenhaft, dann ist die Wahrscheinlichkeit, viele oder gar alle der in der Regel relativ niedrigwahrscheinlichen Security-Schwachstellen aufzudecken, nahezu Null. Mit herkömmlichen Verfahren, die auf einer nichtstrukturierten Zufallsprüfung beruhen, kann Chipkartensoftware mit der heutigen Komplexität nicht hinreichend geprüft werden.

Die Sicherheitsbegutachtung (White-Box-Prüfung) wird in Form eines Walkthrough-Ansatzes durchgeführt, wobei die Analyse der erkannten fundamentalen Sicherheitsprozeduren anhand des Quellcodes vollständig im Einzelschrittverfahren (der Code wird Zeile für Zeile geprüft) erfolgt. Dabei findet eine sogenannte *Prüfung auf vollständige Abdeckung* statt, bei der versucht wird, jede Anweisung, Verzweigung bzw. jeden Pfad mindestens einmal zu durchlaufen. Damit wird sichergestellt, daß alle Programmzeilen und Datenteile während der Prüfung abgedeckt werden. Werkzeuge zur Prüfung auf vollständige Abdeckung zeigen an, welche Anweisungen und Verzweigungen ausgeführt werden, d.h. ob der Code vollständig abgedeckt wird. Die *Pfadabdeckung* untersucht die zahlreichen Einmalpfade, denen ein Programm bei seiner Ausführung folgt. Wegen der enormen Zahl der selbst bei einem kleinen Programm möglichen Kombinationen ist eine komplette Pfadabdeckung nicht praktisch. Jedoch können kritische Pfade wie z.B. häufig benutzte Sequenzen mit sicherheitskritischen Operationen (z.B. Kryptoroutinen) mittels entsprechender Case-Tools hinsichtlich ihres dynamischen Verhaltens geprüft werden. Die *intramodulare* und die *intermodulare* Prüfung auf Pfadabdeckung liefern zwei verschiedene Aspekte von Flußinformation. Die intramodulare Prüfung zeigt die Pfade innerhalb eines Moduls, während die intermodulare Prüfung die Aufrufbeziehungen zwischen Modulen analysiert und sicherstellt, daß alle Verknüpfungen ausgeführt werden und sich der Code entwurfsgerecht verhält.

Es müssen die Sicherheitsanforderungen in allen Spezifikations- und Realisierungsebenen zwischen der vom Hersteller vorzulegenden Feinspezifikation und dem Programmcode eingehalten werden:

- Sämtliche sicherheitsrelevanten Abläufe müssen in der Software ohne Einschränkungen oder Möglichkeiten zur Umgehung umgesetzt sein; Abläufe dürfen nicht vorgetäuscht werden können.

- Weiterhin ist darauf zu achten, daß keine sicherheitssensitiven Daten kompromittiert werden können. Daten, die als sicherheitsrelevant einzustufen sind, müssen in einer sicheren Hardwareumgebung gespeichert und verarbeitet werden. Die Abbildung der Sicherheitskriterien auf die einzelnen Funktionen des Chipkarten-Betriebssystems muß anhand der vorgelegten Dokumentation und Listings leicht nachzuvollziehen sein.

- Alle Algorithmen (z.B. Kryptoroutinen) müssen korrekt und wirksam realisiert worden sein. Zur Unterstützung kann ein Black-Box-Prüfung angewendet werden, z.B. die sog. Burries Cases zur Untersuchung der Wirksamkeit implementierter DES-Algorithmen.

- Auf korrekten Steuerfluß und Schleifenende wird besonders genau geachtet. Der Codeablauf wird auf Programm- und Prozedurebene besonders genau verfolgt, z.B. ob stets die richtigen Pfade eingeschlagen werden. Ein üblicher Programmierfehler am Schleifenende entsteht dadurch, daß eine Wiederholung zu viel ausgeführt wird.

- Mittels der Datenabdeckung wird sichergestellt, daß Variablen und Datenstrukturen ordnungsgemäß initialisiert wurden; die gemeinsame Verwaltung von Speicherzellen bzw. Variablen durch mehrere diverse Routinen ist hierbei besonders kritisch zu betrachten.

- Das Programm sollte entsprechend strukturiert gestaltet sein, z.B. durch eine entsprechende Kommentierung des Sourcecodes.

- Zusätzlich findet eine Untersuchung auf unerlaubte Funktionen, Trapdoors bzw. Sicherheitslücken statt. Es darf z.B. unter keinen Umständen möglich sein, daß ehemalige, im Code verbliebene Testroutinen bzw. Programmwrackteile angesprungen werden könnten.

- Sämtliche Assembler- und Compileroptionen müssen klar erkennbar so parametriert sein, daß das Programm ausschließlich die ihm zugedachte Aufgabe erfüllen kann.

- Parameter müssen spezifikationskonform übergeben bzw. als Completion Codes zurückgegeben werden.

- Ein weiteres Untersuchungspotential bieten der Stack (Tiefe ausreichend, Stackpointer o.k.?), die Verwendung von Schleifen (z.B. Endlosschleifen denkbar?) bzw. das Handling von selbstmodifizierendem Programmcode.

5 Einsatz von Tools

Der **Einsatz von Tools zum dynamischen Testen** des Sourcecodes (z.B. Untersuchung des Stack-Verhaltens) z.B. mittels CASE-Tools (Echtzeit Debug- und Prüfwerkzeug) erfolgt stets parallel zu o.g. Prüfverfahren.

Gegebenenfalls findet noch eine Auditierung der zugehörigen organisatorischen und technischen Maßnahmen bei der Erstellung der Software bzw. bei der Initialisierung, Personalisierung und Komplettierung statt. Hierzu hat die TÜViT einen eigenen Anforderungkatalog („TU4") erstellt.

Soll das System in Deutschland zum Wirkbetrieb zugelassen werden, so ist die Anfertigung eines ZKA-Sicherheitsgutachtens seitens der deutschen Kreditwirtschaft zwingend erforderlich. Anschließend wird das System vom ZKA abgenommen.

6 Referenzbeispiele

Basierend auf o.g. Prüfschemata wurden unter anderem folgende, am Markt befindliche Chipkartensysteme von TÜViT geprüft:

- **STARCOS Chipkartenbetriebssystem S1.1** der Giesecke & Devrient GmbH, München, sowie die darunter laufenden Applikationen „Terminalkarte" und „Kundenkarte" für die elektronische Geldbörse „Quick" in Österreich (ZKA-konforme Sicherheitsuntersuchung).

- **TCOS Chipkartenbetriebssystem V2.0** der Deutschen Telekom AG, Produktzentrum Telesec, Siegen (ITSEC-Evaluation nach Stufe E4 hoch. Befindet sich derzeit Zertifizierungsverfahren durch das Bundesamt für Sicherheit in der Informationstechnik (BSI)).

Literatur

[ITSEC91]   Kriterien für die Bewertung der Sicherheit von Systemen der Informationstechnik (ITSEC), Version 1.2, Juni 1991, ISBN 92-826-3003-X (Bezugsquelle: Bundesanzeiger-Verlag Köln)

[ITSEM93]   Information Technology Security Evaluation Manual (ITSEM), Version 1.0, 10 September 1993, ISBN 92-826-7087-2.

[Rother96]   Rother, Stefan: Aspekte der Sourcecodeprüfung. Vortragsfolien der TÜViT GmbH, Essen, 1996.

[Rother197]   Rother, Stefan: Bedeutung der Informationssicherheit und der Grundsatz der Verhältnismäßigkeit der Kosten. In: Fachkonferenz FinDV97: COMPUTAS Gisaela Geuhs, Köln, 1997.

[Rother297]   Rother, Stefan: Überprüfung, Evaluierung und Akkreditierung von Informationssicherheit durch unabhängige Prüfinstitute. In: Fachkonferenz SIUK97: COMPUTAS Gisaela Geuhs, Köln, 1997.

[ZKA195]   Schnittstellenspezifikation für die ec-Karte mit Chip: Datenstrukturen und Kommandos, Version 2.1, Bank Verlag GmbH, Köln, 17. Mai 1995

[ZKA295]   Schnittstellenspezifikation für die ec-Karte mit Chip: Key-Management, Version 2.1.2, Bank Verlag GmbH, Köln, 08. September 1995

[ZKA395]   Schnittstellenspezifikation für die ec-Karte mit Chip: Das electronic cash-System, Version 2.1, Bank Verlag GmbH, Köln, 16. Juni 1995

[ZKA495]   Schnittstellenspezifikation für die ec-Karte mit Chip: Die elektronische Geldbörse - Börsenkarte -, Version 2.1.1, Bank Verlag GmbH, Köln, 08. August 1995

[ZKA595]   Schnittstellenspezifikation für die ec-Karte mit Chip: Die elektronische Geldbörse - Händlerkarte -, Version 2.1, Bank Verlag GmbH, Köln, 31. Juli 1995

[ZKA695]   Sicherheitskriterien für Chipkarten, herausgegeben vom ZKA in der Version 2.2

[SIEM95]   SLE 44C80S, 8-Bit Security Controller, Siemens Data Sheet 11.95 Version 1.0

[Maier94]   Jürgen Maier-Wolf, 8051 Mikrocontroller erfolgreich anwenden, Franzis-Verlag, Poing 1994

[ISO781695] ISO/IEC 7816-4, Information technology - Identification cards - Integrated circuit(s) cards with contacts - Part 4: Interindustry commands for interchange, First edition 1995-09-01

Chipkartengestützte Sicherheitsfunktionalitäten im europäischen digitalen Mobilfunk

Stefan Pütz[1] · Roland Schmitz[2] · Friedrich Tönsing[2]

[1]T-Mobil
Postfach 300463, D-53184 Bonn
stefan.puetz@t-mobil.de

[2]Deutsche Telekom AG, Technologiezentrum
Am Kavalleriesand 3, D-64295 Darmstadt
{schmitz,toensing}@tzd.telekom.de

Zusammenfassung

Die Authentikation des Nutzers spielt unter den Sicherheitsdiensten im digitalen Mobilfunk eine zentrale Rolle. In neuerer Zeit hat jedoch auch die Authentikation des Netzbetreibers gegenüber dem Nutzer an Bedeutung gewonnen. Der vorliegende Beitrag stellt im Detail die Authentikationsmechanismen vor, wie sie zur Zeit im digitalen Mobilfunk nach den GSM- und DECT-Standards eingesetzt werden. Daneben werden auch die Mechanismen, die zur Zeit in der UMTS-Standardisierung diskutiert werden, behandelt. Insbesondere werden die betrachteten Schemata mittels eines Kriterienkataloges miteinander verglichen und ihre jeweiligen Vor- und Nachteile herausgearbeitet.

1 Einleitung

Die Mobilkommunikation bezweckt, dem Nutzer Zugang zu den Möglichkeiten eines globalen Netzwerks zu gewähren, unabhängig von dessen Standort oder Mobilität. Dieses Ziel kommt den steigenden Wünschen nach Unabhängigkeit und ständiger Erreichbarkeit entgegen, so daß neue Mobilfunknetze aufgebaut und deren Dienste durch Service Provider vermarktet werden. Dabei ist klar, daß die Luftschnittstelle im Mobilfunk einem Angreifer besonders gute Möglichkeiten des Abhörens bietet: Jeder, der einen geeigneten Empfänger besitzt, kann den Verkehr zwischen der Mobil- und der Basisstation abhören, und diese Art des Belauschens kann praktisch weder entdeckt noch verhindert werden. Aus diesem Grund wurde schon früh bei der Standardisierung von GSM (Global System for Mobile Communications), einem digitalen zellularen Mobilfunksystem der zweiten Generation, und DECT (Digital Enhanced Cordless Telephone), einem Standard für die digitale Schnurlostelekommunikation der zweiten Generation, Wert auf die Verschlüsselung der Daten gelegt, die auf der Luftschnittstelle ausgetauscht werden. Weiterhin muß sich im GSM- und DECT-Netz der Nutzer gegenüber dem Netz authentisieren, um so die Entgeltsicherheit für den Nutzer und für den Netzbetreiber zu gewährleisten. Dabei wird sowohl der GSM-Nutzer als auch der

DECT-Nutzer durch eine Chipkarte repräsentiert. Bei GSM bezeichnet man die Chipkarte mit Subscriber Identity Module (SIM), bei DECT mit DECT Authentication Module (DAM). Der Nutzer authentisiert sich gegenüber der Karte durch seine PIN.

Neu hingegen ist die Anforderung, daß auch das Netz sich gegenüber dem Teilnehmer authentisieren sollte. Diese Anforderung ist motiviert durch das zukünftige Aufkommen vieler neuer Netz- und Service-Provider, wie in [Aziz] ausgeführt wird: „Authenticating the base is also necessary, if one considers the situation of competitors located in the same industrial park. The base stations of one competitor should not be able to masquerade as those belonging to the other competitor." Gefragt ist also eine gegenseitige Authentikation von Nutzer und Netzbetreiber. Während eine solche gegenseitige Authentikation bei der schnurlosen Telefonie bereits vorgesehen ist, ist sie im eigentlichen Mobilfunkbereich erst in der dritten Generation, UMTS (Universal Mobile Telephone System, vgl. zum Übergang von GSM auf UMTS z.B. [Aghvami]) genannt, zu erwarten.

Der vorliegende Beitrag beschreibt die heute bei GSM und DECT eingesetzten Authentikationsmechanismen sowie die für den kommenden UMTS-Standard bei ETSI (European Telecommunications Standards Institution) diskutierten Authentikationsschemata. Unter diesen finden sich ein symmetrisch und drei asymmetrisch basierte Verfahren. Sollte eines der asymmetrischen Schemata ausgewählt werden, so würde zum ersten Mal überhaupt ein asymmetrisch basiertes Authentikationsprotokoll im Mobilfunk eingesetzt werden.

In einem abschließenden Abschnitt werden die Mechanismen anhand eines Kriterienkatalogs miteinander verglichen.

2 GSM

GSM ist der bei ETSI erarbeitete europäische Standard für ein digitales zellulares Mobilfunksystem der zweiten Generation, auf dem in Deutschland die Mobilfunknetze D1 der T-Mobil, D2 von Mannesmann Mobilfunk und das von E-Plus (hier genauer das GSM-Derivat DCS-1800) beruhen.

Die wesentlichen Elemente eines Mobilfunknetzes nach GSM, insbesondere unter Berücksichtigung der sicherheitsrelevanten Aufgaben, werden im folgenden kurz vorgestellt (nach [Mich]; siehe Abb. 2.1):

Im Vermittlungssystem (Switching Subsystem, SSS) sind fünf Hauptfunktionen vereinigt. Die Funkvermittlungsstelle (Mobile Switching Centre, MSC) ist eine digitale Vermittlungsstelle hoher Leistungsfähigkeit, die sowohl normale Vermittlungsaufgaben ausführt als auch das Netz verwaltet. Bei ihrer administrativen Arbeit greift sie auf verschiedene Datenbanken zu. Im Heimatregister (Home Location Register, HLR) werden die permanenten Teilnehmerdaten gespeichert. Es enthält auch Informationen über den gegenwärtigen Standort "seiner" Teilnehmer. Das Besucherregister (Visitor Location Register, VLR) dient dem Service für Teilnehmer, die sich vorübergehend im Bereich dieser Funkvermittlungsstelle aufhalten. Dort werden temporäre Daten des Teilnehmers, wie z.B. die temporären Identitäten des Teilnehmers, gespeichert und das Authentikationsergebnis bewertet. Aus der Berechtigungsstelle (Authentication Centre, AUC) kommen die Parameter für die Verschlüsselung und Authentikation. Dort werden die permanenten teilnehmerindividuellen Parameter wie die Benutzeridentität IMSI (International Mobile Subscriber Identity) und der zugehörige Authentikationsschlüssel Ki gespeichert, und dort liegen der Authentikationsalgorithmus A3 und der Schlüsselgenerierungsalgorithmus A8 vor. Das Geräteregister (Equipment Identity Register, EIR) schließlich enthält u.a. Informationen über gestohlene bzw. Fehlerhafte Geräte und kann so deren Gebrauch unterbinden. Das Vermittlungssystem bildet auch die Schnittstelle zu anderen Telekommunikationsnetzen wie z.B. das diensteintegrierende digitale Nachrichtennetz (Integrated Services Digital Network, ISDN) und das öffentliche Telefonnetz (Public Switched Telephone Network).

Das Basisstationssystem (Base Station Subsystem, BSS) umfaßt zwei funktionale Einheiten, den Base Station Controller (BSC) und die Funkfeststation (Base Transceiver Station, BTS). Eine Funkvermittlungsstelle verwaltet einen oder mehrere Base Station Controller, die ihrerseits die Funkfeststationen ansteuern. Der Verschlüsselungsalgorithmus A5 ist in dem BSS untergebracht.

Die Funkfeststation schließlich sorgt für die Funkverbindung zur Mobilstation (Mobile Station MS), dem Endgerät. Die Mobilstation setzt sich aus zwei Einheiten zusammen, dem mobilen Terminal und dem Zugangsgerät des Teilnehmers, der SIM. Die SIM ist eine Chipkarte oder ein Plug-in-Modul und ist das Gegenstück zum AUC auf der Mobilfunkseite. Sie enthält teilnehmerindividuelle Daten wie die Benutzeridentität und den Authentikationsschlüssel Ki. Ki wird u.a. als Input für den auf der SIM implementierten Authentikationsalgorithmus A3 und den auf der SIM implementierten Schlüsselgenerierungsalgorithmus A8 verwendet und verläßt niemals die SIM. Im mobilen Terminal ist der Verschlüsselungsalgorithmus A5 implementiert und die Endgeräteidentität IMEI gespeichert.

Zu erwähnen ist noch das Betriebs- und Unterstützungssystem (Operation and Support System, OSS), welches das gesamte GSM-Netz konfiguriert und kontrolliert - es bildet die Schnittstelle zum Betreiber.

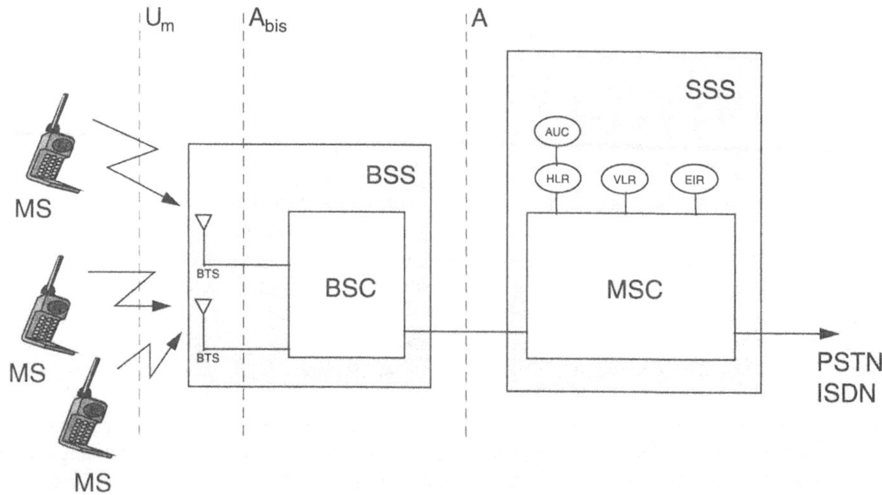

Abb. 2.1: Elemente eines Mobilfunknetzes nach GSM

2.1 Sicherheitsmechanismen nach GSM-Standard

Die für GSM-Systeme vorgesehenen Sicherheitsdienstleistungen sind

- Authentikation von Mobilfunkteilnehmern

- Geheimhaltung der Identität von Mobilfunkteilnehmern auf der Luftschnittstelle

- Geheimhaltung von Signalisierungs- und Nutzdaten auf der Luftschnittstelle

- Überprüfung der Mobiltelefonidentität.

Diese Sicherheitsdienstleistungen werden realisiert mit Hilfe der PIN-geschützten SIM (PIN für Personal Identity Number) und der Verwendung von Einwegfunktionen als Authentikationsalgorithmus und Schlüsselgenerierungsalgorithmus sowie eines Verschlüsselungsalgorithmus.

2.1.1 Authentikation des Mobilfunkteilnehmers

Die Authentikation des Mobilfunkteilnehmers gegenüber dem GSM-Netzbetreiber erfolgt mit dem Ziel der korrekten Rechnungslegung für einen authorisierten Teilnehmer und der Abweisung eines nichtauthorisierten Teilnehmers. Das bei GSM eingesetzte Authentikationsverfahren ist ein Challenge-Response-Verfahren. Die den Teilnehmer repräsentierende SIM authentisiert sich gegenüber dem VLR der Besuchsumgebung bzw. der Heimatumgebung. Das GSM-Authentikationsverfahren basiert auf der Verwendung eines symmetrischen Algorithmus A3, der in der SIM des Teilnehmers und dem AUC des zugehörigen Heimatnetzbetreibers implementiert ist. Der Algorithmus greift auf den teilnehmerindividuellen Schlüssel Ki, der in der SIM und dem AUC gespeichert ist, zu.

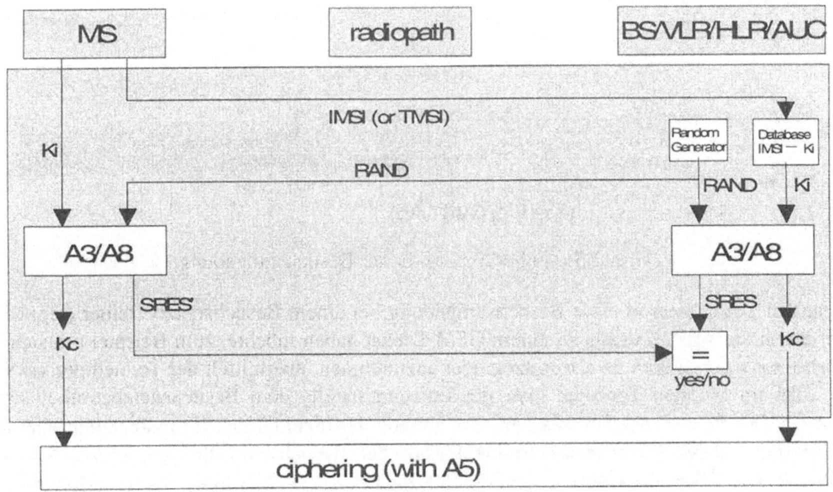

Abb. 2.2: Authentikation in der Heimatumgebung

Das Verfahren arbeitet wie folgt:

Wenn der Teilnehmer in seiner Heimatumgebung bei seinem Heimatnetzbetreiber (repräsentiert durch das HLR) Zugang zu einem GSM-Dienst haben möchte, zum Beispiel um sich zu registrieren oder Gespräche abzusetzen oder anzunehmen, übermittelt die SIM im mobilen Terminal über die Luftschnittstelle dem HLR die Identität des Teilnehmers (siehe auch 2.1.2). Das dem HLR des Heimatnetzbetreiber zugeordnete AUC ermittelt anhand der Identität den zugehörigen teilnehmerindividuellen Schlüssel Ki und produziert die Authentikationsdaten. Diese Daten bestehen aus dem Paar (RAND, SRES). Der Wert SRES wird mittels des Algorithmus A3 unter Einbeziehung des per Zufallszahlengenerator erzeugten Wertes RAND und des teilnehmerindividuellen Schlüssels Ki generiert. Der Heimatnetzbetreiber sendet RAND zur SIM und erwartet eine Antwort SRES'. Diese Antwort wird in der SIM mittels des Algorithmus A3 unter Einbeziehung des empfangenen RAND und des vorhandenen Ki berechnet und dem Netzbetreiber gesendet. Nach Empfang überprüft das VLR der Heimatumgebung, ob SRES und SRES' übereinstimmen. Falls nicht, ist die Authentikation fehlgeschlagen und dem Teilnehmer wird der Zugang verwehrt.

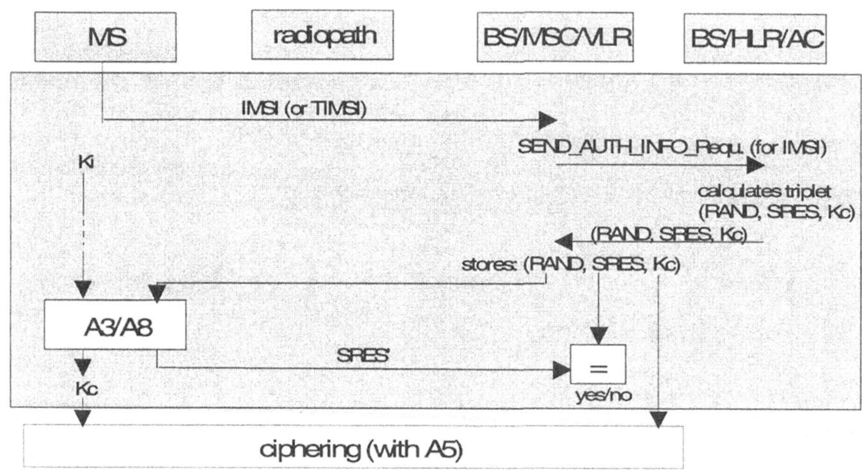

Abb. 2.3: Authentikation in der Besuchsumgebung

Wenn der Teilnehmer in einer Besuchsumgebung bei einem Besuchsnetzbetreiber (repräsentiert durch das VLR) Zugang zu einem GSM-Dienst haben möchte, zum Beispiel um sich zu registrieren oder Gespräche abzusetzen oder anzunehmen, übermittelt der Teilnehmer via seiner SIM im mobilen Terminal über die Luftschnittstelle dem Besuchsnetzbetreiber seine Identität. Der Besuchsnetzbetreiber erkennt anhand der Identität den Heimatnetzbetreiber des Teilnehmers. Beim Heimatnetzbetreiber fordert der Besuchsnetzbetreiber Authentikationsdaten an. Das HLR des Heimatnetzbetreibers instruiert das AUC, die Authentikationsdaten zu produzieren und sendet diese Authentikationsdaten dem Besuchsnetzbetreiber. Diese Daten bestehen aus einem oder mehreren Paaren (RAND, SRES), die wie im vorher beschriebenen Fall der Heimatumgebung berechnet werden. Der Besuchsnetzbetreiber sendet RAND zur SIM und erwartet eine Antwort SRES', die analog zum obigen Fall in der SIM generiert und dann übersendet wird. Nach Empfang überprüft das VLR, ob SRES und SRES' übereinstimmen. Falls nicht, ist die Authentikation fehlgeschlagen und dem Teilnehmer wird der Zugang verwehrt.

Typischerweise fordert der Besuchsnetzbetreiber mehr als ein Paar (RAND, SRES) an, so daß er nicht bei jedem Authentikationsvorgang diese Authentikationsdaten vom Heimatnetzbetreiber anfordern muß. Jedes Paar (RAND, SRES) wird nur einmal verwendet. Die Authentikationsdaten sollen nach Empfehlung der GSM-MoU vertraulich von der Heimat- in die Besuchsumgebung übermittelt werden.

Bei dem Verfahren wird der Authentikationsalgorithmus A3 eingesetzt. Einige Details zu A3:

- der A3-Algorithmus selber ist nicht standardisiert, nur:

Länge von Ki:	128 bits
Länge von RAND:	128 bits
Länge von SRES':	32 bits
Ablaufzeit:	weniger als 500 ms

- jeder GSM Heimatnetzbetreiber kann seinen eigenen Authentikationsalgorithmus verwenden, wodurch unterschiedliche Sicherheitsniveaus zu erreichen sind

- empfohlen wird die Verwendung einer guten Einwegfunktion ("es sollte praktisch unmöglich sein, Ki bei bekanntem RAND und SRES' zu berechnen")

- GSM-MoU hält einen Algorithmus A3 vor, der als MoU-Mitglied bei Unterzeichnung einer Vertraulichkeitserklärung angefordert werden kann.

2.1.2 Geheimhaltung der Identität von Mobilfunkteilnehmern

Die Sicherheitsdienstleistung "Geheimhaltung der Identität von Mobilfunkteilnehmern" sorgt dafür, daß keine Bewegungsprofile der Teilnehmer zu erstellen sind, da aus den Informationen auf der Luftschnittstelle i.allg. die Identität des Teilnehmers nicht erkannt werden kann. Dazu teilt der Netzbetreiber nach einer ersten Authentikation, zu der sich der Teilnehmer noch mit seiner permanenten Identität IMSI identifiziert hat, dem Teilnehmer nach Beendigung des Authentikationsvorganges verschlüsselt die temporäre Identität TMSI (Temporary Mobile Subscriber Identity) mit, mit der sich dann der Teilnehmer vor dem nächsten Authentikationsvorgang zu erkennen gibt. Jeder Authentikationsvorgang wird mit der verschlüsselten Zuteilung einer neuen temporären Identität TMSI abgeschlossen. Die Geheimhaltung der Teilnehmeridentität ist nicht Bestandteil der GSM-Authentikation.

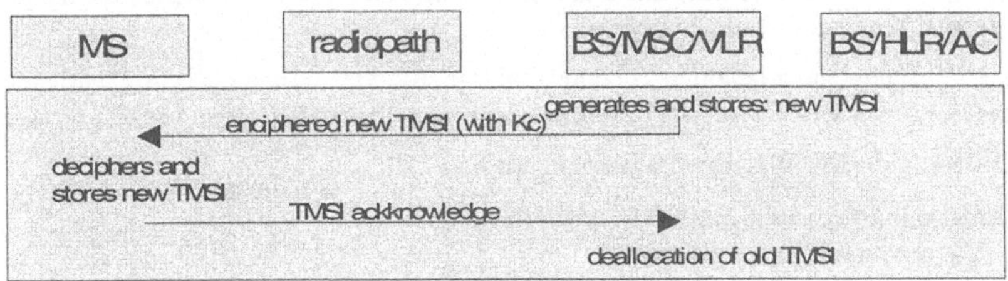

Abb. 2.4: TMSI (re)allocation

2.1.3 Geheimhaltung von Signalisierungs- und Nutzdaten auf der Luftschnittstelle

Da Abhörversuche auf der Luftschnittstelle als wesentlich wahrscheinlicher als Abhörversuche im ortsfesten Teil des GSM-Netzes angesehen werden, wird bei GSM die Sicherheitsdienstleistung "Geheimhaltung von Signalisierungs- und Nutzdaten auf der Luftschnittstelle" vorgesehen. Realisiert wird diese Sicherheitsdienstleistung durch Verschlüsselung auf der Luftschnittstelle. Dabei werden neben den Nutzdaten des Teilnehmers auch die Signalisierungsdaten verschlüsselt. Somit ist die erwähnte verschlüsselte Übertragung der TMSI möglich, denn die TMSI wird über den Signalisierungskanal übermittelt. Die Verschlüsselung erfolgt mit dem symmetrischen Verschlüsselungsalgorithmus A5.

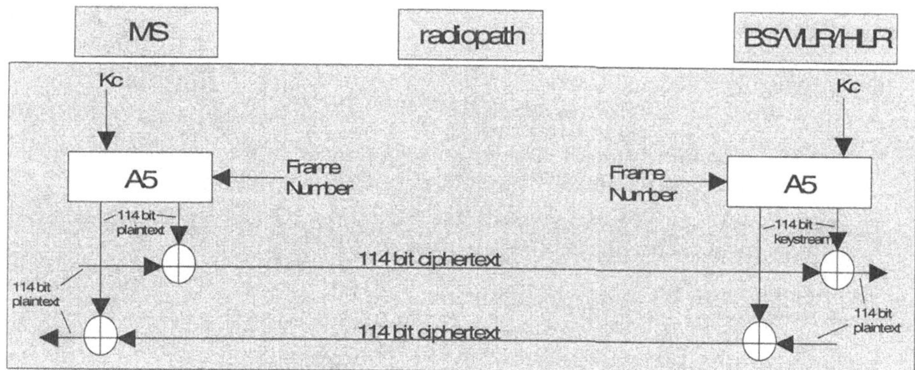

Abb. 2.5: Vertraulichkeit bei GSM

Einige Details zum A5:

- der A5-Algorithmus ist ein standardisierter Stromverschlüsseler und wird von der GSM-MoU vorgehalten

- der A5 ist sowohl in der BSS als auch im mobilen Terminal untergebracht

- Eingabewerte zum A5 sind der 64 bit lange beim Authentikationsverfahren erzeugte Sitzungsverschlüsselungsschlüssel Kc und die zur Synchronisation verwendete TDMA-Rahmennummer

- aus dem Output des A5 werden zwei Blöcke (Länge 114 bit) jeweils zur Verschlüsselung eines 114 bit - Klartextblocks und Entschlüsselung eines 114 bit Ciphertextblocks durch bitweises Addieren modulo 2 verwendet.

Der zur Verschlüsselung benötigte Sitzungsschlüssel Kc wird während des Authentikationsvorganges durch den Schlüsselgenerierungsalgorithmus A8 erzeugt (siehe Abb. 2.3)

Einige Details zum Schlüsselgenerierungsalgorithmus A8:

- der A8-Algorithmus selber ist nicht standardisiert, nur:
Länge von Ki:	128 bits
Länge von RAND:	128 bits
Länge von Kc:	64 bits

- eine Kombination von A3 und A8 ist möglich, was in der Praxis auch umgesetzt wird

- GSM-MoU hält einen Algorithmus A8 vor, der als MoU-Mitglied bei Unterzeichnung einer Vertraulichkeitserklärung angefordert werden kann.

2.1.3 Überprüfen der Mobilfunkidentität

Bei der Sicherheitsdienstleistung "Überprüfen der Mobilfunkidentität" wird die Endgeräteidentität IMEI vom mobilen Terminal zur Netzseite übertragen, wo die IMEI im EIR mit dort gespeicherten IMEIs von z.B. gestohlenen oder fehlerhaften Geräten auf Übereinstimmung verglichen werden kann, um dann geeignete Maßnahmen zu ergreifen. Bei dem Konzept wird ein zentrales EIR verwendet, an das die entsprechende IMEI zum Verteilen auf die EIRs sämtlicher Netzbetreiber übermittelt wird.

2.1.4 Indirekte Authentikation des Netzbetreibers

Neben der in Abschnitt 2.1.1 beschriebenen direkten Authentikation des Mobilfunkteilnehmers kann eine indirekte Authentikation erreicht werden, falls der bei dem Authentikationsvorgang erzeugte Sitzungsverschlüsselungsschlüssel Kc zur Verschlüsselung eingesetzt und alle Übertragungsdaten vertraulich übermittelt werden. Erfolgt die Übertragung hingegen im Klartext oder wird während einer Kommunikationsbeziehung in den Klartext-Modus gewechselt, so erlischt die Authentizität des Netzbetreibers bzw. der Übertragungsdaten sofort, denn von nun an setzt diese Kommunikation die gegenseitige Kenntnis des verwendeten Sitzungsverschlüsselungsschlüssels Kc, auf der die indirekte Authentikation beruht, nicht mehr voraus.

2.1.5 Bewertung der GSM-Authentikation

Als Vorteile des GSM-Ansatzes können angesehen werden:

- Der teilnehmerindividuelle Schlüssel Ki ist nur dem Heimatnetzbetreiber des Teilnehmers, genauer dem AUC des Heimatnetzbetreibers, und der SIM des Teilnehmers bekannt. Er muß und darf keiner anderen Partei zugänglich sein.

- Der Authentikationsalgorithmus A3 kann Heimatnetzbetreiber-spezifisch sein, bzw. spezifisch für den Heimatnetzbetreiber und für eine besondere Gruppe seiner Teilnehmer. Keine andere Partei muß ihn kennen und implementiert haben. Der Heimatnetzbetreiber kann den Algorithmus jederzeit für einen oder mehrere seiner Teilnehmer wechseln, wobei er den Teilnehmern neue SIMs zuteilen muß. Nur die Längen der Parameter RAND und SRES sind standardisiert.

- Der Mechanismus kann zur Ableitung von Verschlüsselungsschlüsseln Kc, die für die Vertraulichkeit von Teilnehmer- und Signalisierungsdaten auf der Luftschnittstelle eingesetzt werden, verwendet werden. Für jedes Gespräch wird dabei ein anderer Verschlüsselungsschlüssel abgeleitet. In der Praxis werden dem Besuchsnetzbetreiber auf seine Anforderung hin vom Heimatnetzbetreiber nicht nur die Paare (RAND, SRES) übermittelt, sondern Triplets (RAND, SRES, Kc).

Nachteile des GSM-Ansatzes:

- Der Besuchsnetzbetreiber muß dem Heimatnetzbetreiber den Zugangswunsch des Teilnehmers signalisieren, um die Triplets zu erhalten. Jedes Triplet ist nur einmal zu verwenden und teilnehmerspezifisch. Das bedingt einen hohen Signalisierungsverkehr.

- Beim GSM-Authentikationsprotokoll handelt es sich lediglich um eine einseitige Authentikation.

- Die GSM-Authentikation setzt einen vertrauenswürdigen Besuchsnetzbetreiber voraus. Insbesondere muß der Besuchsnetzbetreiber die Vertraulichkeit der übermittelten Triplets wahren.

3 DECT

Neben den Mobiltelefonen wurde in der Vergangenheit ein System für schnurlose Kommunikation im Nahbereich entwickelt. Dieses sogenannte DECT-System ermöglicht schnurlose Sprach- und Datenübermittlung bis zu einigen hundert Metern Entfernung von der Basisstation. Es findet daher vor allem im Hausbereich, sowie in Büros, Werkhallen oder auf Firmengeländen Anwendung. Allerdings wird ebenso geprüft, mit DECT die *Letzte Meile* zu überbrücken, also den drahtlosen Zugang (WLL, Wireless Local Loop) an bestehende Kommunikationsfestnetze zu erreichen, um hohe Investitionskosten drahtgestützter Anbindungen im Endbereich zu vermeiden [Arens].

3.1 Sicherheitsmechanismen nach DECT-Standard

Der europäische Standard für digitale Schnurlostelefonie, DECT, beschreibt eine Weiterentwicklung der bestehenden CT (Cordless Telephony)-Standards. Er wurde 1992 vom ETSI verabschiedet, das ebenfalls den GSM-Standard erarbeitet hat und definiert im wesentlichen nur die Luftschnittstelle. Im folgenden werden speziell die Sicherheitsdienste und -mechanismen erläutert, die DECT bietet [ETSI_3].

Im einzelnen stellt DECT die nachfolgend aufgelisteten Sicherheitsdienste bereit.

- Authentikation der Mobilstation

- Authentikation der Basisstation

- Gegenseitige Authentikation

- Teilnehmerauthentikation

- Vertraulichkeit der Übertragungsdaten

Um die Sicherheitsmechanismen zu beschreiben wird zunächst von einer Grundkonfiguration ausgegangen, die aus einem mobilen Endgerät (PP, Portable Part, bzw. PT, Portable Radio Termination) und einer Basisstation (FP, Fixed Part, bzw. FT, Fixed Radio Termination) besteht. Mehrere Basisstationen können zudem zu einem DECT-Netzwerk zusammengefaßt werden, so daß Roaming zwischen den einzelnen Basisstationen möglich ist. Optional werden Mobilstationen mit teilnehmerspezifischen Chipkarten (DAM, DECT Authentication Module) versehen, auf denen Daten gehalten und kryptographische Operationen ausgeführt werden (vgl. SIM bei GSM). Allerdings können diese Funktionen ebenfalls direkt im Endgerät integriert werden. Die Sicherheitsmechanismen von DECT basieren, wie auch bei GSM, auf dem Einsatz symmetrischer Kryptoverfahren und verwenden zur direkten Authentikation Challenge/Response Protokolle. Daher müssen Basisstation und Mobilteil über dieselben Sicherheitsalgorithmen und Authentikationsschlüssel (K, Authentication Key) verfügen.

3.2 Ableitung des Authentikationsschlüssels

Der DECT-Standard legt verschiedene Möglichkeiten dar, um den Authentikationsschlüssel K zu generieren. Abb. 3.1 zeigt dazu drei verschiedene Verfahren auf, wie K von weiteren Ursprungsinformationen abgeleitet werden kann. Die ersten beiden Fälle lassen die Authentikation der Mobil- bzw. Basisstation zu. Der dritte Fall erlaubt zusätzlich die Authentikation des Teilnehmers selbst. Eine eindeutige Teilnehmerkennung (IPUI, International Portable User Identity) ermöglicht dabei in ausgedehnten Netzstrukturen die Identifizierung eines Teilnehmers bzw. seine Zuordnung zu teilnehmerspezifischen Daten, wie z.B. die Zuordnung des Authentikationsschlüssels zu einem Teilnehmer, während dieser roamt.

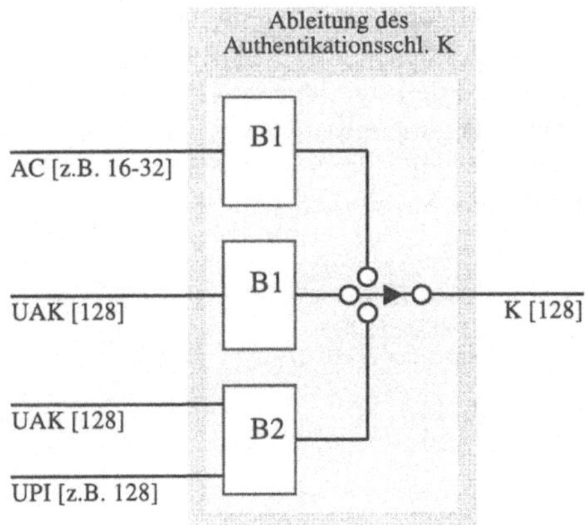

Abb. 3.1: Ableitung des Authentikationsschlüssels K bei DECT [ETSI_3]

- K wird durch den Algorithmus B1 von einem teilnehmerspezifischen Authentikationscode (AC, Authentication Code) abgeleitet, der wiederum über die IPUI systemweit mit dem Teilnehmer assoziiert ist. AC wird teilnehmerseitig entweder innerhalb des PT in einem nicht-flüchtigen Speicher abgelegt oder vom Teilnehmer selbst eingegeben.

- K wird durch den Algorithmus B1 von einem teilnehmerspezifischen Authentikationsschlüssel (UAK, User Authentication Key) abgeleitet, der teilnehmerseitig innerhalb des PT gespeichert und über die IPUI systemweit mit dem Teilnehmer assoziiert ist.

- K wird durch den Algorithmus B2 von einem teilnehmerspezifischen Authentikationsschlüssel (UAK, User Authentication Key) und einer Teilnehmerkennung (UPI, User Personal Identification) abgeleitet, die beide über die IPUI systemweit mit dem Teilnehmer assoziiert sind. UPI ist eine vom Teilnehmer manuell in das PT eingegebene Kennung. Diese explizite Eingabe der UPI erlaubt eine gezielte Teilnehmerauthentikation.

3.3 Authentikation und Vertraulichkeit

Der DECT-Standard unterscheidet zwischen direkter und indirekter Authentikation der jeweiligen Kommunikationspartner. Dabei basiert die direkte Authentikation darauf, in einem separaten Authentikationsprotokoll die Kenntnis des Authentikationsschlüssels K nachzuweisen. Indirekte Authentikation hingegen erfolgt durch die korrekte Ver- bzw. Entschlüsselung der Übertragungsdaten, da somit auf die Kenntnis des Sitzungsschlüssels (CK, Cipher Key) beim jeweiligen Kommunikationspartner geschlossen werden kann, der entweder fest vereinbart ist (SCK, Static Cipher Key) oder aber von K abgeleitet wird (DCK, Derived Cipher Key).

3.3.1 Direkte Authentikation

Zur Durchführung der direkten Authentizitätsprüfung einer Mobilstation sind im FT wie auch im PT die Sicherheitsalgorithmen A11 und A12 implementiert, für eine Authentizitätsprüfung der FT entsprechend die Algorithmen A21 und A22. Diese Algorithmen können sowohl durch DECT-Standardalgorithmen (DSAA, DECT Standard Authentication Algorithm) realisiert werden als auch durch proprietäre Algorithmen. Die Authentikation der Mobilstation verläuft wie folgt (siehe hierzu auch Abb. 3.2).

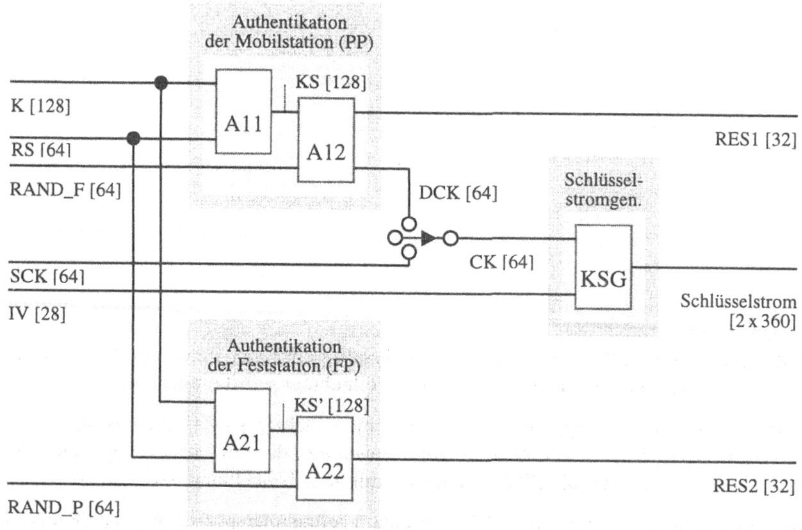

Abb. 3.2: Authentikation und Schlüsselvereinbarung bei DECT [ETSI_3]

Aus dem Authentikationsschlüssel K sowie einer durch die FT (bzw. das Heimatnetz) zufällig gewählten Zahl RS generiert A11 den Sitzungs-Authentikationsschlüssel (KS, Session Authentication Key). A12 berechnet nun aus KS und einer zufälligen Challenge (RAND_F, Random challenge issued by FT) eine Antwort (RES1, Response calculated by PT). RS und

RAND_F werden dem PT gesendet, woraufhin dort die äquivalenten Berechnungen ausgeführt werden. PT sendet RES1' an FT zurück. Stimmen RES1 und RES1' überein, hat sich PT erfolgreich authentisiert (vgl. Authentikation im GSM).

Die Authentikation des FT erfolgt analog mit K, RS, KS' und der durch das PT zufällig erzeugten Challenge RAND_P sowie den Sicherheitsalgorithmen A21 und A22. Als Antworten dienen hier RES2 bzw. RES2'. Der Hauptgrund für die Authentikation des FT ist der, daß bei DECT Applikationen vorgesehen sind, in denen das FT benutzt werden kann, um Daten im PT zu verändern. Die Authentikation des FT basiert im wesentlichen darauf, daß es dem PT gegenüber demonstrieren kann, vom Heimatnetzbetreiber, der den Authentikationsschlüssel K kennt, vertraut zu werden. Eine Gegenseitige Authentikation von PT und FT wird erreicht, indem die beiden Einzel-Authentikationen (Authentikation der Mobilstation und der Basisstation) kombiniert werden, wobei die Authentikation der Mobilstation zuerst erfolgt.

KS bzw. KS' heißen Session Authentication Key und entstehen als Zwischenprodukt bei der Ausführung der Sicherheitsalgorithmen A11/A12 bzw. A21/A22. Die Aufspaltung der Algorithmen ist eingeführt worden, um dem Teilnehmer ein Roaming in heterogenen DECT-Netzen zu erlauben, ohne entweder den Authentikationsschlüssel K oder aber vorausberechnete Einmal-Triplets (vgl. Ki bei GSM) bestehend aus RS, RAND_F, RES1 bzw. RS, RAND_P, RES2 an fremde Netzbetreiber (bzw. DECT-Basisstationen) durchreichen zu müssen. KS bzw. KS' werden durch den Heimatnetzbetreiber (bzw. die DECT-Heimatbasisstationen) generiert und zusammen mit RS an die DECT-Basisstationen bzw. Fremdnetze übertragen, in deren Abdeckungsbereich das entsprechende PT bzw. der Teilnehmer roamt. Somit können nun die FTs dezentral Authentikationen mit den entsprechenden PTs durchführen und umgekehrt. Im Heimatnetz werden teilnehmerbezogene Daten in einer Heimatdatei (HDB, Home Data Base), im besuchten Fremdnetz entsprechend in einer Besucherdatei (VDB, Visitors Data Base) gehalten. Der DECT-Standard beschreibt drei Varianten, das Roaming eines PTs bzw. eines Teilnehmers zu unterstützen, siehe hierzu Abb. 3.3.

- i) Roaming mit Authentikationsschlüssel K
  Auf Anforderung des besuchten Netzes übermittelt der Heimatnetzbetreiber diesem den teilnehmerspezifischen Authentikationsschlüssel K, woraufhin Authentikationen im besuchten Netz ebenso wie im Heimatnetz möglich sind.

- ii) Roaming mit Session Authentication Keys KS, KS'
  Auf Anforderung des besuchten Netzes übermittelt der Heimatnetzbetreiber diesem zwei Session Authentication Keys KS, KS' und die zugehörigen Werte RS, woraufhin Authentikationen im besuchten Netz möglich sind, indem zu den Challenges RAND_F, RAND_P mit den Algorithmen A12, A22 die Werte für RES1 (RES1') und RES2 (RES2') berechnet werden.

- iii) Roaming mit vorausberechneten Triplets
  Auf Anforderung des besuchten Netzes übermittelt der Heimatnetzbetreiber diesem vorausberechnete Triplets bestehend aus RS, RAND_F, RES1 (und optional DCK s.u.) bzw. RS, RAND_P, RES2. Je Trippelt können FT und PT eine Authentikation in der entsprechenden Richtung (Authentikation der Mobilstation bzw. der Basisstation) ausführen.

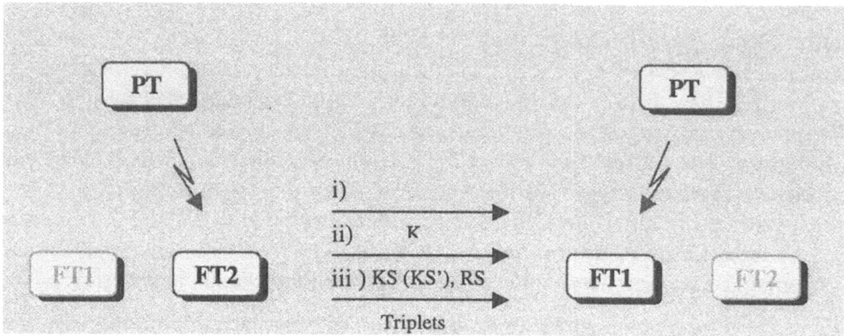

Abb 3.3: Roaming-Varianten bei DECT

Neben RES1 erzeugt der Sicherheitsalgorithmus A12 bei jeder Authentikation des PT einen dynamisch abgeleiteten Sitzungsschlüssel DCK, der zum Schutz der Vertraulichkeit für die Verschlüsselung auf der Luftschnittstelle eingesetzt werden kann. Anderenfalls kann ebenso ein statischer Sitzungsschlüssel SCK Verwendung finden oder aber auf die Verschlüsselung verzichtet werden. Zusammen mit einem Initialisierungsvektor (IV, Initial Vector), der von der TDMA-Rahmennummer abgeleitet wird, fungiert einer der beiden Schlüssel, DCK oder SCK, als Sitzungsschlüssel CK und dient damit als Input für den Schlüsselstromgenerator (KSG, Key Stream Generator). KSG liefert als Output einen Schlüsselstrom mit einer Länge von zweimal 360 bit, die in der jeweiligen Übertragungsrichtung (Uplink oder Downlink) XOR-verknüpft mit dem zu verschlüsselnden Klartext bzw. entschlüsselnden Chiffretext den Chiffretext bzw. Klartext ergeben. Als KSG kann sowohl der DECT-Standardalgorithmus (DSC, DECT Standard Cipher) verwendet werden als auch ein proprietärer Algorithmus.

3.3.2 Indirekte Authentikation

Neben der zuvor beschriebenen direkten Authentikation von PT und FT (durch Ausführen separater Challenge/Response Protokolle) kann eine indirekte Authentikation erreicht werden, falls ein Sitzungsschlüssel CK zur Verschlüsselung eingesetzt und alle Übertragungsdaten vertraulich übermittelt werden. Erfolgt die Übertragung hingegen im Klartext oder wird während einer Kommunikationsbeziehung in den Klartext-Modus gewechselt, so erlischt die Authentizität des Kommunikationspartners bzw. der Übertragungsdaten sofort, denn von nun an setzt diese Kommunikation die gegenseitige Kenntnis des verwendeten Sitzungsschlüssels CK, auf der die indirekte Authentikation beruht, nicht mehr voraus. Für die indirekte Authentikation ergeben sich zwei Fälle.

- Gegenseitige indirekte Authentikation
  Die gegenseitige indirekte Authentikation basiert auf der beiderseitigen Kenntnis des statischen Sitzungsschlüssels SCK. Der Nachweis über die Kenntnis von SCK beruht auf der gegenseitigen korrekten Ver- bzw. Entschlüsselung der ausgetauschten Daten.

- Kombination aus direkter und indirekter Authentikation
  Die Authentikation basiert teilnehmerseitig auf der Ausführung des Authentikationsprotokolls. Die Mobilstation authentisiert sich demnach direkt gegenüber der Basisstation. Allerdings authentisiert sich die Basisstation lediglich indirekt durch die Kenntnis des dynamisch abgeleiteten Sitzungsschlüssels DCK, wobei die Kenntnis von DCK den Besitz des Authentikationsschlüssels K (bzw. KS oder Triplets) voraussetzt. Dieses Authentikationsverfahren entspricht der Teilnehmerauthentikation bei GSM.

Je nach Einsatzbereich des DECT-Systems, ob Telepoint-Anwendungen (PAS, Public Access Service), geschlossener Geschäftsbereich bzw. Firmennetz (BCT, Business Cordless Telecommunications) oder im Privat- bzw. Heimbereich (RU, Residential Use), sind nicht alle Varianten der Authentikation sinnvoll bzw. praktikabel. Beispielsweise ist die Authentikation über den statischen Sitzungsschlüssel SCK bei PAS nicht einsetzbar, da die Überprüfung der Zugangsberechtigung eine Authentikation erfordert, bevor die eigentliche Kommunikation beginnt.

3.4 Bewertung der DECT-Authentikation

Vorteile des DECT-Ansatzes:

- Die Paare RS, KS können vom FT für den speziellen PT so lange verwendet werden, wie der PT bei dem Besuchsnetzbetreiber registriert ist. Da der Heimatnetzbetreiber den Schlüsselquellenwert RS auswählt, können verschiedene FTs, d.h. verschiedene Besuchsnetzbetreiber, oder verschiedene Registrierungen mit unterschiedlichen Schlüsseln für die Authentikation des PT versorgt werden.

- Der teilnehmerindividuelle Schlüssel K ist nur dem Heimatnetzbetreiber des Teilnehmers und der DAM des Teilnehmers bekannt, er muß und darf keiner anderen Partei bekannt gemacht werden.

- Die Schlüsselerzeugungsalgorithmen A11 und A21 können Heimatnetzbetreiber- spezifisch sein, bzw. spezifisch für den Heimatnetzbetreiber und eine besondere Gruppe seiner Teilnehmer. Keine andere Partei muß sie kennen und implementiert haben. Der Heimatnetzbetreiber kann die Algorithmen jederzeit für einen oder mehrere seiner Teilnehmer wechseln, wobei er den Teilnehmern neue Zugangsgeräte verschaffen muß.

- Der Mechanismus kann zur Ableitung von Verschlüsselungsschlüsseln, die für die Vertraulichkeit von Teilnehmer- und Signalisierungsdaten auf der Luftschnittstelle eingesetzt werden können, verwendet werden. Für jedes Gespräch wird dabei ein anderer Verschlüsselungsschlüssel abgeleitet.

Nachteile des DECT-Ansatzes:

- FT, d.h. der Besuchsnetzbetreiber, muß mit dem Heimatnetzbetreiber signalisieren, um den teilnehmerspezifischen abgeleiteten Authentikationsschlüssel KS zu erhalten.

- Algorithmen A12 und A22 müssen allen FTs, d.h. allen Besuchsnetzbetreibern, die Telekommunikationsdienste für den speziellen Heimatnetzbetreiber anbieten, bekannt sein und von ihnen benutzt werden. In der Praxis sollten deshalb standardisierte Algorithmen verwendet werden.

- Der Heimatnetzbetreiber muß dem Besuchsnetzbetreiber vertrauen, daß der Besuchsnetzbetreiber die Vertraulichkeit des abgeleiteten teilnehmerindividuellen Authentikationsschlüssel KS bewahrt.

4 UMTS

Das digitale Mobilfunksystem der dritten Generation wird zur Zeit bei ETSI unter dem Namen UMTS (Universal Mobile Telecommunications System) standardisiert. Die von UMTS bereitgestellten globalen Dienste sollen weit über die von GSM gebotenen hinausgehen: Sprachqualität vergleichbar zu der momentan im Festnetz erreichbaren, eindeutige, von Netzbetreiber und Service Provider unabhängige Teilnehmernummer und Datenraten von bis zu 2 Mbit/s sind nur einige der zur Zeit diskutierten Dienste. Im folgenden sollen zunächst die geplanten Sicherheitsdienste von UMTS vorgestellt werden und danach auf die bei ETSI vorgeschlagenen vier Authentikationsmechanismen eingegangen werden. Die hier gegebenen Beschreibungen der Protokolle A, B und C basieren auf den Ergebnissen des europäischen ASPeCT (Advanced Security for Personal Communications Technologies, s. [ASPeCT]) Projekts. Weitere Details zu Protokoll D findet man in [Pütz_1] und [Pütz_2].

4.1 Sicherheitsdienste von UMTS

Die folgenden, allgemeinen Sicherheitsdienste sollen von UMTS angeboten werden bzw. werden zur Zeit bei ETSI diskutiert.

- *Authentikation*:
  Vorgesehen sind Verfahren, die eine gegenseitige explizite Authentikation liefern. Diese Verfahren können symmetrisch, aber auch asymmetrisch basiert sein.

- *Anonymität/Pseudonymität*:
  Hier werden erweiterte Anonymitäts-/Pseudonymitäts-Mechanismen für den Nutzer angestrebt, die über die Verwendung temporärer Identitäten hinausgehen, z.B. Vertraulichkeit der Teilnehmeridentität.

- *Vertraulichkeit* der über die Luftschnittstelle ausgetauschten Daten

- *Integrität* der über die Luftschnittstelle ausgetauschten Daten
  Diese Funktionalität ist besonders bei über UMTS abzuwickelnden Geldgeschäften erforderlich.

- *Nichtabstreitbarkeit*:
  Bei der Vielzahl der in eine UMTS-Verbindung eingebundenen, sich nicht vertrauenden Parteien sind Nichtabstreitbarkeits-Eigenschaften erforderlich.

Alle im folgenden vorgestellten, vorgeschlagenen Authentikationsmechanismen schließen eine gegenseitige Authentikation des Teilnehmers und des Netzes mit ein, wobei der Teilnehmer wie bei GSM und DECT durch eine Chipkarte repräsentiert wird. Bei UMTS heißt diese Chipkarte User Service Identity Module, kurz USIM. Darüber hinaus wird in allen Protokollen ein Sitzungsschlüssel vereinbart und die Nutzer-Identität geheim gehalten. Unterschiede zeigen sich bei der Authentikation des Service Providers gegenüber dem Nutzer und der Nichtabstreitbarkeit der gesamten Authentikation gegenüber Dritten. Hierbei versteht man in der UMTS-Standardisierung unter der Rolle des Service-Providers einen reinen Diensteanbieter, den wir hier analog zum GSM- bzw. DECT Heimatnetzbetreiber betrachten. Entsprechend ist der UMTS-Netzbetreiber dem GSM- bzw. DECT-Besuchsnetzbetreiber gleich-zusetzen.

4.2 Protokoll A (Royal Holloway College, London)

Protokoll A ist ein am Royal Holloway College in London entwickeltes symmetrisches Verfahren, das mit einen Challenge-Response-Mechanismus arbeitet und daher dem DECT-Verfahren sehr ähnlich ist. Im Gegensatz zum DECT-Verfahren bietet dieser Ansatz jedoch sowohl eine gegenseitige Authentikation des Nutzers und des Netzbetreibers als auch eine Geheimhaltung der Nutzeridentität gegenüber dem Netzbetreiber.

4.2.1 Durchführung

Der Mechanismus besteht im allgemeinsten Fall (d.h. Nutzer und Netzbetreiber besitzen im Vorfeld kein gemeinsames Geheimnis) aus fünf Nachrichten, die zwischen den beteiligten Parteien Nutzer, Netzbetreiber und Diensteanbieter ausgetauscht werden. Für den Fall, daß der Nutzer bereits beim Netzbetreiber angemeldet ist und eine temporäre Identität besitzt, entfallen zwei der fünf Nachrichten und der Diensteanbieter ist nicht beteiligt.

1. Im mobilen Terminal wird eine Zufallszahl RND_U erzeugt und zusammen mit einer vom Service Provider zugewiesenen temporären Identität $TMUI_S$ and den Netzbetreiber geschickt.

2. Der Netzbetreiber leitet diese Daten an den Diensteanbieter weiter. Der Diensteanbieter erzeugt eine neue temporäre Identität $TMUI_{S'}$ und verschlüsselt diese mit Hilfe eines sog. Anonymitätsalgorithmus. Ferner wird ein Authentikationswert $AUTH_S$ aus RND_U, $TMUI_{S'}$ und dem gemeinsamen Geheimnis K_{SU} des Nutzers und des Service Providers errechnet, sowie ein Netzauthentikationsschlüssel K_{NU}, in dessen Berechnung sowohl K_{SU} als auch die Identität des Netzbetreibers eingehen.

3. An den Netzbetreiber übermittelt werden nun die verschlüsselte temporäre Identität $TMUI_{S'}$, K_{NU} und $AUTH_S$, der Authentikationswert des Service Providers. Der Netzbetreiber erzeugt nun seinerseits eine temporäre Identität $TMUI_{N'}$ für den Nutzer und eine Zufallszahl RND_N. Aus diesen beiden Werten sowie dem Netzauthentikationsschlüssel K_{NU} und RND_U wird jetzt ein Netzauthentikationswert $AUTH_N$ und ein Sitzungsschlüssel K_S bestimmt. Außerdem erhält der Netzbetreiber aus RND_U, RND_N und K_{NU} einen User-Authentikationswert $AUTH_U$.

4. Der Netzbetreiber sendet die verschlüsselte temporäre Identität $TMUI_{S'}$, $AUTH_S$, RND_N, $TMUI_{N'}$ und $AUTH_N$ an den Nutzer. Der Nutzer besitzt alle nötigen Informationen, um seine vom Service Provider zugeteilte neue Identität $TMUI_{S'}$ entschlüsseln zu können und um seinerseits die Authentikationswerte $AUTH_S$, $AUTH_N$ und $AUTH_U$ zu berechnen. Durch Vergleich mit den zugesandten Werten kann der Nutzer nun den Service Provider und den Netzbetreiber authentisieren.

5. Der Nutzer sendet $AUTH_U$ an den Netzbetreiber. Durch Vergleich mit dem dort berechneten $AUTH_U$ wird der Nutzer authentisiert.

4.2.2 Bewertung

Durch das Protokoll werden die folgenden Ziele realisiert:

- gegenseitige explizite Authentikation von Nutzer und Netzbetreiber

- Vereinbarung eines Sitzungsschlüssels zwischen Nutzer und Netzbetreiber

- Integrität der Daten durch Einsatz kryptographischer Checkfunktionen

- Geheimhaltung der Nutzeridentität gegenüber dem Netzbetreiber durch temporäre Identitäten.

Nicht realisiert wird hingegen die Nichtabstreitbarkeits-Eigenschaft. Dies ist ohne Einsatz einer dritten vertrauenswürdigen dritten Partei mit symmetrischen Verfahren nicht möglich.

4.3 Protokoll B (Siemens)

Protokoll B ist ein von Siemens entwickeltes, asymmetrisch basiertes Authentikationsprotokoll. Im allgemeinsten Fall, wenn keine authentischen Public Keys des mobilen Terminals bzw. des Netzbetreibers bei der jeweils anderen Seite vorliegen, erfordert das Protokoll fünf Nachrichten, die zwischen Nutzer, Netzbetreiber und einem Zertifikatsserver, der zertifizierte Kopien der benötigten Public Keys vorhält, ausgetauscht werden. Wir stellen hier nur den einfachen Fall vor, daß solche zertifizierten Kopien bereits beim Nutzer bzw. Netzbetreiber vorhanden sind. Dafür werden nur drei Nachrichten benötigt, ohne daß ein Zertifikatsserver beteiligt ist.

4.3.1 Durchführung

Voraussetzung des Protokolls ist, daß man sich im Vorfeld auf eine multiplikative endliche Gruppe G mit erzeugendem Element g geeinigt hat, in dem das diskrete Logarithmusproblem schwierig ist. Der Public Key des Netzbetreibers, der ja beim Nutzer vorliegt, hat also die Form g^s, wobei s der geheime Schlüssel des Netzbetreibers ist.

1. Das mobile Terminal erzeugt eine Zufallszahl RND_U und schickt sie g^{RND_U} dem Netzbetreiber. Dieser bildet $(g^{RND_U})^s$ und errechnet daraus mit Hilfe eines Zufallswerts RND_N und einer Hashfunktion den Sitzungsschlüssel K_S. Aus K_S wird wiederum mit einer zweiten Hashfunktion der Authentikationswert $AUTH_N$ des Netzbetreibers abgeleitet.

2. Der Netzbetreiber schickt an den Nutzer RND_N und $AUTH_N$. Die Chipkarte im mobilen Terminal kann nun ebenso wie der Netzbetreiber K_S und $AUTH_N$ berechnen und so den Netzbetreiber authentisieren.

3. Um sich selbst zu authentisieren, signiert der Nutzer K_S, verschlüsselt seine Unterschrift mittels K_S und schickt sie an den Netzbetreiber. Außerdem verschlüsselt der Nutzer seine Teilnehmerkennung mittels K_S und schickt diese an den Netzbetreiber. Der Netzbetreiber kennt nun die Identität des Nutzers, und kann mittels des zugehörigen Public Keys die Unterschrift des Nutzers verifizieren.

4.3.2 Bewertung

Durch das Protokoll werden die folgenden Ziele realisiert:

- Vertraulichkeit der Daten über der Luftschnittstelle

- Teilnehmeranonymität auf der Luftschnittstelle

- gegenseitige explizite Authentikation von Nutzer und Netzbetreiber

- Vereinbarung eines Sitzungsschlüssels zwischen Nutzer und Netzbetreiber

- Nichtabstreitbarkeit (gegenüber dem Netzbetreiber) wird ebenfalls durch digitale Signaturen erreicht.

Nicht erreicht hingegen wird das Ziel der Pseudonymität des Nutzers gegenüber dem Netzbetreiber. Im Normalfall erfordert das Protokoll nur den Austausch von drei Nachrichten.

4.4 Protokoll C (KPN Research)

Protokoll C ist ebenfalls ein asymmetrisch basiertes Authentikationsprotokoll. Es ist eine Variante des STS (Station-to-Station)-Protokolls [DiOW] und dem oben beschriebenen Protokoll B sehr ähnlich, was den Nachrichtenfluß und den Mechanismus des Schlüsselaustauschs angeht, und wird hier daher nur summarisch wiedergegeben.

4.4.1 Durchführung

Auszutauschende Nachrichten und Berechnungen sind im wesentlichen dieselben wie in Protokoll B. Der wesentliche Unterschied zu Protokoll B liegt in der etwas einfacheren Nutzer-Identifikation: Der Nutzer sendet entweder die Identifikationsnummer eines Zertifikatsservers, der den Public Key des Nutzers zertifiziert hat und später sein eigenes, verschlüsseltes Zertifikat an den Netzbetreiber, oder aber er identifiziert sich durch seine mit dem Sitzungsschlüssel verschlüsselte Nutzeridentität IMUI. In beiden Fällen ist zwar die Vertraulichkeit der Nutzeridentität auf der Luftschnittstelle gewährleistet, nicht jedoch die Anonymität gegenüber dem Netzbetreiber.

4.4.2 Bewertung

Durch das Protokoll werden die folgenden Ziele realisiert:

- Teilnehmeranonymität auf der Luftschnittstelle

- gegenseitige explizite Authentikation von Nutzer und Netzbetreiber

- Vereinbarung eines Sitzungsschlüssels zwischen Nutzer und Netzbetreiber

Nicht erreicht hingegen werden folgende Ziele:

- Pseudonymität des Nutzers gegenüber dem Netzbetreiber

- Nichtabstreitbarkeit (gegenüber dem Netzbetreiber)

Im Normalfall, d.h. wenn bereits der Public Key des Netzbetreibers bzw. Nutzers auf der jeweils anderen Seite vorliegt, sind lediglich drei Nachrichten zwischen mobilem Terminal und Netzbetreiber auszutauschen.

4.5 Protokoll D (T-Mobil/Uni Siegen)

Das Sicherheitsprotokoll basiert auf asymmetrischen, zertifikatsbasierten Kryptoverfahren. Unter der Verwendung zeitvarianter Parameter (Time Variant Parameter, TVP) liefern Digitale Signaturen die Authentikation der Kommunikationspartner. Durch die Einbindung vertrauenswürdiger Zeitstempel, die durch vertrauenswürdige Zeitstempeldienste bereitgestellt werden, erreicht man die Nicht-Abstreitbarkeit der Authentikation selbst sowie die Nicht-Abstreitbarkeit zusätzlicher optionaler Datenfelder. Ein Nachweis, auch Authentikationsnachweis genannt, dokumentiert die erfolgreiche Authentikation gegenüber beliebigen Dritten, beispielsweise gegenüber dem Service Provider des Teilnehmers. Der Zeitstempel garantiert dem Nachweis die erforderliche Einmaligkeit und sichert ihm somit seine Vertrauenswürdigkeit. Die Übertragung der Zertifikate erfolgt, soweit dies erforderlich ist, in einem separaten Setup-Mechanismus, falls der Teilnehmer in einen neuen VLR-Bereich wechselt. Ein vereinbarter Sitzungsschlüssel (Diffie/Hellman unterstützt) dient der Vertraulichkeit der Übertragungsdaten auf der Luftschnittstelle. Teilnehmerzertifikate werden ebenso wie die Identität des Teilnehmers niemals im Klartext übertragen, wodurch sich die Vertraulichkeit der Teilnehmeridentität und des Teilnehmeraufenthaltsortes ergibt. Zudem werden im Setup-Mechanismus Informationen ausgetauscht, die den Aufwand für eine spätere Aushandlung des Sitzungsschlüssels in der Authentikations- und Kommunikationsphase senken.

4.5.1 Durchführung

Das Protokoll umfaßt zur gegenseitigen Authentikation von Teilnehmer und Netzbetreiber insgesamt fünf Protokollschritte M1 bis M5. Davon werden drei Protokollelemente (M1, M4 und M5) über die Luftschnittstelle übertragen und zwei Nachrichten (M2 und M3) zwischen Netzbetreiber und Zeitstempeldienst zur Einbindung des vertrauenswürdigen Zeitstempels ausgetauscht. Ein sechster Schritt (M6) umfaßt den Transport des Authentikationsnachweises zu einem beliebigen Dritten. Allerdings ist dieser Schritt nicht realzeitorientiert.

Die gegenseitige Authentikation erfordert drei Signaturen. In Nachricht M3 signiert der Zeitstempeldienst seinen Zeitstempel und einen Hashwert, der über die Kennungen der Authentikationspartner, deren TVPs und optionale Datenfelder berechnet wurde. Der Netzbetreiber authentisiert sich mit seiner Signatur in Nachricht M4, der Teilnehmer entsprechend mit seiner Signatur in Nachricht M5.

Durch eine Verschachtelung der Signaturen erreicht man gleichzeitig, daß der Teilnehmer mit seiner Signatur in M5 die Authentikation des Netzbetreibers aus Nachricht M4 anerkennt. Mit der insgesamt vierten Signatur in Nachricht M6 bestätigt der Netzbetreiber die Authentikation des Teilnehmers aus Nachricht M5.

Abschließend zeigt sich, daß sich Teilnehmer und Netzbetreiber gegenseitig authentisiert sowie die Authentikation des jeweiligen Kommunikationspartners anerkannt haben. Eben diese gegenseitige Anerkennung der Authentikation erlaubt in Verbindung mit dem vertrauenswürdigen Zeitstempel die Überprüfung des Authentikationsnachweises und damit die Nicht-Abstreitbarkeit des gesamten Authentikationsvorgangs.

4.5.2 Bewertung

Das Protokoll erbringt folgende Ziele:

- Authentizität, Aktualität und Bestätigung des vereinbarten Sitzungsschlüssels KS,
- Perfect forward secrecy [DiOW] des vereinbarten Sitzungsschlüssels KS (bei Diffie/Hellman unterstütztem Schlüsselaustausch),
- Vertraulichkeit der Teilnehmeridentität,
- Gegenseitige explizite Authentikation von Chipkarte und Netzbetreiber,
- Nachvollziehbarkeit und Nicht-Abstreitbarkeit der Authentikation zwischen Chipkarte und Netzbetreiber,
- Generierung eines Authentikationsnachweises, der für Dritte verifizierbar ist,
- Nicht-Abstreitbarkeit optionaler Datenfelder.

5 Vergleich der Systeme

Zum Vergleich der bisher betrachteten Authentikationsprotokolle werden folgende Kriterien herangezogen:

- **Art des Verfahrens**
  - Verfahren symmetrisch oder asymmetrisch?

- **Netz-Authentikation**
  - Findet eine gegenseitige Authentikation, insbesondere Authentikation durch das Netz statt?

- **Vertraulichkeit der Nutzeridentität**
  - Wird die Nutzer-Identität auf der Luftschnittstelle geheim gehalten?

- **Pseudonymität**
  - Kennt der Netzbetreiber die Identität des Nutzers oder nur ein Pseudonym?

- **Nicht-Abstreitbarkeit**
  - Können die an dem Protokoll beteiligten Parteien ihre Beteiligung abstreiten?

- **Verifizierbarkeit**
  - Ist der Authentikationsvorgang durch Dritte verifizierbar?

- **Aufwendigkeit des Verfahrens**
  - Wieviele Nachrichten werden zur Authentikation benötigt?
  - Bei asymmetrischen Verfahren:
    - Wie viele digitale Signaturen müssen geleistet werden?
    - Wie viele Exponentiationen mod p sind notwendig?

Die folgende Tabelle faßt das Ergebnis dieses Vergleichs in übersichtlicher Weise zusammen:

	GSM	DECT	UMTS/ A	UMTS/ B	UMTS/ C	UMTS/ D
symm./asym.	symm.	symm.	symm.	asymm.	asymm.	asymm.
gegens. Auth.	indirekt	direkt oder indirekt	direkt	direkt	direkt	direkt
Vertr. d. Nutzer-Identität	-	-	+	+	+	+
Pseudonymität	-	-	+	-	-	-
Nichtabstreitb.	-	-	-	+	-	+
Verifizierb.	-	-	-	-	-	+
# Nachrichten[1]	5/3	7/4[2]	5/3	5/3	5/3	7/5
# Signaturen[1]	n.a.	n.a	n.a.	3/1	2/2	5/4
# Exponentiationen[1]	n.a.	n.a.	n.a.	4/2	2/2	4/2

(+ : wird durch das Protokoll erreicht, - : wird durch das Protokoll nicht erreicht, n.a.: nicht anwendbar)

---

[1] Angegeben sind die Zahlen für Erst-/laufende Registrierung

[2] Zwei Nachrichten bei indirekter Authentikation

6 Schlußbemerkung

Die Sicherheitsfunktionen für GSM sind unter der Vorgabe entstanden, zum einen die Komplexität dieser Funktionen gering zu halten und zum anderen eine hohe Wirksamkeit der Schutzmechanismen zu gewähren. Vor diesem Hintergrund kann festgehalten werden, daß das GSM-System, bezüglich der Systemsicherheit, einen guten Kompromiß darstellt zwischen dem, was aus Sicht von Teilnehmer und Netzbetreiber wünschenswert ist, und dem, was mit vertretbarem technischen Aufwand machbar ist.

Die DECT-Sicherheitsfunktionen weisen im Vergleich zu GSM folgende wesentliche Erweiterungen statt:

- Verminderung des Signalisierungsaufwandes zwischen Heimat- und Besuchsnetz durch die Verwendung zweier miteinander verknüpfter Authentikationsalgorithmen, allerdings erkauft durch eine erhöhte Mißbrauchsmöglichkeit durch die Besuchsumgebung

- eine Authentikation der Basisstation, die erforderlich ist, da bei DECT Applikationen vorgesehen sind, in denen die Basisstation genutzt werden kann, um Daten in der Mobilstation zu ändern.

Für die dritte Mobilfunkgeneration ist die gegenseitige, explizite Authentikation von Nutzer und Netzbetreiber stärker in den Vordergrund gerückt. Bei ETSI werden derzeit Überlegungen angestellt, statt konkreter Authentikationsverfahren nur ein Authentikationsgerüst für den Ablauf der Authentikation bei UMTS anzugeben. Das prinzipielle Ziel des Authentikationsgerüstes für UMTS soll dabei sein, eine flexible Prozedur für Teilnehmer-Netzbetreiber-Authentikation bereitzustellen. Diese Prozedur soll die Verwendung verschiedener Authentikationsmechanismen, wie etwa in Abschnitt 4 beschrieben, ermöglichen mit der Fähigkeit, von einem Mechanismus zum anderen zu migrieren.

Das Authentikationsgerüst

- soll eine Aushandelprozedur zwischen Nutzer, Netzbetreiber und Service Provider umfassen, wobei ausgehandelt wird, welcher Authentikationsmechanismus verwendet werden kann

- soll die Authentikationsverfahren der Mobilfunksysteme der zweiten Generation wie GSM und DECT umfassen

- soll Migrationskonzepte von SIM/DAM zur USIM aufzeigen

- soll Zeitdauer und Datenumfang als wichtige Parameter für die konkreten Authentikationsverfahren benennen.

Damit können sowohl GSM und DECT als auch die vier in Abschnitt 4 vorgestellten neuen Protokolle als mögliche Authentikationsmechanismen in Mobilfunksystemen der dritten Generation realisiert werden.

Literatur

[Arens]	Arens, Guido: Alles über schnurlose Telefone und Nebenstellenanlagen. Franzis, 1995.

[Aghvami]	Aghvami, Hamid: Future Mobile Communications: Evolution or Revolution. International Journal of Wireless Information Networks, Vol. 4, No.1, 1997, S. 1-5.

[ASPeCT]	http://www.esat.kuleuven.ac.be/cosic/aspect/aspect.html

[Aziz]	Aziz, Ashar; Diffie, Whitfield: Privacy and Authentication for Wireless Local Area Networks. IEEE Personal Communications, First Quarter 1994, S. 25-31.

[DiOW]	Diffie, Whitfield; Oorschot, Paul C. van; Wiener, Michael J.: Authentication and Authenticated Key Exchange. Designs, Codes & Cryptography, Nr. 2, 1992, S. 107-125.

[ETSI_1]	ETSI European Telecommunication Standard: GSM 2.09, Security Aspects, Version 3.0.1

[ETSI_2]	ETSI European Telecommunication Standard: GSM 3.20, Security related network functions, Version 4.3.0

[ETSI_3]	ETSI European Telecommunication Standard: Radio Equipment and Systems (RES): Digital European Cordless Telecommunications (DECT): Common interface – Part 7: Security features. ETS 300 175-7, 1992.

[Michel]	Michel, Uwe: Sicherheitsfunktionen im paneuropäischen Mobilfunknetz, Informatik Fachberichte 271, 1991, S. 133-145

[Pütz_1]	Pütz, Stefan: Zur Sicherheit digitaler Mobilfunksysteme. Datenschutz und Datensicherheit 21/6, 1997, S. 321-327.

[Pütz_2]	Pütz, Stefan: Ein nachweisbares Authentikationsprotokoll am Beispiel von UMTS. In: Müller, G.; Rannenberg, K.; Reitenspieß, M.; Stiegler, H. (Hrsg.): Verläßliche IT-Systeme: Zwischen Key Escrow und elektronischem Geld. Proc. der GI-Fachtagung VIS '97, DuD-Fachbeiträge, Vieweg, Braunschweig, 1997, S. 335-356.

[Walker]	Walker, Michael: Security in Mobile and Cordless Telecommunications. Proceedings Comp. Euro '92, IEEE Computer Society; 1992, S. 493-49

Einsatz von Model-Checking für die Entwicklung von Chipkartensystemen

Rainer Falk · Monika Horak

Lehrstuhl für Datenverarbeitung
TU München
{falk, horak}@e-technik.tu-muenchen.de

Zusammenfassung

Es wird ein Verfahren zur Entwicklung von Chipkartensystemen unter Einsatz der formalen Technik des Model-Checking vorgestellt. Dieses umfaßt die Verifikation von Sicherheitseigenschaften, Szenarioanalysen und die Überführung in eine Implementierung. Das Verfahren wird am Beispiel eines Chipkartensystems aus dem Universitätsbereich erläutert und die gewonnenen Ergebnisse werden dargestellt. Die praktischen Erfahrungen zeigen, daß das erläuterte Verfahren dazu beiträgt, die Qualität der entwickelten Systeme zu erhöhen.

1 Einleitung

Der Entwurf von Chipkartensystemen stellt hohe Anforderungen an den Entwickler, da bei sicherheitskritischen Anwendungen nicht erkannte Fehler große Auswirkungen haben können. Um in kurzer Zeit sichere und robuste Systeme zu entwerfen, ist neben der Wahl vertrauenswürdiger Algorithmen der Einsatz formaler Methoden sinnvoll.

Formale Methoden leisten einen wichtigen Beitrag zum Entwurf sicherheitskritischer Systeme. Sie werden eingesetzt, um die Systemspezifikation unmißverständlich zu formulieren und Aussagen über Sicherheitseigenschaften zu verifizieren. Mit dem Einsatz formaler Methoden ist allerdings ein erhöhter Aufwand verbunden. In der Regel werden zusätzlich zu den Systementwicklern Experten für die zugrundeliegende Logik und Mathematik benötigt, eine Automatisierung ist bei den meisten Verfahren höchstens teilweise möglich. Für einen breiten Einsatz formaler Methoden ist jedoch ein praxisorientierter Ansatz unumgänglich. Die Vorgehensweise bei der Spezifikation und Verifikation muß intuitiv verständlich sein. Ferner ist eine Werkzeugunterstützung für ein automatisiertes Vorgehen wünschenswert.

Die formale Technik des Model-Checking wird heute bereits mit Gewinn in der industriellen Praxis beim Entwurf von Hardware und Steuerungssystemen eingesetzt [Büt97].

Darüber hinaus wurden bereits erste positive Erfahrungen mit dem Einsatz dieser Technik bei der Analyse von Sicherheitsprotokollen [MMS97, MCJ97] erzielt.

In unserem Beitrag geben wir zunächst eine kurze Einführung in die Grundlagen des Model-Checking. Anschließend erläutern wir ein praxisorientiertes Verfahren zur Entwicklung von Chipkartensystemen unter Einsatz von formalen Methoden. Wir veranschaulichen unsere Vorgehensweise anhand eines Chipkartensystems aus dem Universitätsbereich (CAMPUS[1]-System) und berichten über die gewonnenen Ergebnisse und Erfahrungen. Abschließend geben wir einen Überblick über verwandte Arbeiten.

2 Model-Checking

Im Gegensatz zu Theorem-Beweisern (siehe z. B. [ACL97]) erfolgt beim Model-Checking die Verifikation vollautomatisch, d. h. ohne Benutzereingriff. Dadurch können diese Methoden in der industriellen Praxis auch von Entwicklern, die keine Spezialisten auf dem Gebiet der formalen Methoden sind, erfolgreich eingesetzt werden.

Clarke und Emerson [CES86] haben eine spezielle Temporallogik CTL (computation tree logic) sowie einen zugehörigen Verifikationsalgorithmus entwickelt, mit dem zustandsendliche Systeme automatisch auf die Erfüllung einer in CTL formulierten Eigenschaft überprüft werden können. Diese Techniken haben inzwischen eine weite Verbreitung gefunden [Büt97, BP97].

2.1 Grundlagen des Model-Checking

Model-Checking kann auf zustandsendliche Systeme angewendet werden. Die in [CES86] betrachteten zustandsendlichen Systeme sind Tripel $M = (S, R, P)$. S ist dabei eine endliche Menge von Zuständen, R ist eine binäre Relation über S (d. h. $R \subseteq S \times S$), und $P : S \to \wp AP$ weist jedem Zustand die Menge der atomaren Aussagen zu, die in diesem Zustand erfüllt sind.

Ein *Pfad* ist eine (unendliche) Folge von Zuständen $(s_0, s_1, s_2, \ldots)$, wobei $\forall i \cdot (s_i, s_{i+1}) \in R$. Für eine Struktur $M = (S, R, P)$ und einen Startzustand $s_0 \in S$ gibt es einen unendlichen *computation tree* mit der Wurzel s_0, und $s \to t$ ist eine Kante genau dann, wenn $(s, t) \in R$ gilt. Abbildung 1 zeigt eine CTL-Struktur und den zugehörigen Berechnungsbaum, der mit dem Wurzelknoten s_0 beginnt. Im allgemeinen müssen sehr große Zustandsräume untersucht werden. Die Grenze von Verifikationsalgorithmen, die die Zustandsräume explizit durchsuchen, liegt bei ca. 10^9 Zuständen. Beim symbolischen Model Checking [BCM+90] werden die charakteristischen Funktionen der Zustandsmengen und der Transitionsrelation durch BDDs (binary decision diagrams) [Bry86] dargestellt, wodurch wesentlich grössere Zustandsräume untersucht werden können.

---

[1] CAMPUS: Chipkartenbasierte Authentifikation und Meldung zu Prüfungen an Universitäten und höheren Schulen

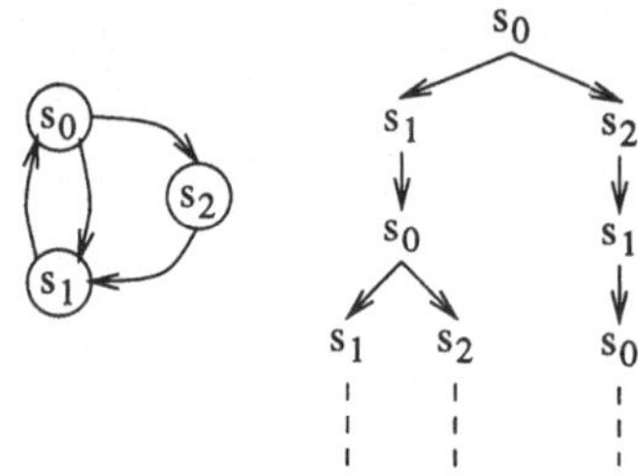

Abb. 1: CTL-Struktur mit Berechnungsbaum für den Startzustand s_0

2.2 Temporale Logik CTL

Systemeigenschaften, die das modellierte System besitzen soll, können in der Temporallogik CTL spezifiziert werden. Die Grundbausteine dieser Formeln sind nicht zeitabhängige Eigenschaften eines Systems, die aussagenlogisch kombiniert werden können. Für solche Aussagen kann dann durch temporale Operatoren deren Gültigkeit im Laufe der Zeit ausgedrückt werden[2]:

$$s_0 \models f_1 \quad gdw \quad f_1 \text{ ist im Zustand } s_0 \text{ erfüllt}$$

$s_0 \models \mathcal{AX} f_1 \quad gdw \quad$ f_1 ist in jedem (direkten) Folgezustand von s_0 erfüllt

$s_0 \models \mathcal{EX} f_1 \quad gdw \quad$ f_1 ist in (mindestens) einem (direkten) Folgezustand von s_0 erfüllt

$s_0 \models \mathcal{AF} f_1 \quad gdw \quad$ in jedem Pfad von Folgezuständen von s_0 gibt es einen Zustand, in dem f_1 erfüllt ist

$s_0 \models \mathcal{EF} f_1 \quad gdw \quad$ es gibt einen Pfad von Folgezuständen von s_0, in dem es einen Zustand gibt, in dem f_1 erfüllt ist

$s_0 \models \mathcal{EG} f_1 \quad gdw \quad$ es gibt einen Pfad von Folgezuständen von s_0 bei dem in jedem Zustand des Pfads f_1 erfüllt ist

$s_0 \models \mathcal{AG} f_1 \quad gdw \quad$ in jedem Pfad von Folgezuständen von s_0 ist in jedem Zustand des jeweiligen Pfads f_1 erfüllt

$s_0 \models \mathcal{A}[f_1 \, \mathcal{U} \, f_2] \quad gdw \quad$ in jedem Pfad von Folgezuständen von s_0 gibt es einen Zustand, in dem f_2 erfüllt ist, und in jedem Zustand zwischen s_0 und diesem Zustand gilt f_1 (until)

$s_0 \models \mathcal{E}[f_1 \, \mathcal{U} \, f_2] \quad gdw \quad$ es gibt einen Pfad von Folgezuständen von s_0, in dem es einen Zustand gibt, in dem f_2 erfüllt ist, und in jedem Zustand auf diesem Pfad zwischen s_0 und diesem Zustand gilt f_1 (until)

Die Syntax und Semantik von CTL sind in [CES86] formal definiert. Dort wird auch der Model-Checking-Algorithmus vorgestellt. Anhand von zwei einfachen Beispielen soll die Bedeutung von CTL-Formeln verdeutlicht werden:

---

[2] *gdw*, genau dann wenn

$\mathcal{AG}(request \rightarrow \mathcal{AF}\,state = busy)$

> Zu allen Zeitpunkten gilt, daß wenn *request* wahr ist, daß dann auch irgendwann (zukünftig) *state = busy* gilt.

$\mathcal{AG}(request \rightarrow \mathcal{A}[request\ \mathcal{U}\ state = busy])$

> Zu allen Zeiten gilt, daß wenn *request* wahr ist, daß dann irgendwann *state = busy* gilt, und daß *request* so lange wahr bleibt, bis *state = busy* gilt.

2.3 SMV

Bei dem hier vorgestellten Projekt wurde der für Forschungszwecke frei verfügbare Model-Checker *SMV* (Symbolic Model Verifier) [McM92] verwendet. Mit diesem symbolischen Model-Checker kann überprüft werden, ob in CTL ausgedrückte Spezifikationen von einem zustandsendlichen System erfüllt werden. Andernfalls wird ein Gegenbeispiel ausgegeben. Ein Zustand ist durch die Werte aller Variablen definiert. Für die Variablen sind nur endliche Typen möglich (Wahrheitswerte, Aufzählungstypen, Felder mit festen Grenzen sowie daraus zusammengesetzte Verbundtypen). Weiterhin wird die Modellierung von nebenläufigen Prozessen unterstützt.

3 Das CAMPUS-System

Die CAMPUS-Card wird derzeit am Lehrstuhl für Datenverarbeitung in Zusammenarbeit mit Giesecke&Devrient entwickelt. Mit ihr werden Studierenden folgende Möglichkeiten angeboten:

- An- und Abmeldung zu Prüfungen und Praktika

- Notenabfrage

- Zugangskontrolle zu den Studierenden-Rechnern

- Zukünftig evtl. Bezahlen (Mensa, Cafeteria, Kopierer, Rückmeldung)

Die CAMPUS-Card stellt gleichzeitig den Studierendenausweis dar. Die Nutzung der Karte soll nicht nur an speziellen Selbstbedienungsterminals erfolgen, sondern auch von allgemein verwendeten Unix-Workstations und PCs sowie von heimischen PCs der Studierenden aus möglich sein. Derzeit wird die erste Version dieses Chipkartensystems entwickelt, das die Funktionen An- und Abmeldung zu Prüfungen und Praktika sowie die Notenabfrage umfaßt. Abbildung 2 zeigt den prinzipiellen Aufbau des CAMPUS-Systems.

4 Methodik

Das Verfahren zur Entwicklung von Chipkartensystemen unter Einsatz formaler Methoden umfaßt folgende Schritte:

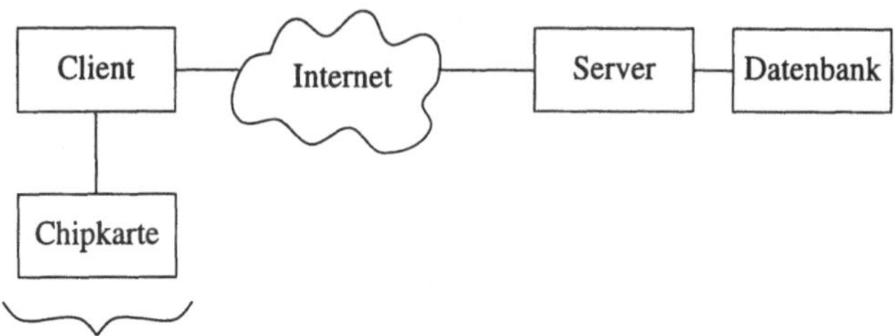

- Selbstbedienungsterminals
- Unix-Workstations
- PCs
- heimische PCs

Abb. 2: Prinzipieller Aufbau des CAMPUS-Systems

1. Verifikation von Sicherheitseigenschaften

- Modellierung des Systems
- Formulierung der Sicherheitseigenschaften
- Verifikation der Eigenschaften

2. Szenarioanalyse

- Erweiterung des Systems durch Angreifer
- Analyse des bedrohten Systems

3. Implementierung

4.1 Verifikation von Sicherheitseigenschaften

4.1.1 Modellierung

Die Komponenten des Systems werden als eigenständige Prozesse modelliert, die über Warteschlangen der Größe 1 kommunizieren. Abbildung 3 zeigt ein Szenario mit den Anwendungen Noteneinsicht und An-/Abmeldung zu Prüfungen und Praktika. Das Modell des CAMPUS-Systems beinhaltet die Prozesse Smartcard, SClient, AClient, Server, SUser und AUser. Über den Prozeß SClient können Studierende die Applikationen des Systems in Anspruch nehmen, AClient bezeichnet den von der Verwaltung verwendeten Prozeß zur Konfiguration, Noteneingabe etc. SUser (Studierende) und AUser (Administration) beschreiben das Verhalten der Systembenutzer. Die Datenbank (Database) wird im Modell nicht als eigenständiger Prozeß formuliert, sondern – auf ihre wesentlichen Eigenschaften abstrahiert – in die Beschreibung des Servers aufgenommen. Der Nachrichtenaustausch zwischen den Prozessen wird in Anlehnung an das in ISO/IEC 7816-4 [ISO94] definierte Nachrichtenformat der Anwendungsschicht durch Befehls- und Antwort-Protokolldateneinheiten (PDUs) modelliert. Eine Befehls-PDU (CMD - PDU) enthält die Komponenten

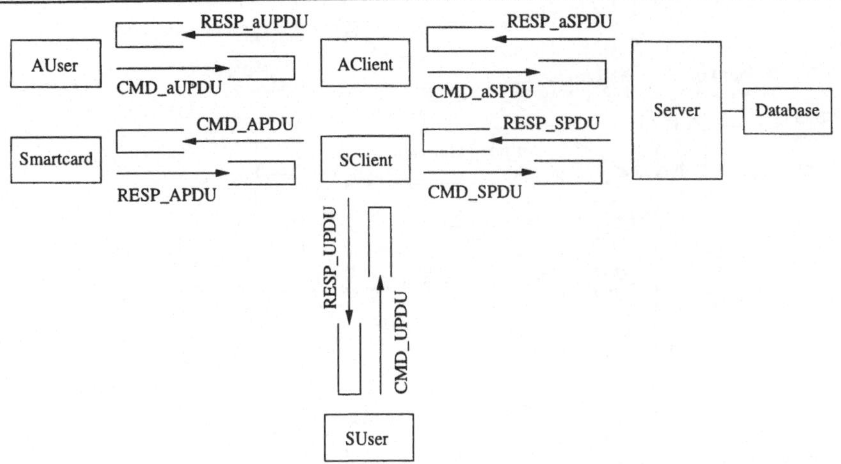

Abb. 3: Modell des CAMPUS-Systems (Ausschnitt)

cmd : eindeutige Identifizierung des Befehls

par : Parameter zur Auswahl der möglichen Funktionen eines Befehls

data : beliebige Daten (Benutzerdaten oder weitere Charakterisierung des Befehls)

Eine Antwort-PDU (RESP - PDU) besteht aus den beiden Komponenten

resp : eindeutige Identifizierung der Antwort

data : beliebige Daten (Benutzerdaten oder Charakterisierung der Antwort)

In dem verwendeten Abstraktionslevel werden weitere Felder der Anwendungsschicht, wie das in Chipkarten-Befehlen enthaltene Class-Byte nicht betrachtet. Ferner wird von den unteren Kommunikationsschichten (Bitübertragungs-, Transportschicht) abstrahiert.

Beim Model-Checking beschreiben Zustandsautomaten das Verhalten eines zustandsendlichen Systems. Jeder Zustand ist durch die Menge der elementaren Aussagen gekennzeichnet, die in diesem Zustand erfüllt sind. Angewandt auf die Problematik kommunizierender Prozesse beschreiben diese elementaren Aussagen einerseits eintreffende Ereignisse, d.h. eingehende Kommandos und die Belegung interner Variablen. Andererseits stellen sie die jeweils ausgelösten Aktionen, also ausgehende Antworten und Veränderungen interner Variablen dar. Abbildung 4 gibt als Beispiel einen Überblick über den Zustandsautomaten der Chipkarte (CAMPUS-Card). Der Prozeß kann die Zustände off, public, client_auth, admin_auth und owner_auth einnehmen. Die Zustandsautomaten der anderen Prozesse sind wesentlich komplexer als der beispielhaft dargestellte Automat der Chipkarte. Auf eine Darstellung dieser Automaten wird hier verzichtet.

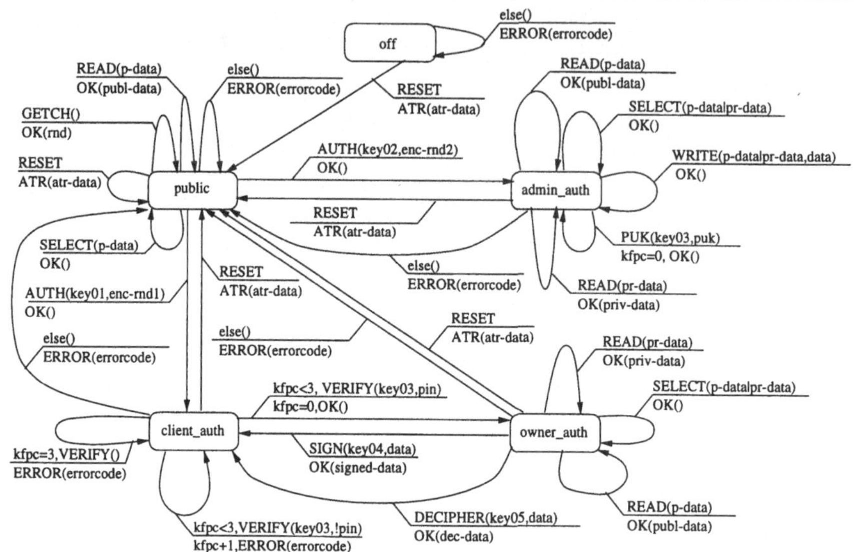

Abb. 4: Zustandsautomat der CAMPUS-Card

4.1.2 Formulierung der Sicherheitseigenschaften

Als Grundvoraussetzung für die Funktionsfähigkeit des Systems werden zunächst elementare Anforderungen wie Deadlock-Freiheit etc. untersucht (siehe Beispiel 1). Für einen sicheren Betrieb des Systems werden hier vor allem Anforderungen betrachtet, die Zugriffe auf persönliche Daten betreffen (Beispiele 2 und 3). Die Sicherheitseigenschaften, die das System erfüllen muß, werden zunächst informell auf einer hohen Abstraktionsebene beschrieben, anschließend schrittweise verfeinert und in der Logik CTL formuliert. Die folgenden Beispiele sollen diese Vorgehensweise erläutern.

Beispiel 1: Anfragen der Studierenden
Ein Benutzer erhält auf eine Anfrage nach einer abgelegten Prüfungsleistung eine Bescheinigung über sein Prüfungsergebnis.

Zu allen Zeiten gilt, daß wenn eine Anfrage nach einer Note ausgeführt wurde, es einen Pfad von Folgezuständen gibt, in dem es einen Zustand gibt, in dem eine Bescheinigung über das Prüfungsergebnis ausgegeben wird.

$$\mathcal{AG}((CMD\_UPDU.cmd = REQUEST \wedge CMD\_UPDU.par = mark)$$
$$\rightarrow \mathcal{EF}(RESP\_UPDU.resp = OK \wedge RESP\_UPDU.data = mark))$$

Beispiel 2: Lesezugriff auf persönliche Daten in der Chipkarte
Die in der Chipkarte gespeicherten persönlichen Daten dürfen nur vom Eigentümer der Karte und von der Verwaltung gelesen werden.

Zu allen Zeitpunkten gilt, daß wenn ein Reset der Chipkarte ausgeführt wurde, solange keine persönlichen Daten aus der Chipkarte ausgelesen werden können, bis eine Benutzerauthentifizierung mit der korrekten PIN durchgeführt wurde oder bis eine korrekte Authentifizierung der Verwaltung anhand einer von ihr mit einem geheimen Schlüssel key02 verschlüsselten Zufallszahl erfolgte. [3]

$$\mathcal{AG}(CMD\_APDU.cmd = RESET \rightarrow$$
$$\mathcal{A}[\neg privd\_read \ \mathcal{U} \ (user\_auth \lor admin\_auth)])$$

$$privd\_read \ \equiv \ RESP\_APDU.data = priv-data$$
$$user\_auth \ \equiv \ CMD\_APDU.cmd = VERIFY \land \ CMD\_APDU.par = key03$$
$$\land \ CMD\_APDU.data = pin$$
$$admin\_auth \ \equiv \ CMD\_APDU.cmd = AUTH \land \ CMD\_APDU.par = key02$$
$$\land \ CMD\_APDU.data = enc-md2$$

Beispiel 3: Schreibzugriff auf in der Chipkarte gespeicherte Daten
Nur die Verwaltung darf die in der Chipkarte gespeicherten Daten schreiben bzw. ändern.

Zu allen Zeiten gilt, daß wenn ein Reset ausgeführt wurde, solange kein Schreibzugriff auf die in der Chipkarte gespeicherten Daten möglich ist, bis eine korrekte Authentifizierung der Verwaltung anhand einer von ihr mit einem geheimen Schlüssel key02 verschlüsselten Zufallszahl erfolgte.

$$\mathcal{AG}(CMD\_APDU.cmd = RESET \rightarrow \mathcal{A}[\neg sc\_write \ \mathcal{U} \ admin\_auth])$$

$$sc\_write \ \equiv \ CMD\_APDU.cmd = WRITE$$
$$\land \ \mathcal{EX}(RESP\_APDU.resp = OK)$$

Bei der Verifikation von Sicherheitseigenschaften wird angenommen, daß nur die rechtmäßigen Instanzen die ihnen zugeordneten geheimen Schlüssel kennen. Bezogen auf die Beispiele 2 und 3 bedeutet dies, daß nur die Chipkarte und die Verwaltung den symmetrischen Schlüssel key02 besitzen und folglich nur diese die verschlüsselte Zufallszahl (Response auf eine Challenge von der Chipkarte) erzeugen können.

4.1.3 Verifikation der Eigenschaften

Mit Hilfe des Model-Checkers SMV werden exemplarisch ausgewählte Eigenschaften überprüft. Abschnitt 5 enthält einen Erfahrungsbericht bei der Modellierung und Verifikation von Systemen mit SMV.

---

[3] $\equiv$, ist logisch äquivalent

4.2 Szenarioanalyse

4.2.1 Erweiterung des Systems durch Angreifer

Szenarioanalysen dienen dazu, Sicherheit darüber zu gewinnen, daß das System die geforderten Eigenschaften nicht nur im Normalbetrieb, sondern auch bei Angriffen erfüllt. Das Systemmodell wird hierfür um den Prozeß Angreifer ergänzt.

Um eine flexible und konfigurierbare Szenarioanalyse zu ermöglichen, werden unterschiedliche Templates für den Angreiferprozeß bereitgestellt. Innerhalb eines Szenarios können beliebig viele Instanzen der unterschiedlichen Angreifer-Prozesse in das Gesamtsystem integriert werden. Folgende Parameter sind konfigurierbar:

- *Ort des Angriffs:* Jede bestehende Kommunikationsverbindung ist als Angriffsort modellierbar (z.B. Client-Chipkarte oder Client-Server). Es können auch mehrere Kommunikationsverbindungen gleichzeitig angegriffen werden.

- *Vermögen des Angreifers:* Angreifer hört nur ab (passiv) oder sendet von sich aus beliebige Nachrichten (aktiv).

- *Wissen des Angreifers:* Der Angreifer kennt nur die öffentlichen Daten oder hat (nach einem bereits erfolgreichen Angriff) auch Wissen über geheime Daten, die er für weitere Angriffsversuche nutzen kann.

Ein aktiver Angreifer wird durch einen Automaten mit nur einem Zustand modelliert. In diesem Zustand werden nichtdeterministisch Nachrichten aus dem vorhandenen Wissensvorrat zusammengesetzt und an den Prozeß gesendet, der Ziel des Angriffs ist. Der Ort des Angriffs ergibt sich durch die entsprechende Belegung der Eingangs- und Ausgangs-Warteschlangen des Zielprozesses.

4.2.2 Analyse des bedrohten Systems

Mit Hilfe des Model-Checkers wird untersucht, ob die geforderten Sicherheitseigenschaften unter verschiedenen Angriffsszenarien erfüllt sind. Dabei geht man zunächst von Angreifern aus, die nur öffentliche Daten kennen. In einer weiteren Analyse werden die Auswirkungen bestimmter erfolgreicher Angriffe ermittelt. Zum Beispiel ist zu untersuchen, welche Folgen das erfolgreiche Knacken eines geheimen Schlüssels für das Gesamtsystem hat.

4.3 Implementierung

Nach erfolgter Verifikation der Sicherheitseigenschaften werden die Beschreibungen der Zustandsautomaten der einzelnen Prozesse in eine Implementierung überführt. Einige Werkzeuge (z. B. StateMate) unterstützen diesen Vorgang durch eine automatische Code-Erzeugung. Andernfalls werden die einzelnen Zustandsautomaten per Hand nachprogrammiert. Die Anwendung auf der Karte kann weitgehend automatisch aus der Zustandsbeschreibung gewonnen werden. Diese Umsetzung ist jedoch stark vom jeweiligen Chipkarten-Betriebssystem abhängig.

5 Verifikation mit SMV

Abbildung 5 zeigt einen Ausschnitt aus der Spezifikation des Chipkarten-Automaten in der SMV-Beschreibungssprache. Die einzelnen Systemkomponenten werden als Module

```
MODULE SmartCard(inapdu,inqueue,outapdu,outqueue)


VAR
state : {off, public, client_auth, admin_auth, owner_auth};


ASSIGN
init(state) := off;
next(state) :=
  case
    inqueue:
    case
      -- (...) states off, public, admin_auth, owner_auth
      state = client_auth:
      case
        inapdu.cmd = VERIFY & inapdu.par = key03
                            & inapdu.data = pin
                            & kfpc < 3             : owner_auth;
        inapdu.cmd = Reset : public;
        inapdu.cmd = VERIFY & kfpc = 3             : state;
        inapdu.cmd = VERIFY & inapdu.par = key03
                            & !(inapdu.data = pin)
                            & kfpc < 3             : state;
                                         1 : public;
      esac;
        1: state;
    esac;
      1 : state;
  esac;
```

Abb. 5: SMV-Eingabefile des Moduls SmartCard (Ausschnitt)

(MODULE) beschrieben. Die Variable state ist ein Skalar, der die symbolischen Werte off, public, client_auth, admin_auth oder owner_auth annehmen kann. Der Initialzustand des Automaten und die Zustandsübergangsfunktion werden durch parallele Zuweisungen bestimmt, die durch das Schlüsselwort ASSIGN eingeleitet werden. In dem Beispiel wird der aktuelle Zustand client_auth betrachtet. Die möglichen eintreffenden Kommandos (VERIFY(key03, pin), VERIFY(key03, !pin) oder Reset) und der Zustand der internen Variable (Fehlbedienungszähler kfpc) lösen jeweils unterschiedliche Zustandsübergänge aus (vergleiche auch Abbildung 4).

In Bezug auf die Leistungsfähigkeit ist das verwendete Werkzeug für die Entwicklung von Chipkartensystemen durchaus geeignet. Für einen industriellen Einsatz ist jedoch eine bessere Eingabemöglichkeit der Systembeschreibung und der Spezifikationen (grafische Eingabe) wünschenswert (vgl. z. B. StateMate). Die für die Verifikation benötigte

Rechenzeit liegt in der Größenordnung von 150 s, für die interne Darstellung der Zustands-automaten und Verifikationsbedingungen werden einige hundert BDD-Knoten benötigt.

6   Verwandte Arbeiten

Sicherheitsanalysen von Authentifizierungsprotokollen mittels BAN-Logik

Burrows, Abadi und Needham definierten eine Logik zur Analyse von Authentifizierungs-protokollen, die auf einem Beweis- oder Regelbasierten Ansatz basiert [BAN90]. Die Logik enthält Relationen zur Beschreibung der Protokollabläufe, der Eigenschaften der verwendeten Schlüssel und Nachrichten und der Annahmen der beteiligten Kommunika-tionspartner (beispielsweise über die Vertrauenswürdigkeit von Partnern). Axiome bzw. Schlußregeln, spiegeln die in Authentifizierungsprotokollen angewendeten Techniken wi-der (z.B. Regeln zur Rückgewinnung des Klartextes) und dienen dazu, Schlußfolgerungen über die o.g. Eigenschaften zu ziehen.

Mit Hilfe der BAN-Logik wurde eine Reihe von Sicherheitslücken in vorher als sicher ein-gestuften Authentifizierungsprotokollen entdeckt. Die Struktur der Logik ist einfach und leicht anzuwenden, dadurch aber in ihren Ausdrucksmitteln beschränkt. Es können daher nicht alle Lücken entdeckt werden. Viele Forscher beschäftigten sich mit der Thematik und führten Schwachpunkte/Einschränkungen aus bzw. brachten Erweiterungsvorschläge vor. Das Verfahren bietet allerdings keine Hilfestellung, wie ein gegebenes Protokoll in die Ausdrucksmittel der Logik übersetzt werden soll. Es besteht Forschungsbedarf hin-sichtlich der Automatisierbarkeit des Verfahrens und der Verfeinerung mit dem Ziel einer Implementierung.

Einsatz von Model-Checking bei der Analyse von Sicherheitsprotokollen

Marrero, Clarke und Jha [MCJ97] setzen Model-Checking zur Verifikation von Sicher-heitsprotokollen ein. Die Modellierung eines Protokolls erfolgt durch einfache Auflistung der Aktionen, die jeder Protokollteilnehmer durchführt. Angreifer werden als spezielle Teilnehmer modelliert, die beliebige Nachrichten abfangen und senden können. Angreifer können ferner Session Keys o.ä. kompromittieren. Ein Protokollauf besteht aus der Inter-aktion zwischen regulären Teilnehmern und dem Angreifer. Durch Analyse eines Durch-laufs können Sicherheitslücken erkannt werden.

Mitchell, Mitchell und Stern wenden das Werkzeug Murφ zur Zustandsraum-Analyse von kryptographischen und sicherheitsrelevanten Protokollen an [MMS97]. Die Modellierung des Protokolls und der gewünschten Eigenschaften erfolgt in einer eigens entwickelten Sprache zur Beschreibung nicht-deterministischer endlicher Zustandsautomaten. Angreifer werden explizit als Teilnehmer des Systems modelliert. Diese können alle Nachrichten abhören, speichern und manipulieren sowie eigene Nachrichten senden.

Verifikation von Chipkartensystemen mit Produktnetzen

In [Neb94] wird ein Verfahren zur Verifikation von Chipkartensystemen mit Produktnetzen beschrieben. Produktnetze sind erweiterte Petrinetze (sogenannte beschriftete Petrinetze mit individuellen Marken). Das Chipkartensystem wird als miteinander kommunizierende, endliche Automaten modelliert. Zustandsübergänge werden durch eingehende Kommandos oder auszusendende Antworten hervorgerufen. Es handelt sich um eine detaillierte Modellierung des Systems: Die Produktnetzspezifikation ist ein identisches Abbild der realen Umgebung. Die Chipkartenanwendung wird durch Belegung der Anfangsmarkierung eingebracht. Die durchgeführten Analysen geben Auskunft über das korrekte Zusammenspiel der Automaten, die Minimalität des Chipkarten-Automaten, Deadlock-Freiheit, Lebendigkeitseigenschaften und Security. Es wurden keine Analysen des Systemverhaltens bei Fehlern oder Manipulationen durchgeführt.

Einsatz von formalen Methoden bei der Entwicklung von Chipkarten-Betriebssystemen

In [AHd96] wird der Einsatz formaler Methoden bei der Entwicklung eines Chipkarten-Betriebssystems vorgestellt. Das entwickelte Betriebssystem enthält einen Interpreter, der dem Java-Bytecode-Interpreter ähnlich ist. Die Semantik eines Teils der verwendeten Programmiersprache wird formal definiert.

Formales Modell des bei der JavaCard verwendeten Byte-Code-Interpreters

Beim Entwurf der Java-Karte wurde ein Teil der virtuellen Maschine (Java Virtual Machine, JVM) formal modelliert [DJV97]. Die verwendete Logik (ACL2) basiert auf Common Lisp. Das Modell beinhaltet Laufzeit-Überprüfungen, um unsichere Ausführungen von Befehlen zu erkennen und zu verhindern. Ferner sind die signifikanten Teile der virtuellen Maschine enthalten: Kreieren von Objekten, Zugriffskontrolle und Methodenaufrufe. Da es sich bei ACL2 um eine ausführbare Logik handelt, kann das erstellte Modell nicht nur ausgeführt werden, sondern es ist auch möglich, die Spezifikation mittels eines Theorembeweisers hinsichtlich der gewünschten Anforderungen zu verifizieren. Das Modell kann als Basis für eine formale Analyse der virtuellen Maschine und des Java-Bytecode-Programms verwendet werden.

7 Zusammenfassung und Ausblick

Unsere praktischen Erfahrungen haben gezeigt, daß Model-Checking eine nützliche Technik bei der Entwicklung von Chipkartensystemen darstellt. Symbolisches Model-Checking ist leistungsfähig genug, um die hohe Komplexität der entwickelten Systeme (Größenordnung 10^{20} Zustände) zu beherrschen. Wir konnten Fehler bereits in frühen Entwicklungsphasen entdecken, in denen sie noch leicht behoben werden konnten.

Die untersuchten Methoden können auch in der industriellen Praxis erfolgreich eingesetzt werden. Der zugrundeliegende Formalismus – endliche Automaten – ist Systementwicklern vertraut. Die untersuchten Eigenschaften können vollautomatisch verifiziert werden, sodaß die Werkzeuge auch von Entwicklern eingesetzt werden können, die keine spezielle Ausbildung in formalen Methoden besitzen. Noch fehlt jedoch eine benutzerfreundliche Schnittstelle zu den Werkzeugen. Formale Methoden sollten entwurfsbegleitend eingesetzt werden können. Dazu muß es möglich sein, ausgehend von der gleichen Systembeschreibung das zu entwickelnde System zu simulieren, soweit möglich automatisch Code zu generieren und Systemeigenschaften formal zu verifizieren. Ziel ist, die beschriebenen Techniken in zukünftige Entwicklungsumgebungen für Chipkartensysteme zu integrieren.

Literatur

[ACL97]   ACL2 Version 2.0, 1997. http://www.cs.utexas.edu/users/moore/acl2/.

[AHd96]   M. I. Alberda, P. H. Hartel und E. K. de Jong Frz. Using formal methods to cultivate trust in smart card operating systems. In P. H. Hartel, P. Paradinas und J-J. Quisquater, Hrsg., *2nd Smart card research and advanced application conference (CARDIS 1996)*, Seiten 111–132. Stichting Mathematisch Centrum, Amsterdam, September 1996.

[BAN90]   Michael Burrows, Martin Abadi und Roger Needham. A Logic of Authentication. *ACM Transactions on Computer Systems*, 8(1):18–36, Februar 1990.

[BCM+90]   J. R. Burch, E. M. Clarke, K. L. McMillan, D. L. Dill, und L. J. Hwang. Symbolic model checking: 10^{20} states and beyond. In *Proceedings of the Fifth Annual IEEE Symposium on Logic in Computer Science*, Seiten 428–439, Philadelphia, Pennsylvania, Juni 1990.

[BP97]   Albert Benveniste und Axel Poigné, Hrsg. *Workshop on Formal Design of Safety Critical Embedded Systems*, Munich, April 1997.

[Bry86]   Randal E. Bryant. Graph-based algorithms for boolean function manipulation. *IEEE Transactions on Computers*, C-35(8):677–691, August 1986.

[Büt97]   Wolfram Büttner. Formale Spezifikation, Verifikation und Synthese zustandsendlicher Systeme. *it + ti – Informationstechnik und Technische Informatik*, 39(3):15–21, 1997.

[CES86]   E. M. Clarke, E. A. Emerson und A. P. Sistla. Automatic Verification of Finite-State Concurrent Systems Using Temporal Logic Specifications. *ACM Transactions on Programming Languages and Systems*, 8(2), April 1986.

[DJV97]   Defensive Java Virtual Machine, 1997. http://www.cli.com/software/djvm/.

[ISO94]   ISO/IEC. *Identification Cards - Integrated Circuit(s) Cards With Contacts - Part 4: Inter-industry Commands for Interchange*, 1994. ISO/IEC 7816-4.

[MCJ97]   Will Marrero, Edmund M. Clarke und Somesh Jha. Model Checking for Security Protocols. Technical Report CMU-SCS-97-139, Carnegie Mellon University, Mai 1997.

[McM92]   K.L. McMillan. *The SMV System*. Carnegie-Mellon University, Februar 1992. http://www.cs.cmu.edu/~modelcheck/smv.html.

[MMS97]   John C. Mitchel, Mark Mitchell und Ulrich Stern. Automated Analysis of Cryptographic Protocols Using Murφ. In *Proceedings of the IEEE Symposium on Security and Privacy*, 1997.

[Neb94]   Markus Nebel. Ein Produktnetz zur Verifikation von Smartcard-Anwendungen in der STARCOS-Umgebung. GMD-Studie 234, Gesellschaft für Mathematik und Datenverarbeitung, Juni 1994.

[Sch96]   Holger Schlingloff. Verification of Finite-State Systems with Temporal Logic Model Checking. In *Proceedings of 3rd WoFACS*, Cape Town, South Africa, Juli 1996.

Trustcenter und Kryptochipkarten für ein sicheres Internet

Volker Bruhn

TC TrustCenter for Security in Data Networks GmbH
bruhn@trustcenter.de

Zusammenfassung

Die Chipkarte mit Kryptokoprozessor als Speichermedium für die geheimen Schlüssel von Public-Key-Verfahren bietet vielfältige Einsatzmöglichkeiten im Bereich der Sicherheit im Internet. Obwohl überall dem Internet-Markt ein enormes Wachstumspotential vorausgesagt wird, ist an vielen Punkten noch ein zögerliches Verhalten zu beobachten. Die Vermarktung von Chipkarten als Werkzeug für Internetsicherheit wird dem Markt neue Impulse geben.

1 Einleitung

Als mögliche Hürden für die Geschwindigkeit, mit der sich in Deutschland die breite Anwendung des Internets als Kommunikationsmedium und kommerziell genutzte Plattform für die verschiedensten Anwendungen entwickeln wird, werden häufig die folgenden beiden Gründe genannt:

- Kosten: Vielen Privatnutzern erscheinen die laufenden Kosten, die bei der Nutzung des Internets schon allein durch die anfallenden Telefongebühren entstehen, zu hoch.

- Sicherheitsbedenken: Das Internet wird allgemein, und zu Recht, als unsicheres Medium angesehen. Alle Daten gehen offen und für weltweit unabsehbar viele Personen und Institutionen mitlesbar über das weltumspannende Datennetz. Daß tatsächlich in großem Stil Daten abgehört werden, belegen aufs neue Meldungen über das systematische weltweite Abhören von E-Mails durch den amerikanischen Geheimdienst NSA mit automatischen Analyseprogrammen [1].

Während wir hier auf den ersten Punkt sicher keinen Einfluß haben, ist der zweite Punkt eng mit der Entwicklung der Chipkartentechnologie verknüpft. Der Einsatz von kryptographischen Verfahren ermöglicht eine sichere Kommunikation auch über das Internet.

Die Sicherheit der grundlegenden kryptographischen Algorithmen und Protokolle wird heute weltweit von Fachleuten als gegeben angesehen. Dies wurde vor allem durch die völlige Offenlegung der Verfahren erreicht. Es sollten nur noch solche Verfahren verwendet werden, deren Sicherheit nicht auf dem Prinzip „Security through Obscurity", d. h. der Geheimhaltung des Verfahrens beruht, sondern darauf, daß das Verfahren weltweit öffentlich bekannt ist, damit intensiv diskutiert und geprüft werden kann und das eigentliche Geheimnis im Schlüssel liegt. Den Durchbruch für die praktikable massenhafte Nutzung von kryptographischen Verfahren zur Datenübertragung bilden dabei die Public-Key-Verfahren.

Zur sicheren Anwendung von Public-Key-Verfahren müssen zwei Dinge gewährleistet sein:

- Die Zuordnung der öffentlichen Schlüssel zu ihren Eigentümern muß zweifelsfrei sichergestellt werden.

- Die geheimen Schlüssel müssen sicher aufbewahrt werden: Der Zugriff auf den geheimen
  Schlüssel darf nur dem Eigentümer des Schlüsselpaares möglich sein. Andererseits muß
  dieser Zugriff bequem und einfach sein, sowie jederzeit und an jedem Ort erfolgen können.

Um den ersten Punkt zu erreichen werden öffentliche Schlüssel in Zertifikaten verteilt. In
diesen Zertifikaten ist die Zuordnung der Schlüssel zu ihren Eigentümern von vertrauenswürdigen Parteien digital signiert. Indem auch die Zuordnung der hierzu verwendeten Signaturschlüssel wieder durch Zertifikate bestätigt wird, entstehen Zertifikatsketten. Bei Ver-
folgung einer Zertifikatskette muß der Anwender irgendwann auf ein Zertifikat stoßen, dem
er „per se" vertraut, etwa, weil er dieses Zertifikat persönlich von einer vertrauenswürdigen
Partei erhalten hat. Zur Gewährleistung dieser Nachprüfbarkeit von Zertifikatsketten gibt es
verschiedene Ansätze. Der heute im kommerziellen Bereich allgemein favorisierte Ansatz ist
die hierarchische Zertifizierungsstruktur. Dabei sind nur bestimmte, in geeigneter Form geprüfte Parteien zur Ausstellung von Zertifikaten berechtigt, die Trustcenter. Die Zertifikatsketten bilden eine flache hierarchische Baumstruktur, in der alle Ketten in einer gemeinsamen Wurzel enden. Wenn in naher Zukunft Internettransaktionen rechtswirksam sein sollen,
ist dieser Ansatz ohne Frage die richtige Wahl. Die Notwendigkeit von Trustcentern für ein
sicheres Internet wird aktuell nicht mehr bestritten.

Für die Aufbewahrung der geheimen Schlüssel wird allgemein die Chipkarte als beste Lösung für viele Anwendungsbereiche favorisiert. Dabei wird vor allem die Verwendung von
modernen Kryptochipkarten diskutiert, da bei diesen schon vom Ansatz her ein größtmögli-
ches Maß an Sicherheit gegeben ist. Dieser Ansatz besteht darin, daß

1. das Schlüsselpaar auf der Chipkarte selber erzeugt werden kann,

2. kryptographische Operationen, die den geheimen Schlüssel verwenden, auf der Karte
   stattfinden, ohne daß der geheime Schlüssel die Karte verläßt, und

3. Zugriff auf den geheimen Schlüssel nur über einen auf der Chipkarte implementierten
   Authentisierungsmechanismus (PIN oder biometrische Merkmale) möglich ist.

Vom unternehmerischen Standpunkt her ist dabei aus unserer Sicht besonders hervorzuhe-
ben, daß die durch diesen Ansatz erzielte Sicherheit auch Nichtfachleuten vermittelbar ist.

2 Neue Impulse für die Marktentwicklung durch Krypto-
chipkarten

Zur Zeit befindet sich die kommerzielle Nutzung des Internets in einer Phase, die vielfach
mit „Henne-Ei-Problem" bezeichnet wird. Die potentiellen Endkunden warten darauf, daß es
mehr Anbieter im Internet gibt, die potentiellen Anbieter warten darauf, daß mehr Teilnehmer im Internet zu erreichen sind.

Dieses Problem gibt es auf mehreren Ebenen. Zum einen existiert es auf der Ebene der Nutzung des Internets überhaupt. Hier spielt auf Kundenseite möglicherweise die Höhe der Telefongebühren sowie die erreichbare Datenübertragungsgeschwindigkeit eine entscheidende

Rolle. Wenn beispielsweise das Herunterladen einer Treibersoftware allein durch die Telefongebühren bereits mehrere DM kostet, während einschlägigen Magazinen zum gleichen Preis ganze Bibliotheken von Treibern auf CD beiliegen, so ist für den Heimanwender kein Vorteil durch die Nutzung des Internets zu erkennen.

Das Henne-Ei-Problem gibt es auch in der Frage der Sicherheit bei der Nutzung von kommerziellen Angeboten im Netz. Während die überwiegende Anzahl von Internetnutzern laut Umfragen WWW-Seiten verlassen, sobald sie um persönliche Daten gebeten werden, haben Anbieter beispielsweise im Versandhausbereich bei Bestellungen über das Internet eine deutlich höhere Fehlbestellungsquote als im klassischen Geschäftsbereich. Mit der Verwendung von SSL (Secure Socket Layer) und Serverzertifikaten läßt sich zumindest den Bedenken der Kunden entgegenwirken. Hier müssen sich – unabhängig von der durch das Stichwort Exportbeschränkung beschriebenen Frage – drei sich gegenseitig beeinflussende Faktoren parallel entwickeln:

1. das Wissen der Kunden über die Bedeutung der SSL-Verbindung.

2. das Angebot von SSL-Seiten.

3. die Etablierung von vertrauenswürdigen Trustcentern.

Die umgekehrte Richtung, die Herstellung von Verbindlichkeit des Kunden gegenüber dem Anbieter, stellt den für den Einsatz von Chipkarten wichtigsten Ansatzpunkt im E-Commerce-Bereich dar.

Zwar ist davon auszugehen, daß hier auch durch Verwendung von nicht auf Chipkarten basierenden Zertifikaten bereits ein deutlicher Fortschritt erzielt werden würde, allein da durch den Zertifizierungsvorgang das Sicherheitsbewußtsein des Kunden erhöht wird, Aussicht auf rechtliche Relevanz vergleichbar einer handschriftlich unterschriebenen Bestellung hat aber zur Zeit nur die auf Chipkarten basierende Lösung.

Zusätzlich würde gerade bei dieser Anwendung im elektronischen Versandhaus der Einsatz von Chipkarten bei den Kunden ein erhöhtes Vertrauen in die Sicherheit bewirken. Da aus Kundensicht mangelnde Sicherheit ein Hauptgrund für Bedenken bei der Nutzung kommerzieller Angebote ist, würde die Chipkarte hier entscheidende Impulse für das Wachstum des Marktes geben.

Wichtig ist dabei, daß dieser Vertrauensgewinn nicht an anderer Stelle wieder zunichte gemacht werden darf, indem man die Schlüssel außerhalb der Karte generiert, so daß der Kunde vermuten kann, daß die Trustcenter seine privaten Schlüssel kennen. Es kann in diesem Zusammenhang gar nicht oft genug betont werden, daß gerade in der Anfangsphase, in der die Akzeptanz durch den Verbraucher ein überlebenswichtiges Kriterium ist, die dem Kunden subjektiv vermittelbare Sicherheit eine entscheidende Rolle spielt. Auch wenn die Lösung, die Schlüssel in einer ITSEC-geprüften [2] Umgebung zu erzeugen und in die Karte zu laden, mit objektiv gleichwertiger Sicherheit durchzuführen wäre wie die Schlüsselerzeugung auf der Karte, wäre damit die Chance vertan, hier einen weiteren Ansatzpunkt für zusätzliches Mißtrauen aus dem Weg zu räumen.

Ebenfalls um ein Henne-Ei-Problem handelt es sich bei der Frage nach der Schlüsselanwendung für die Chipkartentechnologie. Bei allen Vorteilen, die die Verwendung von Chipkarten für den Endanwender bei den verschiedensten Anwendungen haben wird, ist vorher die Hürde des Anschaffungspreises für das Chipkartenlaufwerk zu überwinden. Deshalb bedarf es einer Schlüsselanwendung, für die sich diese Anschaffung lohnt. Es wird im allgemeinen davon ausgegangen, daß diese am ehesten im Homebanking zu suchen sein wird. Eine Möglichkeit könnte dabei sein, daß Banken Ihren Kunden beim Eröffnen eines Kontos ein Chipkartenlaufwerk zur Verfügung stellen.

Dabei sollte aus unserer Sicht besonders auf Kompatibilität geachtet werden. Das für eine spezielle Anwendung erworbene Chipkartensystem sollte möglichst allgemeinen Standards genügen, so daß dieses System auch für weitere Anwendungen zu gebrauchen ist.

Vorteile der Verwendung von Chipkarten

Chipkarten können der Entwicklung des kommerziellen Internetbereiches entscheidende Impulse geben.

Die durch den Einsatz von Chipkarten erzielte Sicherheit ist

- mobil,
- dem Kunden vermittelbar,
- modular,
- prüfbar.

Der größte unmittelbare Vorteil für den Endanwender in vielen denkbaren Anwendungen ist sicherlich die Mobilität der Chipkarte. Eine Chipkarte kann man problemlos ständig bei sich führen, beispielsweise in der Brieftasche. Dies wäre allein von der Bequemlichkeit her schon ein entscheidender Gewinn gegenüber der zur Zeit gängigen Praxis, zum Beispiel PGP-Schlüssel auf Diskette bei sich zu führen. Die Antwort-Mail „Leider habe ich meine Schlüssel gerade nicht bei mir, schick mir die Mail doch noch einmal unverschlüsselt" würde der Vergangenheit angehören.

Für die schnelle Entwicklung des Marktes ist vor allem die Tatsache wichtig, daß die mit Chipkarten erzielte Sicherheit nicht nur objektiv vorhanden ist, sondern auch glaubhaft darstellbar ist. Es leuchtet dem Durchschnittsanwender eher ein, daß eine Chipkarte als speziell für die Speicherung von kryptographischen Schlüsseln entwickeltes Modul diese vor Zugriffen von außen schützt, als daß man die auf Diskette oder auf der Festplatte seines PC's liegenden Schlüssel durch reine Softwaremaßnahmen schützen könnte.

Vom Sicherheitsaspekt her bietet eine Chipkarte vor allem den Vorteil, daß sie ein begrenztes technisches Modul mit einer klar definierten überschaubaren Schnittstelle darstellt. Durch die Chipkarte werden die Sicherheitsfunktionen in einer speziellen technischen Komponente zusammengefaßt. Hierdurch wird ein einheitlicher Sicherheitsstandard für alle Anwendungen, die auf dieses Modul zugreifen, erreicht.

Durch die Modularität werden die Sicherheitsfunktionen zudem in viel höherem Maße prüfbar, als es der Fall wäre, wenn jede Anwendung einzeln geprüft werden müßte.

Zur Erstellung von Signaturen, die dem Gesetz zur digitalen Signatur genügen sollen, scheint heute insbesondere wegen der dort verlangten Prüfung der Sicherheitsfunktionen die Chipkarte die einzig in Frage kommende Lösung zu sein.

Da das Signaturgesetz als Grundlage für eine rechtliche Relevanz von digitalen Signaturen dienen soll, führt bei Anwendungen, die eine rechtliche Relevanz erreichen wollen, offenbar kein Weg an der Chipkarte vorbei.

4 Warum Schlüsselerzeugung auf der Chipkarte?

Die TC TrustCenter GmbH hält die Erzeugung der Schlüsselpaare auf der Chipkarte für den Massenmarkt für absolut notwendig. Nur auf diese Weise kann dem Endanwender vermittelt werden, daß eine Speicherung seines geheimen Schlüssels durch Dritte absolut ausgeschlossen ist. Gerade vor dem Hintergrund der aktuellen Kryptodebatte ist dies ein K.O.-Kriterium für die schnelle Entwicklung des Akzeptanzprozesses.

Werden die Schlüssel nicht auf der Chipkarte erzeugt, so bleiben verschiedene Möglichkeiten:

1. Der Anwender erzeugt das Schlüsselpaar mit einer speziellen Software selber auf seinem eigenen PC.

2. Das Schlüsselpaar wird von einem speziellen Programm in einer sicheren Umgebung erzeugt.

Bei der ersten Möglichkeit wird die objektive Sicherheit, die durch Verwendung der Chipkarte gegen Kompromittierung des Teilnehmerschlüssels durch Schadsoftware erzielt werden soll, wieder ausgehebelt. Zwar ist die Schlüsselerzeugung im Gegensatz zur Schlüsselnutzung ein einmaliger Vorgang, der zudem ohne Internetverbindung durchgeführt wird, und kann unter besonderen Sicherheitsvorkehrungen wie vorherigem Virenscan, Signaturprüfung des verwendeten Programms etc. erfolgen, aber gerade die für die Chipkarte sprechende Modularität, d. h. die Konzentration der sicherheitsrelevanten Funktionen in der Chipkarte, und die damit verbundene Prüfbarkeit der Sicherheitsfunktionen gehen so verloren.

Bei der Erzeugung des Schlüsselpaares in einer sicheren Umgebung stellt sich die Frage, wann dies zu erfolgen hat. Es könnte entweder während des Zertifizierungsvorganges im Trustcenter oder schon vor der Auslieferung beim Hersteller der Chipkarte geschehen. Im Unterschied zu herkömmlichen Karten findet bei Chipkarten zur Aufbewahrung von Public-Key-Schlüsselpaaren der wichtigste Schritt zur Personalisierung gar nicht auf der Karte statt. Die eindeutige Zuordnung der Karte geschieht nämlich im Moment der Zertifizierung des öffentlichen Schlüssels. Auch wenn die objektiv erreichbare Sicherheit in beiden Fällen deutlich über der in der ersten Variante liegen wird, wird immer beim Kunden die Möglichkeit des Mißtrauens gegenüber der schlüsselerzeugenden Stelle bleiben, den privaten Schlüssel absichtlich, beispielsweise zu Strafverfolgungszwecken, aufzubewahren. Es wird sehr schwierig sein, dem Kunden darzulegen, daß einerseits weltweit staatliche Geheimdienste im großen Stil E-Mails mitlesen und er deswegen wichtige geschäftliche Korrespondenz verschlüsselt übertragen sollte, daß aber andererseits von der Möglichkeit, die privaten Schlüssel zu hinterlegen, kein Gebrauch gemacht werden wird.

5 Diskussionspunkte

Soll ein benutzer-authentisiertes Auslesen des geheimen Schlüssels möglich sein? Es ist wichtig, zwischen der Anforderung, daß kryptographische Operationen, die den geheimen Schlüssel verwenden, auf der Karte stattfinden, ohne daß der geheime Schlüssel die Karte verläßt, und der Anforderung, daß das Auslesen des geheimen Schlüssels nicht nur durch Benutzer-Authentisierungsmaßnahmen geschützt ist, sondern von vornherein von der Chipkarten-Software gar nicht vorgesehen ist, zu unterscheiden.

Verließe der Schlüssel die Karte im alltäglichen Gebrauch, so wäre er Angriffen durch trojanische Pferde, Viren und andere Schadsoftware ausgesetzt. Dieses Risiko wird man nur dann eingehen wollen, wenn hohe Anforderungen an die Durchsatzgeschwindigkeit gestellt werden. Dabei wird auf eine hohe Absicherung der Rechnerumgebung durch Firewalls etc. Wert gelegt werden. Die Mobilität der Chipkarte birgt allerdings zusätzlich die Gefahr, daß sie auch in einer nicht entsprechend gesicherten Umgebung verwendet werden könnte. Will man trotzdem auf diese Mobilität nicht verzichten, kann man ebensogut auf eine Speicherchipkarte zurückgreifen. In einer Kryptochipkarte sollte auf keinen Fall das Auslesen geheimer Schlüssel im laufenden Betrieb zum Funktionsumfang gehören.

Wird die Möglichkeit des Auslesens des Schlüssels nur einmalig zum Herstellen eines Backups verwendet, so kann dies unter besonderen Sicherheitsvorkehrungen geschehen, ohne daß die alltägliche Verwendung der Chipkarte dadurch berührt wird.

Befindet sich das Schlüsselpaar auf einer Chipkarte und geht diese Chipkarte verloren, so stellt dies je nachdem ob es sich um Verschlüsselungsschlüssel oder Signaturschlüssel handelt, den Anwender vor unterschiedliche Fragestellungen. Bei Verschlüsselungsschlüsseln muß wiederum unterschieden werden zwischen Verschlüsselung zur Datenübertragung auf unsicheren Übertragungswegen und Verschlüsselung von Daten auf Datenträgern.

Wichtig ist auch, auf welche Weise die Chipkarte verloren ist. Ist sie zerstört, und damit der geheime Schlüssel unwiederbringlich verloren, so ist dies eine andere Situation als die, in der die Gefahr besteht, daß Unbefugte Zugriff auf den geheimen Schlüssel erlangen könnten.

Haben Unbefugte die Chipkarte in den Händen, so scheint es ein zusätzliches Maß an Sicherheit zu sein, wenn das Auslesen des geheimen Schlüssels nicht nur durch Benutzer-Authentisierungsmaßnahmen geschützt ist, sondern von vornherein von der Chipkarten-Software gar nicht vorgesehen ist.

Dies ist abzuwägen gegen die Vorteile, die eine Möglichkeit zum sicheren Backup des geheimen Schlüssels bieten. Weiter sind die technischen Möglichkeiten, eine solche Backupmöglichkeit auf der Karte zu implementieren, zu betrachten. Ideal wäre beispielsweise wenn der geheime Schlüssel nur bei Neuerzeugung des Schlüsselpaares ausgelesen werden könnte. Dabei müßte vorher ein öffentlicher Schlüssel eines bereits vorhandenen Schlüsselpaares an die Karte übergeben werden, mit dem der geheime, neu erzeugte Schlüssel dann vor der Ausgabe verschlüsselt wird.

Bei Signaturschlüsseln und Schlüsseln zur Datenübertragung ist es bei Zerstörung der Karte eine Frage des Aufwandes und der Reputation die der Anwender bei seinen Kommunikationspartnern hat, ob er ein Backup des Schlüssels haben will oder ob er ein neues Schlüsselpaar erzeugen und den öffentlichen Schlüssel an seine Kommunikationspartner verteilen muß. Ob der Anwender in diesem Zusammenhang die Backup-Möglichkeit als notwendig beurteilt, hängt dabei stark von individuellen Faktoren ab. Ist der Anwender beispielsweise Anbieter im Internet oder hat er die Schlüssel in seiner Funktion als Angestellter in einem Unternehmen verwendet, so wird ihm hier die Möglichkeit, seinen guten Ruf bzw. den seines Unternehmens zu waren und seine Kunden oder Kommunikationspartner nicht durch organisatorischen Aufwand zu belästigen als sehr wichtig ansehen.

Ist die Karte nicht zerstört, sondern möglicherweise in Händen unbekannter Dritter, so ist zusätzlich das Risiko in Betracht zu ziehen, daß diese den Benutzer-Authentisierungsmechanismus der Karte überwinden. Nimmt man dies als wahrscheinlich an, so wird man den Schlüssel nicht mehr wiederverwenden wollen, auch wenn man ihn noch als Backup zur Verfügung hat.

Werden Schlüssel zur Verschlüsselung von gespeicherten Daten verwendet, ist grundsätzlich eine Möglichkeit vorzusehen, an diese Daten auch im Falle des Verlustes der Schlüssel wieder heranzukommen. Eine Möglichkeit wäre, ein Backup des geheimen Schlüssels sicher zu verwahren. Eine weitere Möglichkeit ist die, auf einer zweiten Karte ein zweites Schlüsselpaar zu erzeugen, diese sicher zu verwahren, und technisch dafür zu sorgen, daß bei jedem Verschlüsselungsvorgang die jeweiligen Session-Keys des verwendeten symmetrischen Verfahrens mit beiden öffentlichen Schlüsseln getrennt verschlüsselt abgelegt werden.

Allerdings beziehen sich diese Überlegungen nur auf die Methode, besonders sensible Dateien einzeln mit Public-Key-Verfahren zu verschlüsseln. Dies wird im allgemeinen nur als Nebenanwendung eines Public-Key-Systems zur Verschlüsselung von Datenübertragungen angesehen, da ja das Problem der Schlüsselverteilung hier gar nicht vorhanden ist. Auf der anderen Seite kann man auf diese Weise eine differenzierte Zugriffskontrolle implementieren. In der Praxis spielt neben der gezielten Verschlüsselung von einzelnen Dateien die komplette Verschlüsselung ganzer Dateisysteme im laufenden Betrieb eine große Rolle. Dabei werden feste symmetrische Schlüssel verwendet.

Für einige Anwendungen ist die Chipkarte nicht a priori die beste Lösung: Viele Argumente, die für die Verwendung von Chipkarten als Speichermedium für den Endanwender sprechen, gelten für andere Anwendungen nicht, wie die unten beschriebenen Beispiele zeigen. Andererseits ist für viele Anwendungen der Performancegesichtspunkt entscheidend. Grob gesagt, sind dies alle Anwendungen, für die Mobilität keine Rolle spielt, dafür aber eine hohe Verarbeitungsgeschwindigkeit wichtig ist.

Beispiel: Schlüssel zum Aufbau von SSL-Verbindungen. Der Bereich Serverzertifikate stellt heute einen großen Anteil der von TC TrustCenter im Geschäftskundenbereich ausgestellten Zertifikate dar. Es ist anzunehmen, daß hier die Nachfrage in der nächsten Zeit weiter steigt.

Beispiel: Schlüssel für Internet-Shopping-Anwendungen.

In diesen beiden Beispielen ist eine Verarbeitungszeit in der Größenordnung von 1 Sekunde für eine digitale Signatur nicht akzeptabel. Darüber hinaus sind die Schlüssel, auch wenn sie unter Umständen auf den Namen einer Person ausgestellt werden, von der Natur der Sache her eher an den jeweiligen Server gebunden.

Beispiel: Die Zertifizierungsschlüssel eines Trustcenters. Sind diese auf einer Chipkarte gespeichert, so stellt der Mitarbeiter, der die Zertifizierungen durchführt, eine hohe Bedrohung dar. Er kann die Karte über Nacht „ausleihen" und zu Hause in aller Ruhe beliebige Daten signieren. Hier sind zusätzliche Maßnahmen notwendig, um diese Möglichkeit zu unterbinden. Hier ist die Mobilität, die die Chipkarte bietet nicht nur nicht notwendig, sondern sogar unerwünscht. Bei den für den dauerhaften kommerziellen Erfolg notwendigen Anzahl von täglich auszustellenden Zertifikaten wird eine Chipkarte außerdem schnell an ihre Performancegrenzen stoßen. Darüber hinaus scheint uns als Trustcenter-Betreiber die Gefahr, daß die Chipkarte im Dauerbetrieb ausfällt, als ein nicht unerhebliches Risiko. Zwar stellt dies kein Sicherheitsproblem dar, da in dem Fall einfach ein neues Schlüsselpaar generiert werden würde, aber von Marketinggesichtspunkten her wäre dies ein nicht zu unterschätzender Schaden. Läßt man andererseits ein Auslesen des geheimen Schlüssels zur Sicherung zu, hat man keine signaturgesetzkonforme Signatur-Chipkarte mehr vorliegen, und kann im Grunde genau so gut zu anderen Lösungen übergehen.

6 Die Chipkarte in der öffentliche Meinung

Wir vertreten in diesem Aufsatz die These, daß der Einsatz von Chipkarten der breiten Nutzung des Internets für sicherheitssensitive Anwendungen neue Impulse geben würde. Dies begründen wir mit der Annahme, daß auch dem Laien einleuchtet, daß die Chipkarte den geheimen Schlüssel in hohem Maße gegen Angriffe aus dem Internet schützt.

Dem steht entgegen, daß in der öffentlichen Meinungsbildung über Chipkarten vielfach alle möglichen Karten in einen Topf geworfen werden, es wird dabei nicht zwischen einfachen Magnetstreifenkarten und modernen Chipkarten mit Kryptokoprozessor unterschieden. Dadurch gibt es vielfach Ressentiments gegenüber Karten allgemein. Diese emotional begründeten Ressentiments liegen dabei einerseits in Sicherheitsbedenken begründet, die unter anderem auf der Diskussion um die Sicherheit von EC-Karten beruhen, andererseits auf der Angst vor dem Überwachungsstaat, die zum Beispiel auch bei der Einführung der Krankenkassenkarte eine Rolle spielt.

Dieser Angst sollte man auf jeden Fall dadurch begegnen, daß man Karten verwendet, die ihr Schlüsselpaar selbst erzeugen.

Wie den Sicherheitsbedenken zu begegnen ist, ist nicht so einfach zu sagen. Gegenüber der Speicherung des Schlüsselpaares auf der Festplatte birgt gerade der zweite wichtige Vorteil der Chipkarte für den Kunden, die Mobilität, neue Sicherheitsprobleme. Diese entstehen durch die Gefahr des Verlustes der Chipkarte. Eine öffentliches Breittreten der Sicherheitsprobleme von Chipkarten, die die Linie der bei den Verschlüsselungsalgorithmen verfolgten Strategie der totalen Offenlegung auf die Spitze treiben würde, ist sicher nicht der richtige Weg. Das andere Extrem einer totalen Mauerpolitik würde aber unter Umständen zumindest in technisch interessierten Kreisen gerade bei der aktuellen Diskussion um die EC-Karten wieder zu erhöhtem Mißtrauen führen.

7 Trustcenter, Chipkarten und PGP

Der zur Zeit weltweit am meisten genutzte Standard zur sicheren Versendung von E-Mails ist Pretty Good Privacy - PGP [3].

Trustcenter und PGP

Ein Ansatz für Zertifizierungsstrukturen, der nicht hierarchisch ist, ist die heute weltweit in der PGP-Gemeinde verwendete Idee der netzartigen Zertifizierungsstruktur, des „Web of Trust". Dabei stellt jeder Teilnehmer für alle Schlüssel, denen er, z.B. nach persönlicher Übergabe, vertraut, Zertifikate aus. Ziel ist dabei, daß die so entstehenden Zertifikatsketten ein dichtes Netz bilden, in dem sich für jeden jedes Zertifikat bis zu einem vertrauenswürdigen Zertifikat zurückverfolgen läßt. Dieser Ansatz steht und fällt allerdings mit dem Maß an Sicherheitsbewußtsein und technischen Verständnis, daß bei den Teilnehmern vorhanden ist.

Diese Philosophie von PGP ist aber keinesfalls im Programm in irgendeiner Weise fest verdrahtet. Die hierarchische, Trustcenter-basierte Zertifizierungsstruktur läßt sich auch unter PGP zugrunde legen. Dies verringert dabei die Anforderungen an die Anwender, da hier eine eindeutige Richtlinie zur Akzeptanz von Zertifikaten festgelegt wird. Diese ist in eine simple Bedienungsanleitung übersetzbar, bei der der Anwender keinerlei sicherheitsrelevante Entscheidungen zu treffen hat.

Die TC TrustCenter GmbH stellt bereits seit Anfang 1997 PGP-Zertifikate aus, die über die Webseiten http://www.trustcenter.de abrufbar sind.

PGP und Chipkarten

Auch wenn für den privaten Nutzer sicherlich PGP keine Anwendung ist, die die Investition in ein Chipkartenlaufwerk rechtfertigt, wäre die Möglichkeit, Chipkarten zu verwenden sicherlich interessant. Ein wissenschaftlicher Berater von TC TrustCenter hat hierzu bereits auf der Cebit 95 eine modifizierte PGP-Test-Version vorgestellt, die auf eine Philips DX Karte zugreift.

Zur Zeit ist es gängige Praxis, seine Schlüssel auf Diskette bei sich zu tragen. Im Vergleich zu Chipkarten sind Disketten aber relativ groß und empfindlich. Eine Chipkarte hätte in jeder Brieftasche Platz. Eine erhöhte Bequemlichkeit in der Anwendung würde natürlich auch ein erhöhtes Maß an Sicherheit nach sich ziehen, da um so seltener Mails unverschlüsselt geschickt werden würden.

Eine wichtige Frage ist in diesem Zusammenhang die verwendete Zertifizierungsstruktur. Aufgrund der begrenzten Speicherkapazität einer Chipkarte wird ein Anwender seinen Public-Key-Ring im allgemeinen nicht vollständig auf einer Chipkarte unterbringen können. Im Public-Key-Ring sind aber neben den öffentlichen Schlüsseln der Kommunikationspartner auch subjektive Informationen über deren Vertrauenswürdigkeit gespeichert. Zwar lassen sich die öffentlichen Schlüssel überall leicht besorgen, aber mindestens die als vertrauenswürdig eingestuften Zertifikate müssen mit auf der Chipkarte abgelegt werden. Die vollständige Prüfung der Zertifikatskette bis zu einem vertrauenswürdigen Zertifikat ist in einer netzartigen Zertifizierungsstruktur unter Umständen mit erheblichem Aufwand verbunden.

Deshalb eignet sich eine flache, hierarchische Zertifizierungsstruktur mit Trustcentern für einen Einsatz mit Chipkarten sehr viel besser. Im Extremfall, wenn beispielsweise in einer geschlossenen Benutzergruppe nur ein einziges Trustcenter zum Einsatz kommt, hat jeder Teilnehmer nur die öffentlichen Schlüssel von sich und von diesem Trustcenter auf seiner Chipkarte zu speichern.

Anmerkungen

[1]   Heise-Verlag Online, Meldung vom 9.1.1998 http://www.heise.de/newsticker/ data/ae-09.01.98-000/ „US-Geheimdienst fängt europaweit EMails ab . . . Was bislang nur als Gerücht kursierte, wurde von einer Kommission des europäischen Parlaments bestätigt, berichtete der Londoner Telegraph. Gemäß EU-Report läßt die weltgrößte Geheimdienstorganisation, die US National Security Agency (NSA), mit ihrem Echelon-System europaweit EMails, Telefongespräche und Faxe routinemäßig überwachen. . . . Echelon verarbeitet wahllos riesige Informationsmengen, die mittels MEMEX, einem Analyseprogramm der Künstlichen Intelligenz, auf Schlüsselwörter hin untersucht werden.“

[2]   ITSEC – Information Technology Security Evaluation Criteria. Europäischer Standard zur Prüfung von Produkten der IT-Sicherheit.

[3]   Pretty Good Privacy, ein ursprünglich als Freeware entstandenes Programm zur E-Mail-Verschlüsselung. Neueste Informationen zum Thema PGP gibt es neben der Homepage www.pgp.com auch in diversen Newsgroups wie comp.security.pgp.discuss, comp.security.pgp.resource, comp.security.pgp. tech und alt.security.pgp.

MUCK
Multifunktionale Universitäts Chipkarte und integrierte Standardsoftware

Lutz Marten

Universität Würzburg
Informationsmanagement
Lutz.Marten@mail.uni-wuerzburg.de

Zusammenfassung

MUCK ist ein vom Bayerischen Staatsministerium für Unterricht, Kultus, Wissenschaft und Kunst initiiertes Projekt, das zur Zielsetzung hat, eine Effizienzsteigerung der bayerischen Universitäten durch Geschäftsprozessreorganisation zu erreichen. Hierbei sind Geschäftsprozesse in die Studenten, Dozenten und andere Hochschulangehörige involviert sind und die als Selbstbedienungsanwendung implementiert werden können, von besonderem Interesse. Diese werden auf Basis von Standardsoftwarepaketen und der vorhandenen IT-Infrastrukur realisiert. Hieraus und durch die Dezentralisierung der Anwendungsnutzung durch Öffnung der Anwendungen hin zu offenen Kommunikationsnetzen (Internet), resultiert die Problemstellung, den Zugang zu diesen Anwendungen entsprechend dem Sicherheitsbedürfnis aus technischen und rechtlichen Gesichtspunkten (Datenschutzgesetze) gerecht zu werden. Die Chipkarte stellt zur Abdeckung eines Teilbereiches (Träger von schützenswerten Informationen) dieses Problemfeldes ein, aus heutiger Sicht, sinnvoll einsetzbares Mittel dar. Es wird insbesondere angestrebt, als Basis für diesen Chipkarteneinsatz einen nicht unternehmensabhängigen Kartentyp (ZKA-Karte mit Geldkartenchip) einzusetzen, in Verbindung mit einer modularen, anpassbaren und anwendungsunabhängigen Sicherheitssoftware.

1 Einleitung

Als ein Ergebnis des Projektes "Optimierung von Universitätsprozessen" (OUP) des Bayerischen Staatsministeriums für Unterricht, Kultus, Wissenschaft und Kunst, wurde Ende 1996 ein umfassender Abschlußbericht [OUPP96] vorgelegt. Im Verlauf dieses Projektes wurden die Geschäftsprozesse an mehreren bayerischen Universitäten, u.a. an der Universität Würzburg, evaluiert. Es wurden alle Aspekte der universitären Abläufe auf Optimierungsmöglichkeiten hin untersucht. In einem 500-seitigen Geschäftsprozeßhandbuch wurden alle Universitätsprozesse wie Studium und Lehre, Forschung, Personal, Mittelverwaltung und EDV-Leistungen dargestellt. Im Laufe der Analyse wurde neben der integrativen Unterstützung der Abläufe durch DV, auch das Prinzip der Selbstbedienung und der damit verbesserten Kundenorientierung, besonders herausgearbeitet. Als eine Einzelmaßnahme wurde in diesem Zusammenhang auch die Einführung einer *Chipkarte* empfohlen. Basierend auf dieser Empfehlung wurde das Projekt *Multifunktionale Universitäre Chipkarte (MUCK)* definiert. Die

Chipkarte ist im Rahmen dieses Projektes als eine *enabeling-technology* zur Realisierung des Gesamtnutzens aus der Reorganisation der universitären Geschäftsprozesse zu sehen. Der Gesamtnutzen selbst ist ausschließlich aus Reorganisationsgewinnen, die aus der Einführung einer integrierten DV resultieren, zu erlangen. Folgend hieraus ist der Aufbau dieser integrierten DV auf Basis der Anwendungssysteme HISSOS/POS und SAP R/3 der zentraler Punkt des Projektes MUCK. Die Anforderungen, die Aspekte der IT-Sicherheit betreffen, inkl. des Chipkarteneinsatzes, wurden an ein eigenständiges Projekt übertragen (BASILIKA – *Bayerische Sicherheitslösung für Dienstangebote in offenen Kommunikationssnetzen*) [Schu97], da es sich hier primär nicht um universitätsspezifische Problemstellungen und Lösungsansätze handelt, sondern um Sicherheitsaspekte, die im Rahmen jeglichen Dienstangebotes in offenen Kommunikationsnetzen zu betrachten sind. Die im folgenden beschriebenen Konzepte werden zur Zeit in Kooperation zwischen der Bayerischen Julius-Maximilians Universität Würzburg, dem Bayerischen Landesamt für Statistik und Datenverarbeitung, der Firma Siemens AG, Erlangen, dem deutschen Sparkassenverband und der IABG GmbH, Ottobrunn, technisch implementiert [Tron97, IABG97]. Das Endergebnis aus diesen Arbeiten ist dann integraler Bestandteil der Gesamtlösung MUCK [MUCK97].

2 Ziele der Lösung für elektronische Geschäftsabwicklung

Das Ziel von BASILIKA ist es nicht, vorhandene Sicherheitslücken oder Bedrohungsszenarien im Internet zu beseitigen. Vielmehr geht es darum, trotz dieser Lücken eine möglichst modulare Lösung zu suchen, wie zwei Partner (der Nutzer und eine DV-Anwendung) sich gegenseitig, zur Abwicklung von Geschäftsvorfällen, vertrauen können. Damit ist ein enger Zusammenhang mit dem kürzlich vom Bundestag verabschiedeten Signaturgesetz gegeben [SigG97, SigV97]. Dieses deckt aber keineswegs alle Aspekte ab, die für vertrauenswürdige geschäftliche Beziehungen erforderlich sind. Daher ist der konzeptionelle Lösungsansatz weitergehend angelegt und umfaßt technisch wie auch organisatorisch andere Aspekte.

Bei der Definition der Anforderungen im Rahmen der Projektarbeiten, wurde folgende Gliederung vorgenommen:

- Abtrennung der Sicherheitskomponenten

- Logische Zentralisierung der Sicherheitsfunktionen

- Skalierbarkeit, Modularität und Sicherheitsstufen

- Beglaubigung durch eine dritte Instanz

- Minimierung des Administrationsaufwandes

2.1 Abtrennung der Sicherheitstechniken

Die Sicherheit wird von der Applikation weitgehend entkoppelt, da Betriebssystem und Applikation einem häufigen Releasewechsel unterliegen und somit die einzelnen Sicherheitsmechanismen jeweils neu an das Gesamtsystem angepasst werden müssen. Weiterhin ist es organisatorisch wichtig, dass die Administratoren der Anwendung nicht gleichzeitig die Sicherheitskomponenten betreuen.

Es werden keine speziellen Anforderungen an die Clients (z.B. PC) oder an die Netzkonfiguration (TCP/IP) gestellt, die von den Nutzern eingesetzt werden, mit Ausnahme einer Leseeinrichtung am Client für ein Identifikationsmedium (z.B. Chipkarte), wenn auf dieses zugegriffen werden soll.

Es wird nicht auf die Verfügbarkeit von Sicherheitsdienstleitungen eines Netzbetreibers referiert, da diese potentiell unsicher, weder durch den Nutzer noch den Betreiber der Applikation überwachbar sind und durch politische Vorgaben "manipuliert" sein könnten (z.B. U.S. Export Restriktionen).

Die Sicherheit soll möglichst weit als Ende zu Ende Konzept , Nutzer – Applikationsebene und nicht nur Endgerät – Betriebssystem realisiert werden. Dabei kann der Eintritt in den gesicherten Bereich der Applikationen aus einer Domäne erfolgen, zu der die angewählte Applikation nicht gehört.

Damit wird die beidseitige Authentisierung zwischen Nutzer und Applikation in eine Sicherheitsschicht vor der Applikation ausgelagert. Zwischen diesen beiden findet der Nachrichtenaustausch im Sinne eines privaten, virtuellen Kanals (privat: im Sinne des einzelnen Nutzers) statt, unter der Bedingung, daß beide Partner nur sich selbst und erreicht durch eine Dritte Instanz" (z.B. Trustcenter), sich gegenseitig vertrauen. Die Sicherheit verläßt sich nicht auf Techniken der Transportebene, die von niemandem kontrollierbar sind. Der private, virtuelle Kanal ist nutzer-und sessionspezifisch.

Die Herauslösung der sicherheitsrelevanten Aspekte ermöglicht es weiterhin, den Aufwand für eine mögliche Zertifizierung [BSI196], bzw. Validierung ökonomisch günstig zu gestalten. Eine solche Abprüfung auf die Einhaltung bestimmter Sicherheitsstandard, kann insbesondere bei Einsatz proprietärer Standardlösungen wichtig sein.

2.2 Logische Zentralisierung der Sicherheitsfunktionen

Zentralisierung ist hier nicht im Sinne einer physischen, sondern einer logischen und konzeptionellen Zentralisierung definiert.

- Alle sicherheitsrelevanten Funktionen werden logisch konzentriert, so daß sie gleichmäßig über das ganze Unternehmen zugänglich sind.

- Der private, virtuelle Kanal wird durch einen zentralen Kontrollpunkt (Provider Gateway, siehe Abschnitt 3) zeitüberwacht und automatisch geschlossen, wenn er über eine einstellbar lange Zeit inaktiv sein sollte.

- Die Berechtigungsverwaltung (Zugangsberechtigung im Sinne eines Single-Sign-On) und das Auditing des Nachrichtenverkehrs werden zentral administriert und überwacht.

- Die Administration wird an einer Stelle konzentriert, so daß unternehmensweit verbindlich die gleichen Regeln zum Einsatz kommen und das Administrationspersonal optimal ausgebildet sein kann.

- Zentrale Dienste werden realisiert, die von allen Applikationen und Nutzern grundsätzlich und unternehmensweit gleichwertig und einheitlich angesprochen werden können:

  - zweiseitige Authentisierung

  - privater, virtueller Kanal

  - Auditing

- ➢ Berechtigungsverwaltung mit Sperrverwaltung

- ➢ Homogenisierung der Berechtigungsverwaltung und der Nutzeroberflächen über eine heterogene Applikationslandschaft hinweg

- ➢ elektronische Zahlungsfunktionen

- ➢ Signatur und Zeitstempeldienste (Notariatsdienste)

2.3 Skalierbarkeit, Modularität und Sicherheitsstufen

Eine Sicherheitslösung muß sich an den wachsenden Bedürfnissen eines Unternehmens orientiert, inkremental aufbauen lassen. Die Sicherheitslösung wird in sich modular gestaltet werden und in Bezug auf Durchsatz und Ausfallsicherheit skalierbar sein (Clusterlösung), denn die Modularität erhöht die Sicherheit und die Wartbarkeit der Sicherheitsebene. Weiterhin wird durch die Modularität eine Verteilung der Verantwortlichkeiten bei der Administration, sowie eine inkrementale Zertifizierung, ermöglicht.

2.4 Beglaubigung durch eine dritte Instanz

Eine gute Nutzerakzeptanz wird auf breiter Basis nur dann erreicht, wenn beiderseitiges Vertrauen zwischen Nutzer und Applikation besteht. Es bedarf hierzu des Einsatzes vertrauenswürdiger Schlüsselsysteme (im Sinne der Kryptographie) und Software auf Seiten des Clients , wie auch der Applikationen. Es ist also notwendig eine Beglaubigungsinstanz einzuschalten, die als Lieferant für vertrauenswürdige Schlüssel und Software fungiert. Diese Funktion ist wiederum auch eine organisatorische Maßnahme zur Reduzierung des Aufwandes zur Wartung des Gesamtsystems.

2.5 Minimierung des Administrationsaufwandes

Einer der größten Schwachpunkt einer Sicherheitslösung besteht in ungepflegten Datenbeständen und zu weitreichenden Berechtigungen von Administratoren, sowie Personal, das zu wenig geschult oder/und überlastet ist. Hieraus resultiert die Anforderung, daß die Administration der Sicherheitslösung weitestgehend automatisiert sein muß, insbesondere bei der Überwachung des Systems selbst und bei der Dokumentation aller Transaktionen (Auditing). Weiter sollten sind die technischen Sicherheitseinrichtungen externen, zertifizierten Spezialisten installiert und gewartet werden.

Aus diesen Grundsätzen ist die nachfolgend dargestellte Modularisierung und Funktionsaufteilung abgeleitet worden. Dabei wurde natürlich berücksichtigt, welche Marktentwicklungen sich abzuzeichnen beginnen. Ein wesentlicher Grundsatz für erfolgreiche Sicherheitsmaßnahmen ist, daß sie finanzierbar bleiben müssen. Deshalb müssen die Lösungen so konstruiert sein, daß weitestgehend auf dem Markt verfügbare (ggf. zertifizierte [BSI196]) Komponenten zum Einsatz kommen. Diese sind zu kombinieren und nur an wenigen kritischen Punkten durch Sonderentwicklungen so zu ergänzen, daß die gesetzten Ziele erreicht werden können.

Die angesprochenen Anforderungen technischer und organisatorischer Art werden im Sicherheitskonzept zu diesem Projekt zusammengefaßt [IABG97].

3 Technische Realisierung

Die Realisierung der technischen Ebene von MUCK, insbesondere die Integration der Chipkarte als Identifikationsmittel, wird im Rahmen des Projekts BASILIKA auf Basis des Produktes TranSON (Transaction Security for Open Networks) der Siemens AG, realisiert [Tron97]. In den folgenden Abschnitten werden die wesentlichen Komponenten und ihre Funktion dargestellt.

3.1 Identifikation

Zur Identifikation eines Nutzers gegenüber der Sicherheitsschicht können verschiedene Mechanismen in das System integriert werden. Realisierte Beispiele sind Passwörter, Chipkarten und Fingerabdruckscanner. Die Chipkarte ist die zur Zeit vorzuziehende Variante, da sie die Merkmale Portabilität, fortschreitende Standardisierung, Multifunktionalität und hohe Sicherheitsstandards in sich vereinigt. Im Rahmen von MUCK wird zunächst die ZKA-Chipkarte flächendeckend als Identifikationsmittel eingesetzt werden. Folgende Argumente sind hierfür ausschlaggebend:

- Outsourcing des technischen Teils und größerer organisatorischer Teile der Chipkartenorganisation, z.B. bei Nutzung der EC-Karte als elektronischem Ausweis keine Kartenkosten.

- Nutzungsmöglichkeit einer offenen Geldbörse (ZKA-Geldkartenfunktion auf dem Chip), dadurch Outsourcing der Geldwirtschaft, flexible und schnelle Einrichtung von POS (Point-of-Sales) Anwendungen.

- Von der Universität unabhängige Erweiterbarkeit der Multifunktionalität der Karte, z.B. als elektronischer ÖPNV – Ausweis.

- Partizipation an den weiteren Entwicklungen im Bereich der Geldkarte, z.B. asymmetrische Schlüsselsysteme, neue Anwendungen im Bereich Gleitzeiterfassung und Zutrittskontrolle.

- Organisatorisch stellen die Universitäten eine geschlossene Nutzergruppe dar, die nicht unabdingbar komplexe Identifikationsverfahren (z.B. SigG konform [SigG97]), also Kryptoprozessor-Chipkarten einsetzen müssen.

Zur Zeit kann die ZKA-Chipkarte aufgrund Ihrer technischen Ausprägung nur als Identifikationsmittel, jedoch nicht als Träger von hochwertigerer elektronischer Schlüsselinformation (z.B. für asymmetrische Kryptoalgorithmen) dienen. Daher wird neben der ZKA-Chipkarte im Rahmen des Projektes exemplarisch auch der Einsatz einer Krypto-Chipkarte getestet, um den zukünftigen Einsatz solcher Kartentypen auch zu gewährleisten.

Die einzelnen Ebenen und Komponenten der MUCK-Anwendungsarchitektur sind in Abbildung 1 dargestellt.

3.2 Berechtigungsserver und Trustcenter

Die Ablage aller Rechte für Nutzer wird in einem Berechtigungsserver realisiert. Dieser basiert auf einem Directory Service Standardsystem (z.B. NDS) und stellt sowohl für den Provider Gateway, wie auch die Anwendungssysteme, Betriebssysteme und Zutrittskontrollsysteme im abgesicherten Netz, Zugangsinformationen bereit. Das Trustcenter hat neben seinen

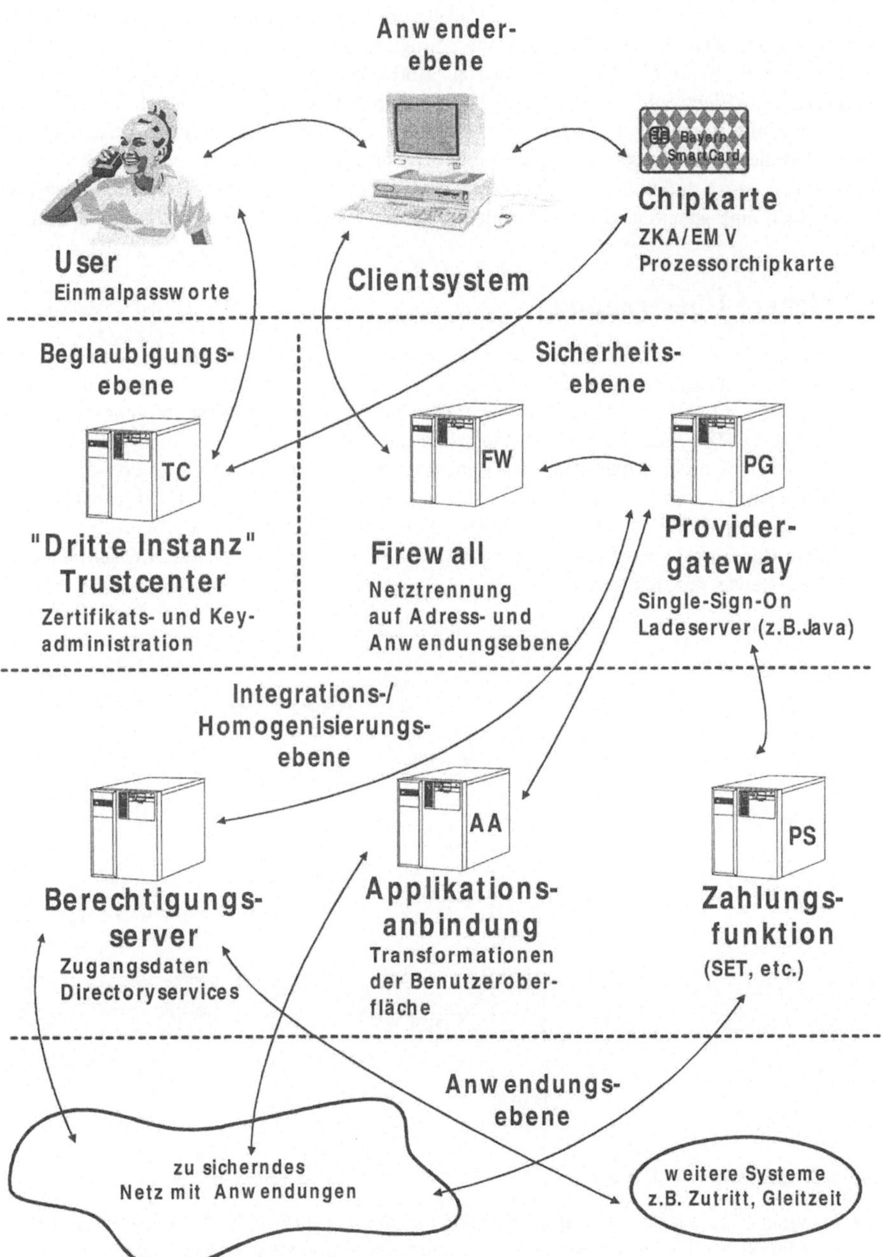

Abbildung 1: Ebenen der MUCK-Anwendungsarchitektur

Aufgaben als CA (Certification Authority) im Rahmen des Einsatzes asymmetrischer Schlüsselsysteme, zertifikatsbasierter Ansätze und Chipkarten, die zusätzlichen Aufgaben der Verteilung digital signierter Software (trusted Software) und Bereitstellung einer ESTS-Funktionalität (Electronic Signature and Time Stamp Services) [LOWR94]. Die Funktionen Trustcenter und Berechtigungsserver wurden getrennt, da das Trustcenter durch Rahmenvorgaben, z.B. dem Signaturgesetz [SigG97] und den daraus resultierenden hohen Sicherheitsanforderungen, auch außerhalb der Universität angesiedelt sein kann. Der Berechtigungsserver stellt jedoch eine spezifische Funktion für die an der Universität betriebenen Anwendungen dar.

3.3 Sichere Übertragung

Zur sicheren Datenübertragung wird der Aufbau eines VPN (Virtual Private Networks) auf Basis der SSL-Spezifikation [SSLV30] bzw. TSL realisiert (siehe Abbildung 2). Dabei werden die Informationen des Identifikationsprozesses durch z.B. die Chipkarte genutzt, um die notwendigen x.509v3 Zertifikate bereitzustellen. Die Informationen kommen entweder von der Clientseite (Chipkartenansatz) oder werden Serverseitig bereitgestellt (Password-, bzw. ZKA-Kartenansatz). Dem so aufgebauten sicheren Kanal sind bestimmte Parameter zugeordnet, so z.B. eine Time-Out Dauer, nach der der Kanal bei Inaktivität automatisch abgebaut wird. Alle den Kanal betreffenden Aktionen, so z.B. Aufbau und Abbau, werden auf dem Provider Gateway protokolliert.

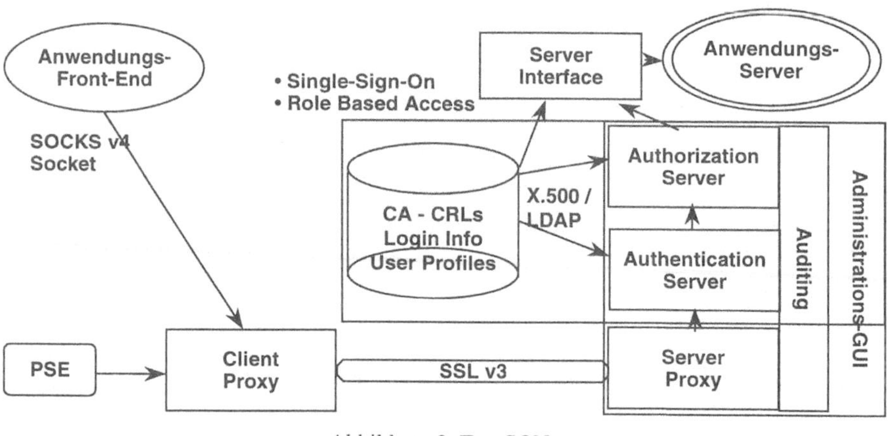

Abbildung 2: TranSON

3.4 Provider Gateway

Das Provider Gateway ist das zentrale Sicherheitsmodul. Das Provider Gateway übernimmt folgende Funktionen, sofern diese nicht schon von einer vorgelagerten Firewall übernommen wurden:

- Paketfilterung

- Applikationsfilterung für Internet-Applikationen, Ports etc.

- Authentisierung des Nutzers mit Chipkarten oder Paßwortmechanismen

- Nachweis der eigenen Identität gegenüber dem Nutzer

- Laden von trusted Software in den Client (mit sessionspezifischer Signatur des Codes)

- vertraulicher, nutzerspezifischer und sessionspezifischer, virtueller Kanal

- Sitzungsüberwachung mit Timeout

- Berechtigungsprüfungen und Zuordnung eines Nutzers zu vordefinierten Nutzerprofilen der Applikationen (Single-Sign-On Funktionalität (SSO)) Homogenisierung der Benutzeroberfläche

- Auslösung von Internet-Zahlungsfunktionen

- Mitführung eines Protokolls des Nachrichtenverkehrs und (verschlüsselte) Speicherung auf einem Auditserver

Das Provider Gateway nutzt zur Aufgabenerfüllung eine Reihe von separierten Hilfsserverfunktionen wie den Berechtigungsserver, das Trustcenter und die Anwendungsadapter (siehe nachfolgende Kapitel. In seiner technischen Ausprägung kann das Provider Gateway auf einem einzelnen Rechner, z.B. jeweils ein Provider Gateway für eine dedizierte Anwendung, installiert sein oder ein System ist zuständig für mehrere Anwendungen gleichzeitig (fanout).

3. 5 Anwendungsadapter

Der Anwendungsadapter stellt den Übergang zwischen dem Providergateway und der Anwendung selbst dar. Dies kann im einfachsten Fall eine vom Anwendungslieferanten bereitgestellte Funktion sein. Ein Beispiel hierfür ist der IST (Internet Transaction Server) der Firma SAP, der den Zugang zum SAP R/3 System basierend auf SSL/TSL und x.509v3-Zertifikaten ermöglicht. Ein Beispiel für einen einfacheren Integrationstypus ist die Simulation eines Passwort basierten Logins durch die Clientseite der Sicherheitschicht, wie sie bei der Anbindung an die HIS-SB Funktionen realisiert ist. Die Anbindung der Netzwerkebene auf Client wie auf Server Seite, wird über die entsprechenden Kommunikationsbibliotheken und Schnittstellen (Socket, SOCKS v4) bereitgestellt (siehe Abbildung 2).

3.5 Zahlungsfunktion

Die Zahlungsfunktion ist, wenn sie auf Basis einer offenen Geldbörse realisiert wird, eine spezielle Anwendung, die nicht ohne kooperative Zusammenarbeit mit den Geldinstituten realisiert werden kann. Im Rahmen von MUCK werden neben konventionellen POS-Installationen, Zahlungsanwendungen über Internet, unter Einsatz der Geldkartenfunktion der ZKA-Chipkarte, realisert. Diese Zahlungsfunktionen werden als Standardanwendung zugekauft und integriert (z.B. IKOSS, G&D, TeleCash).

4 Ausblick

Organisatorisch ist es geplant nach einer MUCK-Pilotphase den Betrieb auf weitere Hochschulstandorte in Bayern auszudehnen. Die Sicherheitslösung auf Basis des Produktes Siemens TranSON befindet sich in ständiger Weiterentwicklung. Es wird am Einsatz anderer Algorithmen für den Schlüsselaustaustausch bzw. die Verschlüsselung selbst (z.B. Diffie-Hellmann, bzw. elliptische Kurven), dem Einsatz von Java-Cards und anderen sich am Markt etablierenden Techniken und Konzepten gearbeitet.

Literatur

[BSI196]   N.N.; "BSI-Zertifizierung Hinweise für Hersteller und Vertreiber", BSI 7138, Bundesamt für Sicherheit in der Informationstechnik, Bonn, 1996

[IABG97]   Spindler, H.; "Sicherheitsvalidierung **BASILIKA**", IABG GmbH, Ottobrunn, 1997

[LOWR94]   Lowry J, Gardiner C, Kimmelmann J; "Quarterly Technical Report: Location Independent Information Object Security", Bolt Beranek and Newmann, ARPA/IOS-1994/04, ARPA Order No. A787, issued by ESC/ENS contract #F198628-93-C-0192, 1994

[MUCK97]   N.N.; "MUCK – Multifunktionale Universitätschipkarte", http://www.zv.uni-wuerzburg.de/muck, Universität Würzburg (Hrsg.), 1997

[OUPP96]   N.N.;"Abschlußbericht der Projektgruppe ‚Optimierung von Universitätsprozessen'", Bayerisches Staatsministerium für Unterricht, Kultus, Wissenschaft und Kunst, München, 1996

[Schu97]   Schuller, G.; Marten, L.: Das **BASILIKA** Konzept (Arbeitspapier), Bay. Julius-Maximilians Universität Würzburg, Würzburg, 1997

[SigG97]   N.N.; "Gesetz zur digitalen Signatur (Signaturgesetz - SigG)", Referentenentwurf, Stand 19. September 1996, Bundesministerium des Inneren, Bonn, 1996

[SigV97]   N.N.; "Verordnung über die Anerkennung von Verfahren zur elekt. Unterschrift (Verordnung über Elektronische Unterschrift - VEU)", Bundesministerium des Inneren, Bonn, 1995

[SSLV30]   Freier A.O., Karlton P., Kocher P.C., "SSL version 3.0" Internet draft, draft-freier-ssl-version3-00.txt, work in progress, IETF, 1995

[Tron97]   N.N.; "TranSON – Transaction Security in Open Networks, Systembeschreibung", Siemens AG, Erlangen,1998

Weitere Informationen und Ansprechpartner: http://www.zv.uni-wuerzburg.de/muck